T0418450

Prions and Diseases

Wen-Quan Zou • Pierluigi Gambetti
Editors

Prions and Diseases

Volume 1, Physiology and Pathophysiology

Editors
Wen-Quan Zou
Department of Pathology
Case Western Reserve University
Cleveland, OH, USA

Pierluigi Gambetti
Department of Neuropathology
Case Western Reserve University
Cleveland, OH, USA

ISBN 978-1-4614-5304-8 ISBN 978-1-4614-5305-5 (eBook)
DOI 10.1007/978-1-4614-5305-5
Springer New York Heidelberg Dordrecht London

Library of Congress Control Number: 2012950411

Printed on acid-free paper

Springer is part of Springer Science+Business Media (www.springer.com)

Contents

Contributors

Jason C. Bartz, Ph.D. Department of Medical Microbiology and Immunology, School of Medicine, Creighton University, Omaha, NE, USA

Ilia V. Baskakov, Ph.D. Department of Anatomy and Neurobiology, Center for Biomedical Engineering and Technology, University of Maryland School of Medicine, Baltimore, MD, USA

David A. Bateman Laboratory of Biochemistry and Genetics, National Institute of Diabetes and Digestive and Kidney Diseases, National Institutes of Health, Bethesda, MD, USA

Emiliano Biasini, Ph.D. Department of Biochemistry, Boston University School of Medicine, Boston, MA, USA

Paul Brown, M.D. CEA Institute of Emerging Diseases and Innovative Therapies, Fontenay-aux-Rose CEDEX, France

Herman K. Edskes, Ph.D. Laboratory of Biochemistry and Genetics, National Institute of Diabetes and Digestive and Kidney Diseases, National Institutes of Health, Bethesda, MD, USA

Anton Gorkovskiy, Ph.D. Laboratory of Biochemistry and Genetics, National Institute of Diabetes and Digestive and Kidney Diseases, National Institutes of Health, Bethesda, MD, USA

David A. Harris, M.D., Ph.D. Department of Biochemistry, Boston University School of Medicine, Boston, MA, USA

Adam C. Kaufman, Ph.D. Cellular Neuroscience, Neurodegeneration and Repair Program, Department of Neurology and Department of Neurobiology, Yale University School of Medicine, New Haven, CT, USA

Amy C. Kelly, Ph.D. Laboratory of Biochemistry and Genetics, National Institute of Diabetes and Digestive and Kidney Diseases, National Institutes of Health, Bethesda, MD, USA

Chaoyang Li, Ph.D. Wuhan Institute of Virology, Chinese Academy of Science, Wuhan, China

Jiyan Ma, Ph.D. Department of Molecular and Cellular Biochemistry, Ohio State University, Columbus, OH, USA

Michael B. Miller, Ph.D. Department of Biochemistry, Dartmouth Medical School, Hanover, NH, USA

Glenn L. Millhauser, Ph.D. Department of Chemistry and Biochemistry, University of California Santa Cruz, Santa Cruz, CA, USA

Fabio Moda, Ph.D. Department of Neurology, Mitchell Center for Alzheimer's disease and Related Brain Disorders, University of Texas Houston Medical School, Houston, TX, USA

Robert B. Petersen, Ph.D. Department of Pathology, Neuroscience, and Neurology, Case Western Reserve University, Cleveland, OH, USA

Sandra Pritzkow, Ph.D. Department of Neurology, Mitchell Center for Alzheimer's disease and Related Brain Disorders, University of Texas Houston Medical School, Houston, TX, USA

Richard Rubenstein, Ph.D. Department of Neurology and Physiology/Pharmacology, SUNY Downstate Medical Center, Brooklyn, NY, USA

Jiri G. Safar, M.D. Department of Pathology and Department of Neurology, National Prion Disease Surveillance Center, School of Medicine, Case Western Reserve University, Cleveland, OH, USA

Hermann M. Schatzl, M.D. Departments of Veterinary Medicine and Departments of Molecular Biology, University of Wyoming, Laramie, WY, USA

Charles R. Schutt, Ph.D. Department of Medical Microbiology and Immunology, School of Medicine, Creighton University, Omaha, NE, USA

Ronald A. Shikiya, Ph.D. Department of Medical Microbiology and Immunology, School of Medicine, Creighton University, Omaha, NE, USA

Claudio Soto, Ph.D. Department of Neurology, Mitchell Center for Alzheimer's disease and Related Brain Disorders, University of Texas Houston Medical School, Houston, TX, USA

Stephen M. Strittmatter, M.D., Ph.D. Cellular Neuroscience, Neurodegeneration and Repair Program, Department of Neurology, Department of Neurobiology, Yale University School of Medicine, New Haven, CT, USA

Surachai Supattapone, M.D., Ph.D., D.Phil. Department of Biochemistry, Dartmouth Medical School, Hanover, NH, USA

Man-sun Sy, Ph.D. Department of Pathology, Case Western Reserve University, Cleveland, OH, USA

Reed B. Wickner, M.D. Laboratory of Biochemistry and Genetics, National Institute of Diabetes and Digestive and Kidney Diseases, National Institutes of Health, Bethesda, MD, USA

Wei Xin, M.D., Ph.D. Department of Pathology, Case Western Reserve University, Cleveland, OH, USA

Wen-Quan Zou, M.D., Ph.D. Department of Pathology and Department of Neurology, National Prion Disease Pathology Surveillance Center, Case Western Reserve University, Cleveland, OH, USA

Chapter 1
Transmissible Spongiform Encephalopathy: From its Beginnings to Daniel Carlton Gajdusek

Paul Brown

Abstract Scrapie was the original member of what has become a family of both animal and human spongiform encephalopathies. Described clearly in the eighteenth century in both England and Germany as a fatal contagious disease of sheep, it was not experimentally transmitted until 1936, and became the subject of wide-ranging research in a number of laboratories in Great Britain. The human analog was first described in 1920 by the German neurologists Creutzfeldt and Jakob, and experimentally transmitted by Gajdusek in 1968, following a similar success in transmitting another analogous human disease (kuru) 2 years earlier. The evolving story of these and other members of the transmissible spongiform encephalopathy family (including "mad cow" disease) has led through a maze of studies involving many unexpected twists and turns, and eventually culminating in the discovery of a new category of infectious disease caused by the misfolding of a normal host protein (PrP^{TSE}).

Keywords Transmissible spongiform encephalopathy (TSE) • Scrapie • Kuru • Creutzfeldt–Jakob disease (CJD) • Transmissible mink encephalopathy (TME) • Chronic wasting disease (CWD) • TSE history

1.1 In the Beginning, …

…there was scrapie. How far back in time is unknown, but it is thought to have originated somewhere in Europe during the late Middle Ages. Whatever the historic beginnings, we know that by the eighteenth century it was prevalent in both England and Germany and that its introduction into England probably came from the importation

P. Brown, M.D. (✉)
Commissariat à l'Énergie Atomique (CEA), Institute of Emerging Diseases and Innovative Therapies, Fontenay-aux-Roses, France
e-mail: Paulwbrown@Comcast.Net

W.-Q. Zou and P. Gambetti (eds.), *Prions and Diseases: Volume 1, Physiology and Pathophysiology*, DOI 10.1007/978-1-4614-5305-5_1,

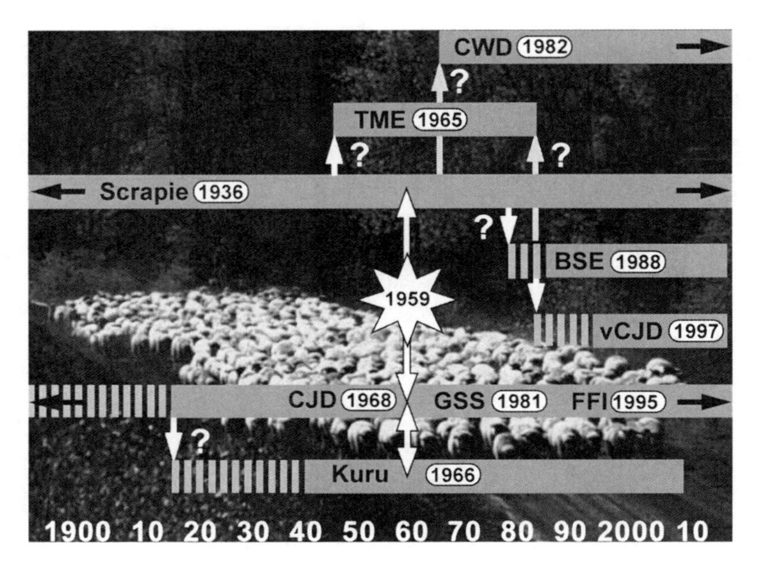

Fig. 1.1 The chronology of TSE. The position and length of the *bars* are keyed to the time line at the *bottom of the figure*. *Striped regions* represent the possible or probable (but unproven) preexistence of the disease. The date of the first reported experimental transmission of each disease is shown within the *bars*. The year 1959 is emphasized to draw attention to its importance as the year in which the kuru–scrapie–CJD connection was made

of Spanish merino sheep that, although highly susceptible to scrapie, had wool of exceptional quality.

At least two centuries elapsed between recognition of the disease and the first attempts to investigate it scientifically. The veterinary literature was limited to its incidence, clinical description, and speculation about its cause until the end of the nineteenth century, when Charles Besnoit and Charles Morel, colleagues in veterinary medicine at Toulouse, France, recognized the regular presence of spongiform change in the spinal cord and adjacent nerves, but considered it to be part of a wider pathology which they thought most likely represented a toxic peripheral neuropathy (Besnoit and Morel 1898). Besnoit also directed a number of transmission experiments in sheep that, unfortunately, were destined to fail because of a surveillance period limited to 9 months (Besnoit 1899), an oversight that a half-century later was also to delay recognition of the transmissibility of the human disease, kuru. Among the younger faculty members at that time was Jean Cuillé, who would later recognize this need for an extended period of postinfection observation, and publish with Paul-Louis Chelle a superb set of experiments between 1936 and 1938 that established beyond any doubt that scrapie was indeed a transmissible disease (Cuillé and Chelle 1936, 1938) (Fig. 1.1).

About the same time that Cuillé and Chelle published their studies, transmissibility was accidentally confirmed when a formalinized louping ill vaccine prepared from sheep CNS tissues was identified as the cause of a mini-epidemic of scrapie in

Scotland (Gordon 1946). Investigation of the outbreak revealed that one batch of vaccine had included material from Cheviot lambs born of ewes that subsequently developed scrapie. These observations laid the groundwork for a flowering of experimental research that was mostly concentrated in Great Britain during the next 30 years, although scrapie was also under study in Iceland, where it had the name "Rida," and in the USA, where it became a growing concern following its diagnosis in Suffolk sheep imported from Great Britain via Canada in 1947.

1.2 Working Out the Biology (in Sheep)

All of the early work on scrapie was conducted in sheep, an extremely inconvenient bioassay animal requiring observation periods of several years in carefully monitored farms, which meant that research remained limited to the very few facilities capable of performing such experiments. Worse still, the unpredictable response of sheep to the same experimental inoculum made it difficult and at times impossible to conduct quantitative titration studies.

Thus, the pioneering work of David R. Wilson at the Moredun Institute in Edinburgh during the 1940s, largely overshadowed by the personalities and careers of the many researchers who followed him, was a remarkable achievement. Conducting experiments almost single-handedly in sheep that had only a 25% transmission rate, he added transmissibility via intradermal and intravenous routes to those reported by Cuillé and Chelle; studied the pathogen's filtration and sedimentation behavior; and discovered its surprising resistance to a variety of chemical and physical treatments, including heat (100 °C for 30 min), exposure to phenol, chloroform, and formaldehyde, and UV irradiation (in retrospect perhaps the most interesting finding). He also documented the survival of infectivity in dried brain tissue after a 2-year storage. A great deal of experimental work published during the next several decades built upon the foundation laid down by Wilson.

The fact that scrapie was of lesser concern to the sheep industry than several other diseases, and was not known (then or now) to be a human pathogen, resulted in little governmental interest in the disease. That indifference changed when, in the early 1950s, North America, Australia, and New Zealand placed embargos on the importation of British sheep in response to the existence of undiagnosed scrapie in their exported sheep. (Never underestimate the power of commercial interests on the funding of scientific research, which recently surfaced again when "mad cow disease" appeared on the scene). Increased funding from the UK expanded the program at Moredun under the continuing direction of Wilson, and later John Stamp, and at Agriculture Research Council (ARC) facility at Compton, England, under the direction of William Gordon.

Gordon conceived and executed a massive study using over 1,000 sheep to investigate the breed susceptibility to scrapie (the "twenty-four breed experiment"), leading to the selection for experimental purposes of two flocks of the Herdwick breed: one highly susceptible and the other relatively resistant. He also put together a very active group of scientists, including Gordon Hunter, Geoffrey Millson, Richard Kimberlin, Carol Walker, and Iain Pattison, who produced a flood of research papers during the 1960s to the 1980s dealing with genetic susceptibility, pathogenesis, and the nature of the scrapie agent.

Meanwhile, at Moredun, Stamp and Alan Dickinson began a wide-ranging study of scrapie strains in Cheviot sheep, producing, for the first time, sound experimental evidence for the maternal transmission of infection and spread of disease through close contact, and in a remarkable set of classical genetic analyses established that a single gene (*Sip*) with two alleles controlled the incubation period in sheep. Dickinson later became the founding Director of the ARC and MRC Neuropathogenesis Unit, also in Edinburgh, where he was soon joined by Kimberlin, Hugh Fraser, Moira Bruce, and David Taylor (and later by Jim Hope, Nora Hunter, and Jean Manson)—who as a group with wide-ranging expertise in pathogenesis, disinfection, molecular biology, and molecular genetics would advance knowledge in each of these areas in the years that followed.

1.3 The Mouse that Roared

In 1961, at Compton, Richard Chandler succeeded in adapting sheep scrapie to the mouse (Chandler 1961). This accomplishment immediately opened the door to studies that would have been prohibitive if limited to bioassays in sheep, and later made possible all of the genetic engineering that is crucial to so much work being done today. Pattison describes the event with his customary flair (Pattison 1972):

> I still feel the urge to genuflect as I pass the spot at our Institute (Compton) beside the boiler house, where my colleague R.L. Chandler paused 1 day in 1960 to suggest to me that he might inoculate three strains of mice (C57, CBA and Swiss) with brain material from two clinical types of goat scrapie (drowsy and scratching). Chandler had already found that the three strains of mice had different susceptibilities to *M. johnei*. He subsequently injected the two strains of scrapie i/c and he transmitted the drowsy strain in 7 months in the Swiss strain and to the other two strains a few weeks later. These mouse strains of scrapie bred true with an incubation period of 4 months. Thus occurred the greatest single advance in scrapie research since experimental transmission of the disease by Cuillé and Chelle in 1936.

This technical advance nearly, but not quite, extinguished all further experimental studies in sheep: the exceptions being studies in which non-rodent species are used to confirm the results in mice, or where there is a need for large amounts of tissues or fluids (blood, for example), or most recently, in studies designed to explore the behavior of bovine spongiform encephalopathy (BSE) infection in sheep. Three of the most important early studies in mice were conducted at the following laboratories:

- At the NIH Rocky Mountain laboratory in Montana, Carl Ecklund and William Hadlow initiated an exhaustive study of the distribution and level of infectivity in a wide variety of tissues and fluids in Chandler's strain of mouse-adapted scrapie, and in mice inoculated with material from naturally and experimentally infected sheep and goats.

- At Compton, Kimberlin and Walker extended these pathogenesis studies to the dynamics of peripheral infection, implicating lymph nodes and spleen along a pathway through visceral sympathetic nerves to the thoracic spinal cord and thence to the brain.
- At the ARC unit in Edinburgh, Dickinson's group applied the same classical genetic approach they had used in sheep, discovering that a similar gene (*Sinc*) controlled the incubation period in mice. They also showed that distinctive patterns of brain lesion distribution were reproducibly associated with different scrapie strains. The conjunction of these two observations led to a method of TSE strain identification that would later serve as the most persuasive evidence for a close strain similarity between BSE and vCJD (Bruce et al. 1997).

1.4 The Nature of the Beast

Amidst all of this work, two crucial questions stood out: what was the relative importance of an infectious versus genetic origin of the naturally occurring disease and, assuming the existence of an infectious agent, what were its biochemical components? The first question was a major topic of discussion at a 1964 meeting convened by the USDA in Washington DC. After listening to 3 days of heated debate, novitiates in the audience were left wondering if all medical meetings were going to be similarly confrontational (they would not be disappointed). Two participants were in almost diametrical opposition: H.B. (James) Parry, an Oxford veterinarian who argued for genetics as the exclusive cause of the naturally occurring disease, and Dickinson, who argued that scrapie was caused by an infectious agent that was influenced by genetic susceptibility. In due course, Dickinson's position would be fully validated. In fact, the *Sip* and *Sinc* genes that Dickinson had identified by classical genetics were none other than the prion-encoding *Prnp* alleles later identified by molecular genetics.

The other question—biochemical characterization of the infectious agent—was (and continues to be) a subject of intense research interest and importance. Although the burden of evidence for different strains of the scrapie agent clearly implied the existence of a nucleic acid genome, there were indications as early as the 1960s that nucleic acid was not only unlikely to be the sole constituent of the scrapie pathogen but, based on radiation resistance data, unlikely even to be present. The first clue came from the early inactivation studies by Wilson, noted above, that included a resistance to standard sterilizing doses of UV radiation. Then came the set of inactivation studies by Hunter, Millson, and Kimberlin that, in conjunction with their demonstration of a firm association of infectivity with cell membranes, led Gibbons and Hunter to propose that the infective entity was a modified glycoprotein subunit of membranes that multiplied by inducing similar chemical or conformation changes in newly "infected" cell membranes (Millson et al. 1976).

The "coup de grace" came from a set of rigorously controlled irradiation studies published by Tikvah Alper and colleagues between 1966 and 1971, in which both the resistance of scrapie brain extracts to very high doses of ionizing and UV radiation and the UV inactivation profile were inconsistent with any known virus or nucleic acid. One paper in particular began with the following point-blank abstract: "Scrapie is a slowly developing disease of the nervous system. Experiments on the effects of ultra-violet irradiation of suspensions of infected mouse brain extracts confirm that the agent responsible for it does not depend on a nucleic acid for its ability to replicate. No evidence is obtained, however, to indicate whether the agent is associated with a protein" (Alper et al. 1967).

No one doubted the validity of Alper's radiation resistance work, but no one knew how to deal with it—in other words, how to accommodate a clear indication of the absence of nucleic acid in the pathogenic agent, and still satisfy the dogma of nucleic acid-directed replication. Explanations invoking protection or repair of nucleic acid eased acceptance of her data, but her conclusions remained in a kind of limbo for years.

1.5 The Transition from Biology to Molecular Biology

In 1967, the mathematician John Stanley Griffith suggested three ways by which a protein might self-replicate, remarking that "there is no reason to fear that the existence of a protein agent would cause the whole theoretical structure of molecular biology to come tumbling down" (Griffith 1967). He presented free energy equations for the polymerization of protein subunits on preexisting dimerized molecules, i.e., a template mechanism, as had been suggested by Gibbons and Hunter. He went on to say that "there is an obvious analogy between the idea presented here and the idea that a gas can only condense on nuclei which are already present: many of the more general schemes could be summed up by saying that the subunits can only polymerize by utilizing condensation nuclei of polymers which are already there." He concluded that scrapie could be "a protein or a set of proteins which the animal is genetically equipped to make, but which it either does not normally make or does not make in that form. It may be passed between animals but be actually a different protein in different species. Finally, in either case there is the possibility of spontaneous appearance of the disease in previously healthy animals."

Credit for the discovery of the first disease-specific structure in a transmissible spongiform encephalopthy (TSE) goes to Patricia Merz, working at the Institute for Basic Research in Developmental Disabilities in Staten Island, New York, who in the late 1970s began to study extracts of scrapie-infected mouse brains under the electron microscope. She identified fibrillar structures very similar to the those seen in Alzheimer's disease, which she named "scrapie-associated fibrils" (SAF), and in further studies also found them in the brains of humans and experimental animals infected with CJD (Merz et al. 1981; Merz and Somerville 1983).

What all of these experiments lacked was a molecule that specifically co-purified with infectivity, but this was finally rectified by 1982 in Stanley Prusiner's laboratory, using the 263 K hamster model of scrapie that had been developed by Kimberlin and Walker in 1977 (Kimberlin and Walker 1977). This model proved to have exceptionally high concentrations of infectivity in the brain (10^{10} LD_{50}/g) after an incubation period of only 2 months, a fortuitous combination that made it possible to undertake the purification of a sufficiently large amount of highly infectious fibrils (renamed "prion rods" by Prusiner) to isolate a peptide subunit that could then be subjected to the tools of modern molecular biology.

The overall contribution of scrapie to the field of TSE was aptly summarized by Pattison in (1972), who concluded his reflections with the statement that "Scrapie is one of four closely similar diseases, the others being kuru, Jakob–Creutzfeldt disease, and transmissible mink encephalopathy. Research on scrapie was responsible for recognition of this group of diseases, to which others may be added in due course, and knowledge of the vagaries of scrapie has been of great value in planning research on them all, for in planning a complicated journey it is reassuring to know that similar ground has already been covered."

1.6 The Discovery of Kuru

In the mid-1950s, a young pediatrician turned research scientist named Carleton Gajdusek was stationed at the Walter Reed Army Medical Center where, in 1954, he was assigned to spend a year in Australia to study the immunology of liver disease in the laboratory of Sir MacFarlane Burnet. Ever the explorer, he traveled widely during his stay, including a trip to Papua New Guinea to satisfy what would become a lifelong interest in primitive cultures, and there met Vincent Zigas, a charming if somewhat eccentric Lithuanian physician who was working as a Medical Officer in the Eastern Highlands. Zigas told him about a strange neurological disease (kuru) that was decimating the Foré-speaking peoples in his area of practice, and invited him to the Highlands to see for himself. He did so and was intrigued by the high incidence, age and sex distribution, and neurological characteristics of the disease (Gajdusek and Zigas 1957). His journals and letters detail the heroic efforts needed to establish a beachhead in Okapa, the administrative center of the Foré region, including a dedicated hospital that for many years operated under the direction of Dr. Michael Alpers, and a native personnel network to identify and transport the continuing stream of new patients to and from Okapa.

He experienced many difficulties with the Australian colonial authorities (Papua New Guinea was then a dependency of Australia), which sometimes resented his dramatic intrusion into their territory. He once remarked that the US government would not be pleased in the converse situation of an Australian research team studying a new disease on an Indian reservation. In fact, one of Gajdusek's most remarkable and generous traits was, with a single exception, his acceptance of people and events that would depress or anger almost anyone else, as part of the "comédie humaine." He was simply incapable of feeling offended or bitter, and never looked back.

He was also an authentic genius, whose interests spanned physics, anthropology, medicine, music, and literature, and his early career was spent in the laboratories of a number of Nobel Laureates. It did not take him long to join their ranks: in 1976 he was awarded a Nobel Prize for his demonstration that kuru, a neurodegenerative disease, had an infectious cause. Kuru had been recognized for decades by the affected population (who considered it to be due to sorcery) and by European locals—everyone from missionaries to bush pilots—who attributed the disease to cannibalism. The difficulty was proving it, as is evident from the innumerable failures to find the cause in toxic, hormonal, nutritional, and infectious causes during the first several years of study.

1.7 The Kuru–CJD–Scrapie Triangle

The year 1959 was a banner year for TSE (Fig. 1.1). Since his encounter with kuru, Gajdusek had been spending a good part of each year in the field, establishing a kuru hospital in Okapa, the administrative center of the region, organizing the care of kuru patients, doing autopsies, trying to discover the cause of the disease, and conducting preliminary therapeutic trials based on all the possible causes under study. During this time, he sent brains from a dozen kuru cases to Igor Klatzo, a neuropathologist working at the NIH. In 1959 he published his findings, noting widespread neuronal degeneration (including vacuolation), myelin loss, astroglial and microglial proliferation, scattered perivascular cuffing, and, in half the cases, a predominantly cerebellar location of amyloid plaques. He did not mention spongiform change, and attributed the neuronal vacuolation to postmortem artifact. However, in his discussion comparing kuru to other diseases, he concluded that "Creutzfeldt–Jakob disease appears to be closest in resemblance" (Klatzo et al. 1959).

This astute observation by Klatzo was all the more remarkable because the diagnostic criteria for Creutzfeldt–Jakob disease had been in disarray since its initial description in 1920 and remained so through the late 1960s. Creutzfeldt's original case was described as a "new and unusual type of neurological disease" in a 22-year-old woman with a 1-year illness characterized by tremors, spasticity, pyramidal signs, nystagmus, ataxia, myoclonus, and dementia (Creutzfeldt 1920). Neuropathology showed diffuse neuronal loss and astrogliosis, but vacuolation was neither mentioned nor illustrated. A year later, in 1922, Jakob reported four cases that he thought resembled Creutzfeldt's case (Jakob 1921). A review of the slides from Jakob's cases was undertaken by Colin Masters in 1982 (Creutzfeldt's slides had not survived), who concluded that only one of the cases (a 42-year-old male) satisfied the criteria for what we now call Creutzfeldt–Jakob disease: the histopathology included neuronal loss, astrogliosis and a diffuse spongiform change throughout the cerebrum and cerebellum (Masters and Gajdusek 1982).

Over the next several years, Jakob and his students gradually acquired a fuller appreciation of spongiform encephalopathy as a pathological entity, including the first case of familial CJD, and somewhat later, in the mid-1930s, Gerstmann,

Straüssler, and Scheinker reported the first family with the disease that now carries their names (GSS) (Gerstmann et al. 1936). Nevertheless, the clinical and neuropathological characteristics of CJD remained elusive until the bedrock criterion of transmissibility allowed its clear separation from a host of other neurodegenerative diseases of unknown etiology.

Hadlow's recollection of events that led him to make the kuru–scrapie connection was recounted in a reminiscence published in 2008:

> The unlikely linkage of these two diseases came about fortuitously while I was an employee of the USDA studying the pathology of scrapie at Compton. William Jellison, a friend and colleague from Rocky Mountain Laboratory, Hamilton, Montana, where I had worked before coming to England visited me in Compton and casually mentioned an exhibit he saw the previous day at the Wellcome Medical Museum in London. It had to do with a strange brain disease affecting the primitive people in Papua New Guinea. He thought I might like to see it owing to my interest in neuropathology. Five days later I saw the exhibit in London. Neuronal degeneration and intense astrocytosis likened kuru to scrapie. The likeness was made even more so by the single and multilocular vacuoles in the perikaryon of large neurons. From the start I was drawn to them for they were so much like those in scrapie (Hadlow 2008).

In his letter to Lancet, Hadlow recalled that "scrapie can be induced experimentally in the sheep and in the closely related goat but not in other species so far tested...," and he concluded that "It might be profitable, in view of veterinary experience with scrapie, to examine the possibility of the experimental induction of kuru in a laboratory primate, for one might surmise that the pathogenetic mechanisms involved in scrapie—however unusual they may be—are unlikely to be unique in the province of animal pathology" (Hadlow 1959). He had recognized the twin needs for extended observation periods and the use of a species closely related to humans (Bjorn Sigurdsson, working in Iceland, had in 1954 set out criteria for "slow infections" that included species specificity).

1.8 Experimental Transmission of Kuru

At the NIH, brain tissue had already been inoculated into numerous laboratory rodents, observed for periods of up to several months, with negative results, but now Gajdusek went about organizing a primate colony at the Patuxent Wildlife Center in Laurel, MD, under the able direction of Clarence J. (Joe) Gibbs, Jr., who had served with him at the Walter Reed Army Medical Center. By 1963 all was in readiness, but Gajdusek decided to wait until new autopsy specimens could be obtained under optimal conditions for survival of any infectious agent before initiating a chimpanzee inoculation program. The author well remembers being sent to New Guinea only a few months after joining the laboratory in July 1963 with instructions to get autopsies on any kuru patients who died during his month-long stay. Only one patient died, and in a hut under the flickering light of a hurricane lantern, with the deceased woman's husband hovering nearby, it was necessary to barter for each organ that was taken (coffee, canned goods, flashlights, knives, etc.), and also satisfy his very sharp eye for reassembling the body to its pre-autopsy condition. Gajdusek had set

Table 1.1 Animal species used in TSE experiments

Primates	
Apes	**Chimpanzee**, Gibbon
Prosimians	Bushbaby, Lemur, Slow Loris
Old World monkeys	**African green**, Baboon, Bonnet, **Cynomolgus**, Langur, Mangabey, Patas, **Rhesus**, Pig-tailed, Stump-tailed, Talapoin, Vervet
New World monkeys	**Capuchin**, Marmoset, Owl, **Spider**, **Squirrel**, Wooly
Non-primates	
Rodents	**Guinea pig**, **Hamster**, **Mouse**
Carnivores	Mink, Ferret
Ungulates	Horse
Felines	**Domesticated cat**
Avians	Chicken, Duck, Turkey
Suidae	Domesticated pig
Caprinae	Sheep, goat

The most frequently used species are shown in bold type

up an elaborate logistical system to preserve the viability of any infectious agent that might be present, including canisters of liquid nitrogen at the autopsy site, Land Rovers and Piper Cubs on call, and way-station reservoirs of additional liquid nitrogen at each airport between the middle of New Guinea and Washington DC. As it turned out, the brain from this case was among the first three to transmit kuru to chimpanzees (the two others having been collected by Gajdusek himself). Little did we then know that the transmissible agent could have withstood boiling, standard sterilizing chemicals, and burial in the ground for 3 years and still have remained infectious!

The publication in 1966 (Gajdusek et al. 1966) of the first experimental transmission of kuru from three of seven patients, whose brain tissue homogenates had been inoculated intracerebrally into chimpanzees 18–21 months earlier, was followed by an explosive decade of activity in Gajdusek's NIH laboratory, and as Pattison had said, the earlier studies of scrapie provided a valuable road map for this new exploration of kuru. The first order of business was to validate transmissibility of the disease and, if successful, begin to characterize the properties of what appeared to be a "slow" or "unconventional" virus. Chimpanzee to chimpanzee passage of kuru was accomplished in 1967 (Gajdusek et al. 1967), and a large series of experiments in a variety of primate species was carried out to determine the physical/chemical resistance, filtration size, host range, and pathogenesis of this new "virus" (Table 1.1).

1.9 The Expanding Horizon of Transmissible Spongiform Encephalopathy

The other pressing need, in view of Klatzo's observation of the neuropathological similarities between kuru and CJD, was to find a case of CJD to inoculate, which was not an easy task considering the rarity of the disease and its confusion with other dementia syndromes. However, a fully typical neuropathologically verified case was soon provided by Peter Daniel and Elizabeth Beck at the Maudsley Hospital in London, England, which transmitted disease to a chimpanzee 13 months after intracerebral inoculation, in 1968 (Gibbs et al. 1968). Ironically, that same year Kirschbaum published a comprehensive review of all known cases of CJD, favoring an etiology of vascular origin (Kirschbaum 1968).

Although interest shifted dramatically from scrapie to CJD in the years following its experimental transmission, two animal diseases, transmissible mink encephalopathy (TME) and chronic wasting disease (CWD) of deer and elk, were recognized as belonging to the TSE family by Dieter Burger and Hartsough (1965) and by Elizabeth Williams and Stuart Young (1980), respectively (Burger and Hartsough 1965; Williams and Young 1980; Williams et al. 1982). Both diseases may have originated from exposure to scrapie-infected sheep that had been present in the USA since the late 1940s, but that epidemiologically plausible hypothesis will never be proven. In fact, one of the more interesting features of TME is its association with the consumption of cattle rather than sheep carcasses on two US mink ranches in 1963 and 1985, leading to speculation about an early undetected occurrence of BSE in the USA (Marsh et al. 1991). No further incidents have occurred in the USA since the second outbreak (TME has also been diagnosed in Canada, Finland, and Russia as late as 1986). In contrast, CWD has assumed more and more importance as it spreads from its origin in Colorado mule deer to other species of deer in regions of the USA that now include the Midwest and both US coastlines. It poses an obvious risk to the comparatively small number of humans who hunt and/or consume venison and other vital organs, and a potentially greater future threat via cross-contamination of wild predators (the cat family is highly susceptible), and eventually to captive animals and livestock. The unique attribute of CWD that makes it important is its presence in free-ranging animals that cannot be subjected to the kinds of preventive or destructive measures applied to animals in captivity.

The most recent addition to the TSE family—BSE—appeared on the scene in 1986 in the UK as a new disease of cattle, and spread through most European and a few non-European countries within the next few years. Strictly speaking, it qualifies for discussion in this historical account, but as its occurrence extends well beyond the era when Gajdusek was actively engaged in the field, and it is sufficiently important to deserve a detailed discussion in a chapter of its own, we will instead return to the human diseases with which Gajdusek was most involved.

As news of the transmissibility of CJD spread through the neurological community, the NIH laboratory became a global clearinghouse of case referrals including hundreds of cases of possible or suspected CJD, all of which were inoculated into primates. The early use of chimpanzees rapidly gave way to a variety of monkeys (Table 1.1), and as features of the disease came to be defined in each species, the squirrel monkey became the preferred assay animal because of a susceptibility greater than 90% (nearly equal to the chimpanzee) combined with a comparatively short mean incubation period of 24 months (Table 1.2; Fig. 1.2). However, the observation that the same inoculum could sometimes produce disease after widely spaced incubation periods in replicate monkeys signaled caution in accepting incubation period length as a measure of infective dose in

Table 1.2 Characteristics of CJD transmissions in the most frequently used primate species

		New World monkeys			Old World monkeys	
	Chimpanzee	Squirrel	Spider	Capuchin	Rhesus	Cynomolgus
No. animals inoculated	29	211	31	45	28	23
Transmission rate (%)	97	93	97	80	68	22
Mean incubation period (months)	17	25	32	40	64	61
Mean duration of illness (months)	1.7	1.3	1.6	2.4	3.2	2.1

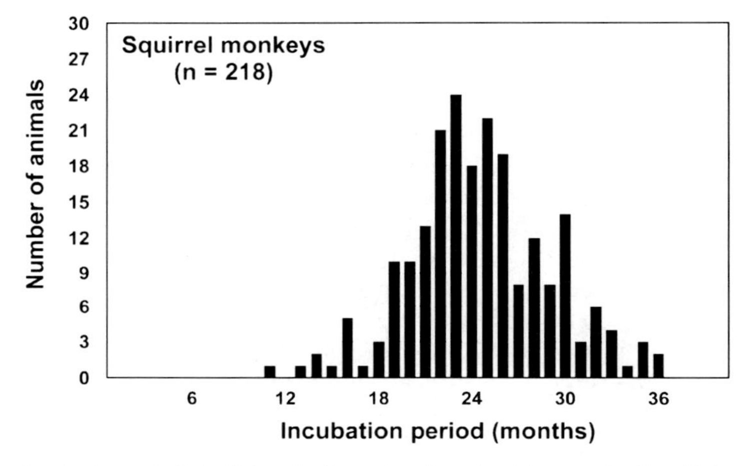

Fig. 1.2 Incubation periods in 218 squirrel monkeys inoculated intracerebrally with human CJD brain homogenates

any experiment using only two or three animals, a point that is sometimes forgotten in current research studies (Fig. 1.3).

The search for additional cases suspected of having CJD or diagnoses of other neurodegenerative diseases, and the laborious task of characterizing the transmissible agent, including its host range and pathogenesis, consumed a much larger number of animals and a much longer period of time, lasting well into the 1980s. Consider the simple matter of estimating the mean lethal dose (LD_{50}) of infectivity in a given tissue. Working with mice or other rodents, the usual technique would be to inoculate groups of 5–6 animals with a spread of dilutions large enough to bracket an unknown end point, typically totaling 40–50 animals, which would be unthinkable when using primates. Even a "stripped down" titration using pairs of animals at successive 100-fold dilutions would require at least eight animals. Add to this the need for observation periods of at least 5 years, and the difficulty of obtaining even the most basic information becomes formidable.

Over the years, the NIH laboratory bought, bred, and housed thousands of monkeys and hundreds of apes used in primary isolation and passage attempts, species

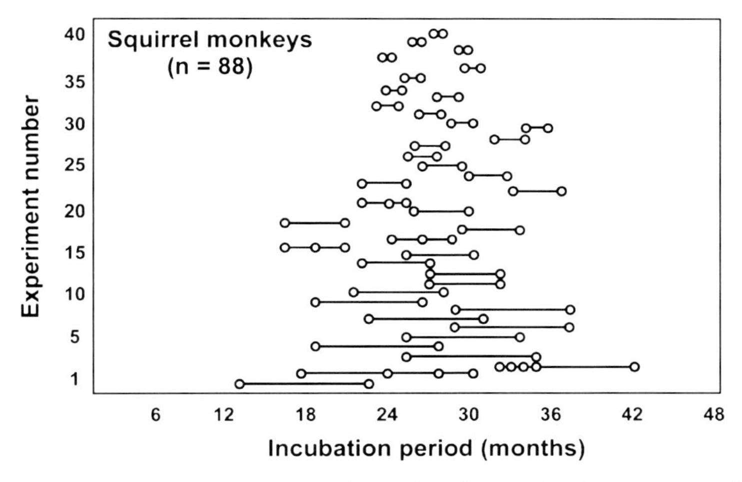

Fig. 1.3 Incubation periods in 40 experiments in which replicate (or in a few cases, more than two) squirrel monkeys were inoculated intracerebrally with the same human CJD brain homogenate

susceptibility experiments, and pathogenesis bioassays, located at various sites in California, Hawaii, Louisiana, New Mexico, New York, Texas, and Virginia, as well as overseas in Paris and Marseille. Eventually, all primate research was consolidated to Gulf South in the middle of Louisiana Cajun country, and Fort Detrick, about 30 miles north of the NIH in Frederick, MD. Transmission experiments on non-primate species were mostly conducted at a spacious farm-like facility in Otisville in southern New York State. It is to the everlasting credit of Dr. Joseph Smadel, NIH Associate Director who had earlier been Gajdusek's chief at the Walter Reed Army Institute of Research, and Dr. Richard Masland, Director of the NIH Institute of Neurological Diseases and Blindness, to have at its inception approved and assisted in this gigantic undertaking.

1.10 Clinical and Epidemiological Precisions

During the 1970s, the unassailable criterion of transmissibility led to an appreciation of the range of clinical syndromes associated with CJD, and made it possible, finally, to define the essential features with a precision that had hitherto been impossible. This evolving understanding was recorded in several papers based on larger and larger numbers of cases culminating in a synthesis based on 300 transmitted cases of transmissible spongiform encephalopathy published in 1994 (Brown et al. 1994a). During this period, the two remaining members of the quartet of human spongiform encephalopathies were also found to be transmissible: GSS in 1981 (Masters et al. 1981) and fatal familial insomnia (FFI) in 1995 (Tateishi et al. 1995). However, the need for diagnostic verification of cases by transmission studies was, in most

Table 1.3 Disease categories of referrals to the NIH laboratory for transmission studies

Disease category	Number of cases	Number of animals	Observation period (years)	Number of transmissions
TSE	440	1,914	1–21	291
Alzheimer's disease	105	240	1–24	0
Other neurodegenerations	115	224	1–30	0
Other neurological diseases	453	1040	1–26	0
Non-neurological Diseases	53	76	1–30	0
Total	1,113	3,418	–	–

instances, abolished by the twin discoveries of a high level of protein kinase inhibitor (14–3–3) in the spinal fluid with a diagnostic specificity >90%, and of a specific pathognomonic amyloid protein (PrP^{TSE}) in brain tissue that could be detected by ELISA or Western blot.

In stark contrast to the multiple transmissions of each of the spongiform encephalopathies, not a single transmission followed similar inoculations of any non-spongiform neurological disease (including Alzheimer disease, Pick's disease, Parkinson's disease, Huntington's disease, amyotrophic lateral sclerosis, and multiple sclerosis) or a wide variety of non-neurological diseases of unknown etiology like sarcoidosis, lupus erythematosus, Crohn's disease, and rheumatoid arthritis (Table 1.3). It is sometimes forgotten in the present-day impulse to demonstrate transmissibility of Alzheimer's disease using various "seeding" techniques and genetically altered susceptible mice that over 100 cases of neuropathologically verified Alzheimer's disease have been inoculated into primates with uniformly negative results (Brown et al. 1994a). Thus, whatever the similarities between the two diseases (and there are many), inoculation of host species closely related to humans under conditions typically used to demonstrate infectivity simply does not transmit disease, and any claim that Alzheimer's disease is infectious must contend with these consistently negative results. Stated another way, facilitating or accelerating disease in animal models of Alzheimer's disease should not be confused with causing disease in humans.

Given the experimental transmissibility of sporadic CJD, and the increasing repertory of cases referred to the NIH, it was not long before the question of human contagion arose, which led to a burgeoning series of epidemiological studies beginning in 1971 with Giovanni Alemà's search for cases of CJD in Italy (Alemà 1971). This was really only a "sketch" that served to inaugurate the much larger canvases to come, but Alemà deserves credit for first recognizing the need to look at epidemiology, a fact that is almost never cited. Brian Matthews and Robert Will substantially extended the epidemiological exploration of CJD in a systematic 5-year retrospective study in England and Wales (Will and Matthews 1986), and Françoise Cathala and the author followed with an even more intensive 10-year investigation of CJD in France (Brown et al. 1987). With the appearance of variant CJD (vCJD) in 1996, the entire European community, together with individual countries elsewhere in the world (e.g., Argentina, Australia, Canada, and Japan), established a coordinated program of surveillance that continues to this day. The most important results with respect to sporadic CJD, as everyone now knows, are that it occurs worldwide in a random distribution at an average annual incidence of about 1 case per million population, with no natural source or person-to-person spread of disease yet identified.

However, beginning in the 1970s, some cases of apparently sporadic disease began to be recognized as having an iatrogenic origin, at first in operative procedures involving contamination of a corneal graft and a neurosurgical stereotactic electrode, and later on a much larger scale from cadaveric sources of human growth hormone and dura mater grafts. A more recent cause of iatrogenic disease has come from secondary infections in recipients of packed red cell donations from individuals incubating vCJD in a pre-symptomatic stage of disease. Three primate transmission experiments contributed to knowledge about iatrogenic disease by demonstrating infectivity (1) on the "sterilized" stereotactic EEG needle; (2) in one lot of human growth hormone; and (3) in leukocytes during the preclinical phase of disease of an experimentally inoculated chimpanzee.

1.11 Therapeutic Essays

Studies having potential relevance to therapy may be said to have begun with early experiments on the resistance of the scrapie agent to physical and chemical treatments. Unfortunately, the pathogen was far more resistant than its host to heat, radiation, and chemicals, and the most effective treatments (now used for ex vivo disinfection) such as autoclaving, or exposure to strong solutions of NaOH (lye) or NaOCl (bleach), although obviously irrelevant for therapeutic considerations, provided a clue to the challenges that lay ahead. Furthermore, in the era of pre-molecular biology, when the etiology of TSE was thought to be an unconventional virus, all such trials were mere shots in the dark, in the hopes that something that worked on viruses might work equally well on TSE infections. All failed to qualify as practical therapeutic agents, although two categories—polyanionic compounds and polyene antibiotics—were found to prolong the incubation period of scrapie-infected laboratory rodents when given at or near the time of infection. Because this is almost never known in human infections, even the occasional promising results in experimental animals could not be realized in humans (Brown 2010).

With the development of ex vivo infectivity assays, a few such chemical agents were found to reverse or even abolish cell culture infections, and it became tempting to move from these successes directly to human trials without the intermediate step of animal experiments. The recent experience with quinacrine dramatically illustrates the error of this haste, as subsequent experiments in animals confirmed its failure to affect the disease in humans.

Some elegant strategies involving genetic manipulations or prophylactic vaccines in mice are unfortunately either impractical or of limited use in humans. Another conceptual approach of targeting chaperones rather than the prion protein itself is in its infancy. Whatever the prospective treatment, it is first going to have to pass the stringent test of efficacy in already symptomatic experimental animals before any human therapeutic trial is undertaken, unless a reliable assay for preclinical infection in humans becomes available.

1.12 The End of an Era

If the 1970s were about biology, and the 1980s saw a transition to molecular biology, the 1990s can be considered the decade of molecular genetics. Since the time of Jakob and Gerstmann, it had been known that CJD could in rare cases also assume a familial form, and that the even rarer occurrence of GSS was always restricted to families. With the discovery in the 1980s of a host gene that encodes the normal "prion" protein, the time had come to search for mutations responsible for familial forms of human spongiform encephalopathy. Identification of the first such mutation was reported by Karen Hsiao et al. in 1989—at codon 102 in a family with GSS (Hsiao et al. 1989).

By the turn of the century, over 30 different mutations had been identified (there are now more than 40), and here again, Gajdusek played a major role because of his extensive global contacts and the efforts of a small research team led by Lev Goldfarb, which first identified the polymorphism at codon 129 (Goldfarb et al. 1989), then what were to become the two most common PRNP mutations worldwide at codons 200 and 178 (Goldfarb et al. 1991a, 1992), as well as several other more restricted mutations among the many being identified in other laboratories. In collaboration with Robert Petersen in Pierluigi Gambetti's laboratory, they also discovered the determining influence of codon 129 on whether the codon 178 mutation would result in the clinical syndromes of CJD or FFI (Goldfarb et al. 1991b) and, of historical interest, identified the codon 178 mutation in the original CJD family reported by Jakob (Brown et al. 1994b), and the codon 102 mutation in the original GSS family reported by Gerstmann (Hainfellner et al. 1995).

As the decade progressed, and the NIH primate program wound down, molecular research—both biological and genetic—overtook the dwindling number of "classical" transmission experiments in both quantity and importance, and genetically embellished mice came to be the preferred method for exploring a number of remaining issues related to host susceptibility and pathogenesis. There is currently an understandable tendency to equate the detection of PrP^{TSE} by protein amplification methods, or transmissibility in humanized transgenic mice, with a risk of "real-life" transmission. Until this assumption is confirmed by transmission to normal animals under natural experimental conditions, this risk remains speculative, and the most appropriate animals for such confirmatory experiments are primates.

Gajdusek retired from the NIH in 1996, and most of the laboratory staff either found other employment or retired. Gibbs stayed on until his death in 2001, and the author remained until 2004, bringing to a close the largest, longest, costliest, and possibly most fruitful experimental animal study ever undertaken in the field of medical science. Gajdusek died sometime after 4 p.m. on December 11, 2008, at the age of 86. The last page of his journal contains the following two entries:

1.13 December 11, 2008

10 a.m. Psychology and Law Library, University of Tromsø
I am at my library office trying to sort out my life. I'm much better placed than my crowded hotel room desk. I have most of my mailing done. Now I can concentrate on getting a recorder to play my CDs. What luxury I live in!

To bring 2008 to a close is my current goal. I dare not contemplate much further. I would like to finish some further journals, but that is appearing unlikely. To have lived into my 86th year is much more than I ever anticipated or planned on. Now, I wonder what I should do. My life is essentially finished.

I've mailed a check to Yavine and hope all is well with him. The only outstanding payment is my lost check to Magame. I will attend to that shortly. Now, to get off these mundane matters, and back to serious thoughts. To start listening to the Gregorian Chants and early Baroque music I have on hand is my first priority. That should bring me back to this world.

4 p.m. Clarion Hotel, Bryggen Tromsø
Returned from the University where I copied pp. 120–164 of ledger XVIII for the last 11 of the individual archivists, which is a prodigious sharing of my current journal with 38 individuals.

These last "mundane matters" nevertheless bear witness to an abiding generosity towards the Oceanic family he had nurtured, an undiminished range of intellectual and esthetic sensibility, and a clear presentiment of mortality, aware of what he had accomplished and what he was leaving behind. His journal, begun during childhood, grew to more than 70 volumes containing over 10 million words and was still growing at the time of his death, bringing to a close the daily record of one of the most distinguished scientific careers of the twentieth century.

Acknowledgments The author apologizes to the many scientists whose names and contributions were omitted from this review due to the constraints of covering a very large subject in a very small space. He expresses his gratitude to Drs. O. Andeoletti, D.M. Asher, R. Bradley, K. Haynes, and R.H. Kimberlin for help in assuring factual accuracy, and especially to Drs. Bradley and Kimberlin for succumbing to the author's plea for critical readings of the manuscript.

References

Alemà G (1971) Aspectos epidemiológicos de la enfermedad de Jakob-Creutzfeldt (consideraciones sobre los casos italianos). In: De la Fuente R, Weisman MN (eds) Proceedings of the 5th world congress of psychiatry. Excerpta Medica, International Congress Series, Amsterdam, pp 1221–1227

Alper T, Cramp WA, Haig DA, Clarke MC (1967) Does the agent of scrapie replicate without nucleic acid? Nature 214:764–766

Besnoit C (1899) La tremblante ou névrite périphérique enzootique du mouton VI. Etiologie. Rev Vét (Toulouse) 21:333–343

Besnoit C, Morel C (1898) Note sur les lesions nerveuses de la tremblante du mouton. Rev Vét (Toulouse) 23:397–400

Brown P (2010) An historical perspective on efforts to treat transmissible spongiform encephalopathy. CNS Neurol Disord Drug Targets 8:316–322

Brown P, Cathala F, Raubertas RF et al (1987) The epidemiology of Creutzfeldt–Jakob disease: conclusion of a 15-year investigation in France and review of the world literature. Neurology 37:895–904

Brown P, Gibbs CJ Jr, Rogers-Johnson P et al (1994a) Human spongiform encephalopathy: the National Institutes of Health series of 300 cases of experimentally transmitted disease. Ann Neurol 35:513–529

Brown P, Cervenáková L, Boellaard JW et al (1994b) Identification of a PRNP mutation in Jakob's original Creutzfeldt-Jakob disease family. Lancet 344:130–131

Bruce ME, Will RG, Ironside JW et al (1997) Transmissions to mice indicate that 'new variant' CJD is caused by the BSE agent. Nature 389:498–501

Burger D, Hartsough GR (1965) Encephalopathy of mink II. Experimental and natural transmission. J Infect Dis 115:393–399

Chandler RL (1961) Encephalopathy in mice produced by inoculation with scrapie brain material. Lancet 1(7191):1378–1379

Creutzfeldt HG (1920) Über eine eigenartige herdförmige Erkrankung des Zentralnervensystems. Z Ges Neurol Psychiatr 57:1–18

Cuillé J, Chelle P-L (1936) La maladie dite "tremblante" du mouton; est-elle inoculable? CR Acad Sci 203:1552–1554

Cuillé J, Chelle P-L (1938) Investigations of scrapie in sheep. Vet Med 34:417–418

Gajdusek DC, Zigas V (1957) Degenerative disease of the central nervous system in New Guinea. N Engl J Med 257:974–978

Gajdusek DC, Gibbs CJ Jr, Alpers M (1966) Experimental transmission of a kuru-like syndrome to chimpanzees. Nature 209:794–796

Gajdusek DC, Gibbs CJ Jr, Alpers M (1967) Transmission and passage of experimental "kuru" to chimpanzees. Science 155:212–214

Gerstmann J, Straüssler E, Scheinker I (1936) Über eine eigenartige hereditär-familiäre erkrankung des zentralnervensystems Zugliech ein Beitrag zur frage des vorzeitigen lokalen alterns. Z Neurol 154:636–762

Gibbs CJ Jr, Gajdusek DC, Asher DM et al (1968) Creutzfeldt–Jakob disease (spongiform encephalopathy): transmission to the chimpanzee. Science 161:388–389

Goldfarb LG, Brown P, Goldgaber D et al (1989) Patients with Creutzfeldt–Jakob disease and kuru lack the mutation in the PRIP gene found in Gerstmann-Sträussler syndrome, but they show a different double-allele mutation in the same gene. Am J Hum Genet 45(Suppl):A189

Goldfarb LG, Brown P, Mitrova E et al (1991a) Creutzfeldt–Jakob disease associated with the PRNP codon 200^{Lys} mutation: an analysis of 45 families. Eur J Epidemiol 7:477–486, Lancet 337:425

Goldfarb LG, Petersen RB, Tabaton M et al (1991b) Fatal familial insomnia and familial Creutzfeldt–Jakob disease: disease phenotype determined by a DNA polymorphism. Science 258:806–808

Goldfarb LG, Brown P, Haltia M et al (1992) Creutzfeldt–Jakob disease co-segregates with the codon 178[Asn] PRNP mutation in families of European origin. Ann Neurol 31:274–281
Gordon WS (1946) Advances in veterinary research. Vet Rec 58:516–520
Griffith JS (1967) Self-replication and scrapie. Nature 215:1043–1044
Hadlow WJ (1959) Scrapie and kuru. Lancet ii:289–290
Hadlow WJ (2008) Kuru likened to scrapie: the story remembered. Philos Trans R Soc B 363:3644
Hainfellner JA, Brantner-Inthaler S, Cervenáková L et al (1995) The original Gerstmann-Stäussler-Scheinker family of Austria: divergent clinicopathological phenotypes but constant PrP genotype. Brain Pathol 5:201–211
Hsiao K, Baker HF, Crow TJ et al (1989) Linkage of a prion protein missense variant to Gerstmann-Sträussler syndrome. Nature 338:342–344
Jakob A (1921) Über eine eigenartige erkrankung des zentralnervensystems mit bemarkenswerten anatomischen Befunde (Spastische Pseudosklerose-Encephalomyelopathie mit disseminierten Degenerationsherden). Z Ges Neurol Psychiatr 64:147–228
Kimberlin RH, Walker C (1977) Characteristics of a short incubation model of scrapie in the golden hamster. J Gen Virol 34:295–304
Kirschbaum WR (1968) Jakob–Creutzfeldt disease. Elsevier, New York
Klatzo I, Gajdusek DC, Zigas V (1959) Pathology of kuru. Lab Invest 8:799–847
Marsh RF, Bessen RA, Lehman S, Hartsough GR (1991) Epidemiological and experimental studies on a new incident of transmissible mink encephalopathy. J Gen Virol 72:589–594
Masters CL, Gajdusek DC (1982) The spectrum of Creutzfeldt–Jakob disease and the virus-induced subacute spongiform encephalopathies. In: Recent advances in neuropathology. Julius Springer, Edinburgh
Masters CL, Gajdusek DC, Gibbs CJ Jr (1981) Creutzfeldt–Jakob disease virus isolations from the Gerstmann-Sträussler syndrome. Brain 104:559–588
Merz PA, Somerville WHM (1983) Scrapie-associated fibrils in Creutzfeldt–Jakob disease. Nature 306:474–476
Merz PA, Somerville RA, Wisniewski HM, Iqbal K (1981) Abnormal fibrils from scrapie-infected brain. Acta Neuropathol 54:63–74
Millson GC, Hunter GD, Kimberlin RH (1976) The physicochemical nature of the scrapie agent. In: Kimberlin RH (ed) Slow virus diseases of animals and man. North Holland, Amsterdam
Pattison IH (1972) Scrapie – a personal view. J Clin Pathol 25(Suppl):110–114
Prusiner SB (1982) Novel proteinaceous infectious particles cause scrapie. Science 216:136–144
Tateishi J, Brown P, Kitamoto T et al (1995) First experimental transmission of fatal familial insomnia. Nature 376:434–435
Will RG, Matthews WB (1986) A retrospective study of Creutzfeldt–Jakob disease in England and Wales 1970–1979. II. Epidemiology. J Neurosurg Pschiatry 49:749–755
Williams ES, Young S (1980) Chronic wasting disease of captive mule deer: a spongiform encephalopathy. J Wildl Dis 16:89–98
Williams ES, Young S, Marsh RF (1982) Preliminary evidence of transmissibility of chronic wasting disease of mule deer (Abstract No. 22). In: Proceedings of the wildlife disease association annual conference. Madison, Wisconsin, 19 Aug 1982

Chapter 2
The Rich Chemistry of the Copper and Zinc Sites in Cellular Prion Protein

Glenn L. Millhauser

Abstract Research over the last decade clearly demonstrates that the function of the cellular form of the prion protein, PrP^C, is related to its ability to bind copper and zinc. Zinc (Zn^{2+}) coordination is homogeneous and localized to the octarepeat domain, with participation of the histidine side chains. In contrast, copper uptake is complex and dependent on the oxidation state of the metal ion (Cu^+ or Cu^{2+}), and its concentration. This chapter will cover a brief history of PrP^C–metal interactions leading to the current structural models, a recently recognized relationship between Cu^{2+} coordination and inherited prion disease arising from octarepeat inserts, and new findings that suggest an electrochemical basis for PrP^C neuroprotection and transmembrane signaling.

Keywords Copper • Zinc • Octarepeat domain • Transmembrane signaling • Octarepeat inserts • Electrochemistry • Familial prion disease

2.1 Introduction

Research over the last decade continues to find remarkable functional roles for the normal cellular form of the prion protein (PrP^C). PrP^C supports myelin development (Bremer et al. 2010), influences sleep–wake cycles (Tobler et al. 1996), is upregulated at sites of ischemic injury (McLennan et al. 2004), promotes neuron development (Kanaani et al. 2005) and protects nerve cells against chemical and oxidative assaults (Rachidi et al. 2003; Klamt et al. 2001). Although one cannot yet assign a sole function to PrP^C as, say, a signaling molecule, enzyme or transporter, it is clear that the

G.L. Millhauser, Ph.D. (✉)
Department of Chemistry and Biochemistry, University of California Santa Cruz, 1156 High Street, Santa Cruz, CA 95064, USA
e-mail: glennm@ucsc.edu

W.-Q. Zou and P. Gambetti (eds.), *Prions and Diseases: Volume 1, Physiology and Pathophysiology*, DOI 10.1007/978-1-4614-5305-5_2,

protein is required for normal neurological function. Most functional investigations link PrP^C to metal ion binding, specifically to copper and zinc. This link was recently emphasized in an elegant X-ray fluorescence study that examined the spatial location and relative levels of iron, copper, and zinc in mouse brain (Pushie et al. 2011). Comparison of wild-type, PrP knockouts (KO) and 20X overexpressers revealed remarkable differences in specific brain regions, with each metal ion exhibiting a unique PrP-dependent profile. For example, PrP appears to drive copper levels near the ventricles and thalamus, whereas zinc is upregulated in cortical regions. And while there is scant evidence suggesting that PrP directly binds iron, levels are nevertheless influenced by PrP expression, perhaps suggesting a relationship between distinct metal transporters, as established in yeast (Bleackley and Macgillivray 2011).

This chapter will begin with a brief historical review of the PrP metal ion literature, with emphasis on works that frame current thinking. Next, I will describe the biophysical features of the copper and zinc sites in PrP^C. Unlike most other metal binding proteins that present a unique high affinity site, PrP responds dynamically with a rich variation of coordination modes that depend on metal concentration and the presence of competing species. Recognition of these distinct coordination modes provides new insight into inherited disease resulting from octarepeat inserts. Finally, I will describe new electrochemical work that not only provides a detailed characterization of PrP–copper redox properties, but also suggests a mechanism for PrP-mediated signaling.

2.2 Brief History

PrP^C is able to bind both copper and zinc, but most studies emphasize the specific interaction with Cu^{2+}. (Copper possesses two common, biologically relevant oxidation states: Cu^+ and Cu^{2+}.) Hornshaw et al. recognized that the histidine-rich octarepeat domain, containing four tandem PHGGGWGQ segments, would likely bind Cu^{2+}, and demonstrated this directly with mass spectrometry (Hornshaw et al. 1995a). Moreover, they showed a persistent 1:1 complex, although it was also noted that the OR region could take up additional equivalents. Next, using circular dichroism (CD), which detects conformational changes, and fluorescence quenching, they estimated a Cu^{2+} dissociation constant in the low micromolar range (Hornshaw et al. 1995b).

In 1997, Brown et al. published a remarkable study that clearly identified a physiological connection between PrP and copper (Brown et al. 1997). First, using a peptide corresponding to the PrP N-terminal domain, PrP(23–98), they showed that the protein takes up multiple Cu^{2+} equivalents with positive cooperativity, described by an unusually high Hill coefficient. Estimated affinity was higher than initially found by CD, as reflected in a low, submicromolar dissociation constant. Brown and colleagues further compared brain copper levels between wild-type and KO mice, and reported a severe reduction in brain copper in the transgenics. Many aspects of this work have been revisited in the last 15 years, but there is little doubt that this initial publication firmly established PrP^C as a copper metalloprotein.

The lowered copper content in the mouse KO suggested that perhaps PrP^C functions as a transporter. PrP^C is attached to membrane surfaces through a GPI anchor and is cycled from the extracellular space to early endosomes through endocytosis, with approximately 90% of the protein returned to the surface by exocytosis. As monitored in N2a mouse neuroblastoma cells, Pauly and Harris showed that addition of 200 μM copper stimulated rapid PrP^C internalization, while removal of the metal ion allowed the protein to redistribute back to the membrane surface (Pauly and Harris 1998). Elimination of the octarepeats, or the His residues within the repeats, fully disrupts these copper-dependent processes (Perera and Hooper 2001). Similarly, certain mutations in the octarepeat domain that give rise to familial prion disease also interfere with copper-stimulated endocytosis (Perera and Hooper 2001). Collectively, these findings suggest that PrP^C may play a key role in copper trafficking. However, early examinations of tissue copper, and copper protein activity, in brain fractions derived from wild-type and transgenic mice possessing different levels of PrP^C, failed to find a correlation between PrP^C expression and copper content (Waggoner et al. 2000). Consequently, this promising line of research did not progress. However, the X-ray fluorescence imaging work described in the Introduction, certainly motivates a renewed look at the role of PrP^C in neuronal copper distribution.

In parallel to cellular assays were several notable structural and biophysical investigations (Stöckel et al. 1998; Garnett and Viles 2003; Viles et al. 1999; Valensin et al. 2004; Aronoff-Spencer et al. 2000; Burns et al. 2002, 2003; Chattopadhyay et al. 2005; Van Doorslaer et al. 2001). Early work focused primarily on the octarepeat domain, although newer research finds copper sites outside this region. Viles et al. performed a wide array of spectroscopic experiments including CD, nuclear magnetic resonance (NMR), and electron paramagnetic resonance (EPR) (Viles et al. 1999). This work demonstrated a 1:1 stoichiometry between each histidine (His) containing repeat segment and Cu^{2+}, and suggested a micromolar dissociation constant. Moreover, they identified a strong pH dependence, with tight copper binding only at pH 6.0 and above. These findings have endured many follow-up studies. To account for cooperative uptake, they proposed a ring-like structure of alternating His imidazole side chains and Cu^{2+} ions. While there is precedence for this type of structure in the inorganic chemistry literature, it is now considered unlikely to be a significant biological conformation.

Most copper binding proteins exhibit a very high affinity, reflected by a low dissociation constant (K_d). For example, the K_d for copper at the active site of superoxide dismutase is approximately 10^{-14} M. Early work with PrP N-terminal peptides pointed to a much weaker affinity, suggesting that perhaps PrP might not take up copper in vivo. This was addressed with detailed MS and fluorescence assays to carefully assess copper binding thermodynamics in full-length PrP (Kramer et al. 2001). Analysis of the observed fluorescence quenching revealed both affinity and detailed stoichiometry, with five Cu^{2+} per protein. Copper uptake showed positive cooperativity with the last equivalent exhibiting a K_d of ~2 μM, well below the level of Cu^{2+} in blood estimated at 18 μM. It is not clear, though, how relevant the comparison to blood copper levels is, given that high levels of PrP are localized to extracellular presynaptic surfaces in the CNS (Herms et al. 1999). As will be discussed, more recent analyses find specific binding modes that display very high affinity, below 1.0 nM, and thus eliminate doubt that PrP takes up Cu^{2+} in vivo.

2.3 Features of Cu^{2+} and Zn^{2+} Coordination in PrP

Copper binds within PrP's N-terminal region, with the relevant segment from the human sequence shown below:

PrP(51–111) PQGGGGWGQ<u>P**H**GGGWGQP**H**GGGWGQP**H**GGGWGQP**H**G GGWGQ</u>GGGT**H**SQWNKPSKPKTNMK**H**

There are five tandem eight-residue repeats, each with the canonical sequence PXGGGWGQ, but in the first repeat, a Gln fills the X position. Since histidine, with its imidazole side chain, is required for copper uptake, the first repeat does not participate. Thus, from a sequence or genetics perspective, there are five N-terminal octarepeats, but from a metal ion coordination perspective, there are four (underlined in the sequence). Beyond the octarepeat domain, copper also interacts with high affinity at the His residues at positions 96 and 111 (Walter et al. 2009; Jones et al. 2005). The current consensus is that all copper coordination is within the segment PrP(61–111) (human) bounded by the histidines (His, bold H) in the sequence shown above.

A number of early investigations used peptide design, NMR, mass spectrometry, circular dichroism, Raman spectroscopy, molecular modeling, and related biophysical approaches to develop insight into the structure of the Cu^{2+}–octarepeat complex. Ultimately, though, EPR provided the essential insights leading to the current models. EPR is sensitive to the chemical environment at paramagnetic Cu^{2+} centers and, through hyperfine couplings to copper's unpaired electron, can directly reveal nearby nuclei and atomic features of the coordination environment. Details of the relevant EPR techniques have been reviewed elsewhere (Millhauser 2004, 2007); a summary of the coordination features is given in Fig. 2.1. The copper coordination environment depends critically on the ratio of copper to protein. At low copper concentration, the four octarepeat His imidazole side chains bind simultaneously to a single Cu^{2+}, as shown in the figure and inset (Chattopadhyay et al. 2005). This is often referred to as the low occupancy binding mode or "component 3," based on component analysis of the EPR spectra. The affinity for this mode is very high, with a dissociation constant of approximately 0.10 nM (Walter et al. 2006).

At intermediate Cu^{2+} concentration, the octarepeats take up two copper equivalents, with each coordinated by two His side chains (not shown) (Chattopadhyay et al. 2005). At high copper concentrations, the octarepeat domain saturates at 4 equiv., with each His binding to a single Cu^{2+}, as shown in Fig. 2.1 (Aronoff-Spencer et al. 2000; Burns et al. 2002, 2003; Chattopadhyay et al. 2005). This high occupancy binding mode is referred to as "component 1." The copper affinity for this state is lower than that of component 3, with a dissociation constant of approximately 10 μM (Walter et al. 2006). The specific coordination features of this high occupancy

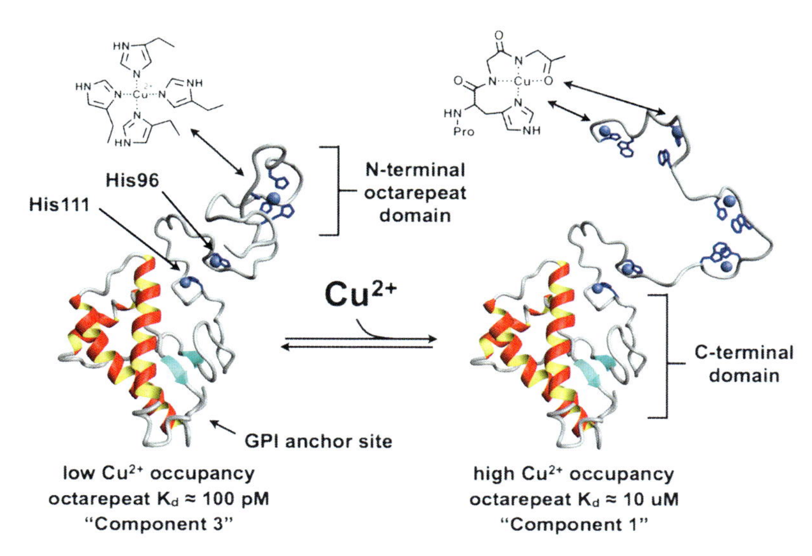

Fig. 2.1 Structural features of PrP^C at low and high Cu^{2+} concentrations. The C-terminal domain is helical, whereas the N-terminal domain is flexible and able to restructure to accommodate different copper coordination modes. At low $[Cu^{2+}]$, the metal ion coordinates to sites localized to His96 and His111. In addition, a single equivalent of Cu^{2+} binds within the octarepeat domain, coordinated by the four His imidazole side chains ("component 3," details shown in the inset). The affinity in the octarepeat domain is high, as characterized by a low K_d of approximately 100 pM. At high $[Cu^{2+}]$, the octarepeat domain restructures to take up four copper equivalents, each coordinated to single His side chain and backbone nitrogens ("component 1," *inset*). The affinity for this coordination mode is lower than that of component 3

site, shown in the inset, were determined by isotopic labeling, in combination with a range of EPR techniques (Aronoff-Spencer et al. 2000), and confirmed by X-ray crystallography of the Cu^{2+}–HGGGW complex (Burns et al. 2002).

The specific features of the component 1 site are unusual compared to previously characterized protein copper sites. In most copper metalloproteins, the metal ion is coordinated to His or Cys side chains. For example, copper superoxide dismutase contains the metal ion with four tetrahedrally placed His imidazoles. As seen in the inset, the Cu^{2+} ion coordinates to the His side chain, the deprotonated amide nitrogens of the two Gly residues that immediately follow the His, and a Gly carbonyl. In addition, there is an axially coordinated water molecule that hydrogen bonds to the Trp indole hydrogen (not shown). A coordination sphere with deprotonated amides has been seen previously with the N-terminal copper binding segment of albumin (Harford and Sarkar 1997), and also in peptides, but not in the interior polypeptide segments of a protein. The involvement of amide nitrogens confers significant pH sensitivity since an increase in the H^+ concentration (lower pH) protonates at the nitrogen and competes with copper complexation. Consequently, high occupancy copper binding is unstable below pH ~ 6.0. It has been proposed that this might provide a chemical mechanism for release of Cu^{2+} in the endosomal compartments (Burns et al. 2002).

In addition to Cu^{2+} uptake in the octarepeats, there are two additional binding sites localized to His96 and His111 (human PrP numbering), and these also exhibit sub-nanomolar affinity. These two sites are often referred to as the "5th sites," since early studies suggested only the involvement of His96, beyond that of the four sites in the octarepeat domain (Burns et al. 2003). We prefer to label these as "non-octarepeat" coordination sites, thus underscoring their distinct location and chemical properties (Walter et al. 2009). At both of these non-octarepeat sites, copper coordinates to the imidazole side chain, the His backbone nitrogen, and two additional backbone nitrogens from the residues on the N-terminal side of the His (Burns et al. 2003). Affinity at these sites is high, with a K_d that is similar to that found for the multi-His component 3 mode in the octarepeat domain. Titration studies show that these non-octarepeat sites take up copper simultaneously with component 3 (Walter et al. 2009). Once PrP^C is saturated with Cu^{2+}, the octarepeat domain restructures to component 1 coordination thus enabling additional binding equivalents, as shown in Fig. 2.1.

Like copper, zinc also binds to PrP^C and stimulates endocytosis (Pauly and Harris 1998). Because this metal ion is found only as diamagnetic Zn^{2+}, EPR is of limited use in directly evaluating its coordination features. To address this, we applied two different approaches (Walter et al. 2007). First, using an octarepeat peptide, as well as full-length PrP^C, we competed Zn^{2+} against Cu^{2+} and monitored by copper EPR. Interestingly, we found that regardless of concentration, Zn^{2+} was not able to displace Cu^{2+}, which shows that copper has a much higher affinity than zinc. However, Zn^{2+} was able to influence the Cu^{2+} coordination mode, shifting the distribution to favor component 1 binding. Next, we tested Zn^{2+} coordination to a range of octarepeat-derived peptides and monitored binding with the reagent diethylpyrocarbonate (DEPC). DEPC chemically modifies free imidazole groups, but only if they are not involved in metal ion coordination. Analysis by mass spectrometry showed protection against DEPC modification only with the full octarepeat domain. Collectively, these experiments demonstrate that Zn^{2+} coordinates to the four octarepeat His imidazoles, equivalent to that observed for Cu^{2+} in its low occupancy mode. With a K_d of approximately 200 μM, the affinity is substantially lower than any of the coordination modes found for Cu^{2+}. However, because Zn^{2+} competes with Cu^{2+}, it is able to influence copper coordination in a concentration-dependent fashion. These results, summarized in the scheme in Fig. 2.2, show that when copper levels are low, PrP can simultaneously bind both copper and zinc. At higher copper levels, the protein accommodates the zinc by shifting to the high occupancy binding mode that minimizes the ratio of histidines to copper. However, when no rearrangement can accommodate both zinc and the available copper, it is the zinc that is displaced, not the copper. These results are consistent with previous screens that identified copper and zinc as the sole biologically relevant metal ions that coordinate to PrP^C, and perhaps suggest mechanisms by which both metal ions may stimulate endocytosis. What is also clear is that copper exhibits a substantially higher coordination affinity, thus arguing against zinc as the dominant species in PrP metallobiochemistry.

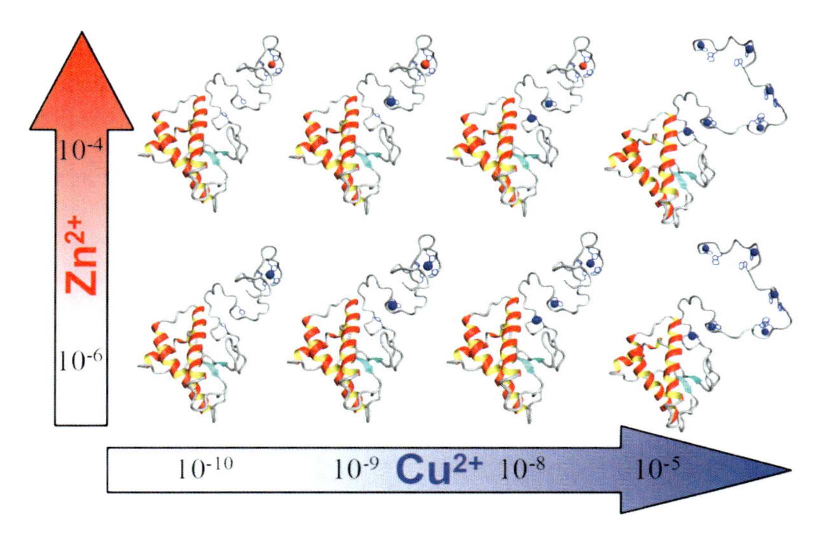

Fig. 2.2 Models representing metal binding in the N-terminal domain of PrP. *Top row* (high zinc); zinc (*red*) is bound by the octarepeat region (*left*), while non-octarepeat sites (H96 and H111) are available for copper binding (*blue*, *middle*). Copper at high concentration will displace zinc from octarepeats to form up to 4 equiv. of component 1 (*right*). *Bottom row* (low zinc); copper (*blue*) is bound by the octarepeats in component 3 when copper is low (*left*), with increasing copper loads in the non-octarepeat sites (*middle*). High copper (*right column*) results in component 1 copper binding by the octarepeats. Approximate molar metal concentrations are shown in the *arrows*

2.4 A Role for Altered Copper Coordination in Octarepeat Expansion Disease

Approximately 10–15% of human TSE cases are inherited and arise from mutations in the *PRNP* gene (Prusiner 2004). Of these, most are missense mutations in the folded C-terminal domain. For example, the E200K mutation causes midlife development of CJD, with most patients dying 6–24 months after onset (Colombo 2000). In addition to these point mutations are insertional mutations of one to nine PHGGGWGQ segments in the octarepeat domain (Goldfarb et al. 1991). This class of mutations is enigmatic insofar that they modify a region of the protein that is not essential for propagating prion disease. Treatment of PrP^{Sc} with proteinase K cleaves the protein at approximately residue 90, thereby removing the octarepeat domain, but the remaining protease resistant aggregate retains infectivity. Despite these results, early studies with transgenic mice showed that the PrP octarepeats modulate the disease process. Specifically, inoculated mice expressing a modified PrP^{C} lacking residues 32–93 develop disease with longer incubation times than wild type, produce tissues with lower prion titers, and exhibit a reduced presentation of prion plaques (Flechsig et al. 2000).

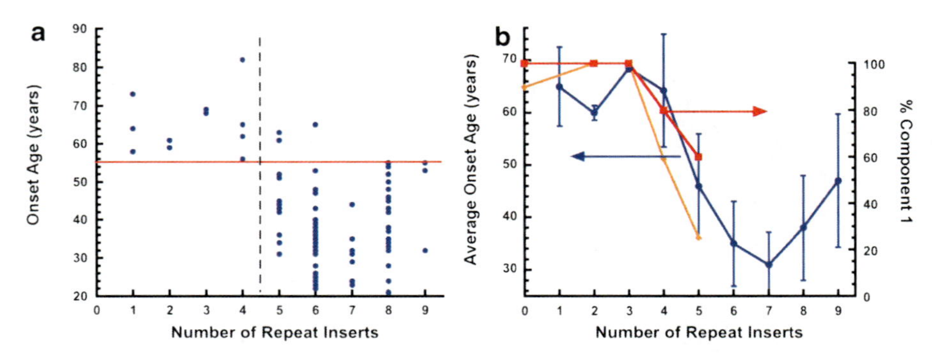

Fig. 2.3 The relationship between onset age for familial prion disease resulting from octarepeat inserts and copper coordination modes. (**a**) Onset age for individual cases as a function of extra octarepeat inserts. Note that wild type corresponds to four repeats, so three inserts correspond to seven total repeat segments. The *horizontal red line* is at 55.5 years and represents a statistically defined separation between late and early onset. (**b**) Average onset age, with standard deviation (*blue circles, left axis*), and component 1 coordination (*orange diamonds* and *red squares*, right axis, for 3.0 and 4.0 equiv. Cu^{2+}, respectively) as a function of extra octarepeat inserts. At both copper concentrations, component 1 coordination drops suddenly at approximately the same OR length threshold as average onset age

Disease progression in individuals with octarepeat expansions depends on the number of inserts. Individuals with one to four extra octarepeats develop disease with an average onset age of 64 years, whereas five to nine extra octarepeats result in an average onset age of 38 years, a difference of almost three decades (Croes et al. 2004; Kong et al. 2004). A number of previous studies examined the biophysical properties of expanded octarepeat domains with emphasis on either the rate of amyloid production or its uncomplexed backbone conformation (Leliveld et al. 2006, 2008; Dong et al. 2007). However, none of these identified a quantitative link between octarepeat length and age of disease onset.

Given the profound influence of octarepeat domain length on expansion disease, we explored whether the domain's response to copper is altered by insertion number (Stevens et al. 2009). We also reevaluated all known cases of human prion disease resulting from octapeptide insertions, and compared the findings to biophysical studies that examined the balance between component 1 and component 3 coordination, as a function of octarepeat domain length. Beginning with statistical data from two existing studies (Croes et al. 2004; Kong et al. 2004), we surveyed the recent clinical literature, pooled the data, and established a new data set covering approximately 30 families and 108 individuals. Onset age for individual cases is shown in Fig. 2.3a. The red line is drawn at 55.5 years. All cases up to four octarepeat inserts (eight repeats total) are above this line and 96% of the cases of five or more octarepeat inserts are below the line. Although there is significant scatter in reported onset age for each specific octarepeat length, the dramatic shift to early onset disease between four and five inserts is apparent. A detailed statistical analysis shows that the results are indeed consistent with the presence of two groups, one composed of individuals with one to four OR inserts and another of individuals with five to eight inserts.

We then performed EPR analysis on a series of PrP-derived constructs from four to nine repeats, corresponding to zero to nine insertions. The experiments showed that domains with four to seven repeats (i.e., zero to three insertions) behave much like the wild type. However, constructs of eight or nine repeats exhibit persistent component 3 coordination. Moreover, these constructs take up approximately twice as much copper as wild type. Equivalent trends were observed with full-length recombinant protein, where we compared wild type with mutant PrP^C containing five repeat inserts. To underscore these findings, we compared the average onset age and standard deviation, as a function of octarepeat length, to Cu^{2+} binding properties. The longest OR expansions favor component 3 coordination and resist component 1. Thus, component 1 coordination serves as a convenient measure of altered Cu^{2+} binding properties. Figure 2.3b shows the relative population of component 1 coordination for each OR construct superimposed on the average age of onset. For wild type and expansions involving up to seven repeats (three inserts beyond wild-type), component 1 coordination is dominant for both 3.0 and 4.0 equiv. Cu^{2+}. However, at eight and nine ORs (four and five inserts, respectively), the population of component 1 coordination drops precipitously.

These data reveal a remarkable relationship, where decreased onset age and persistent component 3 coordination take place at a threshold of eight or more total repeats. In turn, our findings suggest an important protective role for component 1 coordination that may be lost in cases of octarepeat expansion disease with four or more inserts. Together, these findings motivate a careful examination of the distinct chemical properties and reactivity of component 1 vs. component 3 copper coordination.

2.5 Electrochemical Properties of the PrP Copper Sites

Copper's ability to cycle between the Cu^{+} and Cu^{2+} oxidation sites is essential for life. For example, cellular respiration relies on cytochrome c oxidase, a copper-dependent enzyme that converts molecular oxygen to water, ultimately leading to the production of ATP. Since the earliest studies connecting PrP^C to copper uptake, there has been interest in understanding reduction–oxidation (redox) cycling at the copper sites. One line of enquiry suggests that PrP^C functions as a superoxide dismutase (SOD), which inactivates toxic O_2^-, converting it to the more benign hydrogen peroxide (H_2O_2). This hypothesis has been controversial, and is reviewed elsewhere (Brown 2009; Daniels and Brown 2002). The connection between copper coordination mode and onset age for octarepeat expansion disease, discussed above, certainly motivates an evaluation whether component 1 and component 3 coordination sites give rise to distinct redox properties.

Initial electrochemical studies used cyclic voltammetry to evaluate short single repeat peptides as models of component 1 coordination (Bonomo et al. 2000). Reduction of Cu^{2+} to Cu^{+} was found to be energetically unfavorable, leading to the possibility that PrP^C may stabilize copper in its oxidized form. From a neuroprotective perspective, this could be important since weakly complexed copper readily cycles between oxidation states, resulting in the production of reactive oxygen species that are often cytotoxic. By stabilizing copper in a single oxidation state, PrP^C may quench this deleterious chemistry.

Component 3 coordination, with four His residues, appears somewhat similar to the active site in SOD and initially suggested that it might readily undergo redox cycling. Redox kinetics, as measured by bathocuproine absorbance, suggested that indeed component 3 was more easily reduced than component 1 (Miura et al. 2005). Building from these results, it was proposed that PrP^C might function in concert with endocytosis as a copper reductase. In this scenario, extracellular Cu^{2+} binds to PrP^C with component 1 coordination, and the complex is internalized by endocytosis. Next, the low pH drives rearrangement in the octarepeat domain to favor component 3 coordination, leading to reduction to Cu^+. Finally, the copper is released and internalized through a copper transporter.

In a collaborative work with Zhou and coworkers, we recently revisited the detailed electrochemical features of the component 1 and component 3 coordination modes (Liu et al. 2011). The full octarepeat domain with 1 equiv. of Cu^{2+} served as a model for component 3 coordination. Cyclic voltammetry performed in the presence of ascorbate, with and without oxygen, and under nearly reversible conditions showed facile reduction to Cu^+, along with a significant increase in affinity. Thus, as opposed to cycling copper, these data suggest that Cu^+ is very stable in this low occupancy mode, and unlikely to be reoxidized back to Cu^{2+}. Next, we used the same conditions to examine component 1 coordination and found reduction potentials consistent with a copper center that supports cycling between its oxidation states. However, when we compared the findings to free copper, or simple copper–peptide complexes like those found in blood or cerebral spinal fluid, we observed that the reaction was controlled and less likely to produce cytotoxic species such as hydroxyl radicals. Additional assays demonstrated that copper bound to PrP with component 1 coordination, under reducing conditions by ascorbate, gently converts dissolved oxygen to hydrogen peroxide. A summary of these findings is shown in Fig. 2.4.

The ability to bind copper and facilitate redox cycling is shared with the Aβ peptide and α-synuclein, which are causative in Alzheimer's and Parkinson's disease, respectively. Unlike PrP^C, however, these species exhibit only a single binding mode and, therefore, a single profile for producing hydrogen peroxide. Comparing these two neurodegenerative species with PrP^C, we find that component 3 is by far the least reactive, producing hydrogen peroxide at the lowest rate, whereas component 1 is the most reactive (Liu et al. 2011). Thus, PrP^C exhibits vastly different electrochemical profiles, depending on copper occupancy. Both modes are neuroprotective with component 3 coordination completely inhibiting copper redox activity, and component 1 regulating activity with the controlled formation of hydrogen peroxide.

Together, these findings support a role for PrP^C in suppressing copper's inherent redox activity that would otherwise be very damaging to cellular components. However, the discovery that high copper occupancy PrP^C produces hydrogen peroxide suggests additional biochemical control. Similar to nitric oxide, hydrogen peroxide is now considered a signaling species of particular importance in the immune

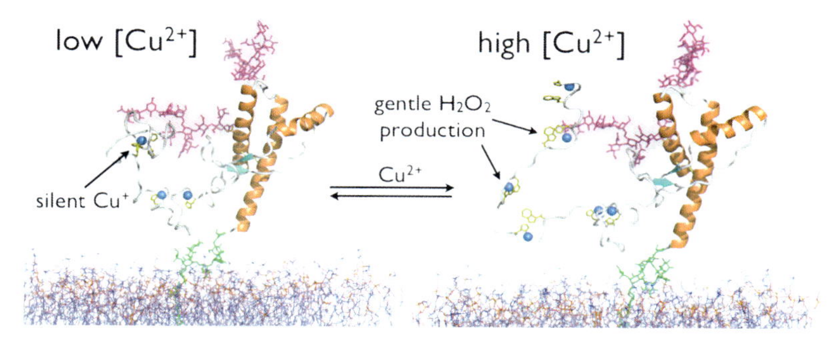

Fig. 2.4 Schematic representation of the possible roles of PrP^C–Cu^{2+} complexes in quenching the Cu^{2+} redox cycling or gradual production of H_2O_2 for signal transduction. PrP is tethered to cell membrane via the GPI anchor (*green*) with its α-helices in the C terminus shown in *orange*, *N*-linked carbohydrates in *purple*, and the N-terminal copper binding segment depicted in *white*. When [Cu^{2+}] is at a low level (nM or lower), Cu^{2+} (*blue sphere*) remains bound in the component 3 mode (*left*), quenching the Cu^{2+} redox cycling. At higher [Cu^{2+}] (μM), the binding mode transitions to component 1 (*right*), leading to a gradual and controlled production of H_2O_2"(Coordinates for the PrP^C-terminal domain, along with carbohydrates, GPI anchor, and membrane, were kindly provided by Professor Valerie Daggett (U. Washington))

system and also in protein localization (Veal et al. 2007). There are likely several possible mechanisms for H_2O_2 action. However, PrP^C has been linked to transmembrane signaling (Mouillet-Richard et al. 2000) and it is noteworthy that hydrogen peroxide readily crosses membrane bilayers and inactivates phosphatase and kinase active sites by reaction with catalytic residues.

The cumulative findings reviewed here emphasize the complex connection between zinc and copper uptake, and the variability in copper binding as controlled by concentration. The relationship between copper coordination modes and onset age for prion disease, resulting from octarepeat expansion, suggests that metal ion regulation may also factor into the development of disease. New electrochemical findings provide a foundation for understanding how PrP^C protects cells against oxidative assaults, and also reveal a possible mechanism for transmembrane signaling. Further refinement of these concepts is sure to lead to a precise function for PrP^C and perhaps insight into how the loss of function contributes to neurodegenerative disease.

Acknowledgments This work was supported by NIH grant GM065790. The author wishes to thank Professor F. Zhou of California State University, Los Angeles, for review and insightful comments on the PrP–copper electrochemistry section.

References

Aronoff-Spencer E, Burns CS, Avdievich NI, Gerfen GJ, Peisach J, Antholine WE, Ball HL, Cohen FE, Prusiner SB, Millhauser GL (2000) Identification of the Cu^{2+} binding sites in the N-terminal domain of the prion protein by EPR and CD spectroscopy. Biochemistry 39:13760–13771

Bleackley MR, Macgillivray RT (2011) Transition metal homeostasis: from yeast to human disease. Biometals 24(5):785–809. doi:10.1007/s10534-011-9451-4

Bonomo RP, Impellizzeri G, Pappalardo G, Rizzarelli E, Tabbi G (2000) Copper(II) binding modes in the prion octarepeat PHGGGWGQ: a spectroscopic and voltammetric study. Chemistry 6:4195–4202

Bremer J, Baumann F, Tiberi C, Wessig C, Fischer H, Schwarz P, Steele AD, Toyka KV, Nave KA, Weis J, Aguzzi A (2010) Axonal prion protein is required for peripheral myelin maintenance. Nat Neurosci 13(3):310–318. doi:nn.2483 [pii]10.1038/nn.2483

Brown DR (2009) Brain proteins that mind metals: a neurodegenerative perspective. Dalton Trans 21:4069–4076. doi:10.1039/b822135a

Brown DR, Qin K, Herms JW, Madlung A, Manson J, Strome R, Fraser PE, Kruck T, von Bohlen A, Schulz-Schaeffer W, Giese A, Westway D, Kretzschmar H (1997) The cellular prion protein binds copper in vivo. Nature 390:684–687

Burns CS, Aronoff-Spencer E, Dunham CM, Lario P, Avdievich NI, Antholine WE, Olmstead MM, Vrielink A, Gerfen GJ, Peisach J, Scott WG, Millhauser GL (2002) Molecular features of the copper binding sites in the octarepeat domain of the prion protein. Biochemistry 41:3991–4001

Burns CS, Aronoff-Spencer E, Legname G, Prusiner SB, Antholine WE, Gerfen GJ, Peisach J, Millhauser GL (2003) Copper coordination in the full-length, recombinant prion protein. Biochemistry 42(22):6794–6803

Chattopadhyay M, Walter ED, Newell DJ, Jackson PJ, Aronoff-Spencer E, Peisach J, Gerfen GJ, Bennett B, Antholine WE, Millhauser GL (2005) The Octarepeat domain of the prion protein binds Cu(II) with three distinct coordination modes at pH 7.4. J Am Chem Soc 127(36): 12647–12656

Colombo R (2000) Age and origin of the PRNP E200K mutation causing familial Creutzfeldt–Jacob disease in Libyan Jews. Am J Hum Genet 67(2):528–531. doi:10.1086/303021

Croes EA, Theuns J, Houwing-Duistermaat JJ, Dermaut B, Sleegers K, Roks G, Van den Broeck M, Van Harten B, Van Swieten JC, Cruts M, Van Broeckhoven C, Van Duijn CM (2004) Octapeptide repeat insertions in the prion protein gene and early onset dementia. J Neurol Neurosurg Psychiatry 75(8):1166–1170. doi:10.1136/jnnp. 2003.020198

Daniels M, Brown DR (2002) Purification and preparation of prion protein: synaptic superoxide dismutase. Methods Enzymol 349:258–267

Dong J, Bloom JD, Goncharov V, Chattopadhyay M, Millhauser GL, Lynn DG, Scheibel T, Lindquist S (2007) Probing the role of PrP repeats in conformational conversion and amyloid assembly of chimeric yeast prions. J Biol Chem 282(47):34204–34212

Flechsig E, Shmerling D, Hegyi I, Raeber AJ, Fischer M, Cozzio A, von Mering C, Aguzzi A, Weissmann C (2000) Prion protein devoid of the octapeptide repeat region restores susceptibility to scrapie in PrP knockout mice. Neuron 27(2):399–408

Garnett AP, Viles JH (2003) Copper binding to the octarepeats of the prion protein. Afffinity, specificity, folding and cooperativity: insights from circular dichroism. J Biol Chem 278(9):6795–6802

Goldfarb LG, Brown P, McCombie WR, Goldgaber D, Swergold GD, Wills PR, Cervenakova L, Baron H, Gibbs CJ Jr, Gajdusek DC (1991) Transmissible familial Creutzfeldt–Jakob disease associated with five, seven, and eight extra octapeptide coding repeats in the PRNP gene. Proc Natl Acad Sci USA 88(23):10926–10930

Harford C, Sarkar B (1997) Amino terminal Cu(II)- and Ni(II)-binding (ATCUN) motif of proteins and peptides - metal binding, DNA cleavage, and other properties. Acc Chem Res 30(3): 123–130

Herms J, Tings T, Gall S, Madlung A, Giese A, Siebert H, Schürmann P, Windl O, Brose N, Kretzschmar H (1999) Evidence of presynaptic location and function of the prion protein. J Neurosci 19:8866–8875

Hornshaw MP, McDermott JR, Candy JM (1995a) Copper binding to the N-terminal tandem repeat regions of mammalian and avian prion protein. Biochem Biophys Res Commun 207:621–629

Hornshaw MP, McDermott JR, Candy JM, Lakey JH (1995b) Copper binding to the N-terminal tandem repeat region of mammalian and avian prion protein: structural studies using synthetic peptides. Biochem Biophys Res Commun 214(3):993–999

Jones CE, Klewpatinond M, Abdelraheim SR, Brown DR, Viles JH (2005) Probing copper^{2+} binding to the prion protein using diamagnetic nickel^{2+} and 1 H NMR: the unstructured N terminus facilitates the coordination of six copper^{2+} ions at physiological concentrations. J Mol Biol 346(5):1393–1407

Kanaani J, Prusiner SB, Diacovo J, Baekkeskov S, Legname G (2005) Recombinant prion protein induces rapid polarization and development of synapses in embryonic rat hippocampal neurons in vitro. J Neurochem 95(5):1373–1386

Klamt F, Dal-Pizzol F, Conte DA, Frota ML Jr, Walz R, Andrades ME, Gomes DA, Silva E, Brentani RR, Izquierdo I, Moreira JCF (2001) Imbalance of antioxidant defense in mice lacking cellular prion protein. Free Radic Biol Med 30:1137–1144

Kong Q, Surewicz WK, Petersen RB, Zou W, Chen SG, Gambetti P, Parchi P, Capellari S, Goldfarb L, Montagna P, Lugaresi E, Piccardo P, Ghetti B (2004) Inherited prion diseases. In: Prusiner SB (ed) Prion biology and diseases. Cold Spring Harbor Laboratory Press, Cold Spring Harbor, NY, pp 673–775

Kramer ML, Kratzin HD, Schmidt B, Romer A, Windl O, Liemann S, Hornemann S, Kretzschmar H (2001) Prion protein binds copper within the physiological concentration range. J Biol Chem 276:16711–16719

Leliveld SR, Dame RT, Wuite GJ, Stitz L, Korth C (2006) The expanded octarepeat domain selectively binds prions and disrupts homomeric prion protein interactions. J Biol Chem 281(6): 3268–3275

Leliveld SR, Stitz L, Korth C (2008) Expansion of the octarepeat domain alters the misfolding pathway but not the folding pathway of the prion protein. Biochemistry 47(23):6267–6278

Liu L, Jiang D, McDonald A, Hao Y, Millhauser GL, Zhou F (2011) Copper redox cycling in the prion protein depends critically on binding mode. J Am Chem Soc 133(31):12229–12237. doi:10.1021/ja2045259

McLennan NF, Brennan PM, McNeill A, Davies I, Fotheringham A, Rennison KA, Ritchie D, Brannan F, Head MW, Ironside JW, Williams A, Bell JE (2004) Prion protein accumulation and neuroprotection in hypoxic brain damage. Am J Pathol 165(1):227–235

Millhauser GL (2004) Copper binding in the prion protein. Acc Chem Res 37(2):79–85

Millhauser GL (2007) Copper and the prion protein: methods, structures, function, and disease. Annu Rev Phys Chem 58:299–320

Miura T, Sasaki S, Toyama A, Takeuchi H (2005) Copper reduction by the octapeptide repeat region of prion protein: pH dependence and implications in cellular copper uptake. Biochemistry 44(24):8712–8720

Mouillet-Richard S, Ermonval M, Chebassier C, Laplanche JL, Lehmann S, Launay JM, Kellermann O (2000) Signal transduction through prion protein. Science 289(5486): 1925–1928

Pauly PC, Harris DA (1998) Copper stimulates endocytosis of the prion protein. J Biol Chem 273:33107–33119

Perera WS, Hooper NM (2001) Ablation of the metal ion-induced endocytosis of the prion protein by disease-associated mutation of the octarepeat region. Curr biol 11(7):519–523

Prusiner SB (2004) Prion biology and diseases, 2nd edn, Cold Spring Harbor Monograph Series. Cold Spring Harbor Laboratory Press, Cold Spring Harbor, NY

Pushie MJ, Pickering IJ, Martin GR, Tsutsui S, Jirik FR, George GN (2011) Prion protein expression level alters regional copper, iron and zinc content in the mouse brain. Metallomics 3(2):206–214. doi:10.1039/c0mt00037j

Rachidi W, Vilette D, Guiraud P, Arlotto M, Riondel J, Laude H, Lehmann S, Favier A (2003) Expression of prion protein increases cellular copper binding and antioxidant enzyme activities but not copper delivery. J Biol Chem 278(11):9064–9072

Stevens DJ, Walter ED, Rodriguez A, Draper D, Davies P, Brown DR, Millhauser GL (2009) Early onset prion disease from octarepeat expansion correlates with copper binding properties. PLoS Pathog 5(4):e1000390. doi:10.1371/journal.ppat.1000390

Stöckel J, Safar J, Wallace AC, Cohen FE, Prusiner SB (1998) Prion protein selectively binds copper(II) ions. Biochemistry 37:7185–7193

Tobler I, Gaus SE, Deboer T, Achermann P, Fischer M, Rulicke T, Moser M, Oesch B, McBride PA, Manson JC (1996) Altered circadian activity rhythms and sleep in mice devoid of prion protein. Nature 380(6575):639–642. doi:10.1038/380639a0

Valensin D, Luczkowski M, Mancini FM, Legowska A, Gaggelli E, Valensin G, Rolka K, Kozlowski H (2004) The dimeric and tetrameric octarepeat fragments of prion protein behave differently to its monomeric unit. Dalton Trans 9:1284–1293

Van Doorslaer S, Cereghetti GM, Glockshuber R, Schweiger A (2001) Unraveling the Cu^{2+} binding sites in the C-terminal domain of the murine prion protein: A pulse EPR and ENDOR study. J Phys Chem 105:1631–1639

Veal EA, Day AM, Morgan BA (2007) Hydrogen peroxide sensing and signaling. Mol Cell 26(1):1–14. doi:10.1016/j.molcel.2007.03.016

Viles JH, Cohen FE, Prusiner SB, Goodin DB, Wright PE, Dyson HJ (1999) Copper binding to the prion protein: structural implications of four identical cooperative binding sites. Proc Natl Acad Sci USA 96:2042–2047

Waggoner DJ, Drisaldi B, Bartnikas TB, Casareno RLB, Prohaska JR, Gitlin JD, Harris DA (2000) Brain copper content and cuproenzyme activity do not vary with prion protein expression level. J Biol Chem 275:7455–7458

Walter ED, Chattopadhyay M, Millhauser GL (2006) The affinity of copper binding to the prion protein octarepeat domain: evidence for negative cooperativity. Biochemistry 45(43): 13083–13092

Walter ED, Stevens DJ, Visconte MP, Millhauser GL (2007) The prion protein is a combined zinc and copper binding protein: Zn2+ alters the distribution of Cu2+ coordination modes. J Am Chem Soc 129(50):15440–15441. doi:10.1021/ja077146j

Walter ED, Stevens DJ, Spevacek AR, Visconte MP, Dei Rossi A, Millhauser GL (2009) Copper binding extrinsic to the octarepeat region in the prion protein. Curr Protein Pept Sci. doi:CPPS-13 [pii]

Chapter 3
Role of Cellular Prion Protein in the Amyloid-β Oligomer Pathophysiology of Alzheimer's Disease

Adam C. Kaufman and Stephen M. Strittmatter

Abstract Alzheimer's disease (AD) is the most common form of dementia affecting millions worldwide. The primary histopathological features of AD are amyloid-beta (Aβ) plaques and neurofibrillary tangles. Aβ oligomers (Aβo) are believed to be essential mediators of the synaptotoxicity and cell death that are characteristic of this illness. For decades, the exact mechanism for how Aβ exerted its toxic effect remained unknown. Recently, it has been shown that the cellular Prion Protein (PrP^C) acts as a high-affinity binding partner for Aβo. Moreover, it has been demonstrated that PrP^C is necessary for memory loss, impaired long-term potentiation, and neuronal dysfunction in transgenic mouse models of AD. Antagonizing PrP^C in AD mouse models has also been shown to reverse memory deficits, so targeting PrP^C is a potential avenue for treatment. This chapter will review the evidence connecting PrP^C to Aβo pathophysiology.

Keywords Alzheimer • Amyloid beta peptide • Oligomer • Neurodegeneration • Signal transduction • Transgenic • Spatial memory • Long-term potentiation • Synaptic plasticity

---

A.C. Kaufman
Cellular Neuroscience, Neurodegeneration and Repair Program, Department of Neurology and Department of Neurobiology, Yale University School of Medicine, New Haven, CT, USA

S.M. Strittmatter, M.D., Ph.D. (✉)
Cellular Neuroscience, Neurodegeneration and Repair Program, Department of Neurology and Department of Neurobiology, Yale University School of Medicine, 295 Congress Ave, New Haven, CT 06536, USA
e-mail: stephen.strittmatter@yale.edu

W.-Q. Zou and P. Gambetti (eds.), *Prions and Diseases: Volume 1, Physiology and Pathophysiology*, DOI 10.1007/978-1-4614-5305-5_3,

3.1 Introduction

Alzheimer's disease (AD) is a chronic and progressive neurodegenerative disease estimated to affect approximately 35 million individuals worldwide (Prince et al. 2009). AD is responsible for 50–70% of all cases of dementia. As the population continues to age, its prevalence is expected to quadruple by the year 2050 (Brookmeyer et al. 2007). The classical clinical manifestations of AD are an amnestic memory impairment, language deterioration, and visuospatial deficits, eventually leading to death (Cummings 2004). Patients with AD have a post-diagnosis median survival ranging from 3 to 8 years (Helzner et al. 2008). It is now the sixth most common cause of death in the USA (Thies and Bleiler 2011). Current treatment options for AD are limited to partial efficacy and to symptomatic control. There is no disease-modifying therapy for AD in clinical practice today. Due to these factors, AD places a tremendous burden on individuals and families, with societal costs of 100 billion dollars each year (Meek et al. 1998).

The disease was first described in 1907 as a condition with progressive memory loss, atrophic brain, visible plaques, and intraneuronal fibrils (Alzheimer et al. 1995). The specific histological pattern is, to this day, the definitive way to diagnose AD (1997). The National Institute of Aging has proposed a criterion based on biomarkers that may broaden diagnoses (McKhann et al. 2011). The classical histological lesions have since been determined to be composed of extracellular insoluble plaques of polymeric beta-amyloid (Aβ) peptide (Glenner and Wong 1984) and intraneuronal fibrillary tangles of the hyperphosphorylated microtubule-associated protein, tau (Kosik et al. 1986). Efforts to understand the pathophysiology of AD focus on these proteins and lesions.

3.2 Amyloid Hypothesis

Over the past decade, there has been a growing consensus that the key mediator of the memory loss associated with AD is the 38–43 amino acid peptide Aβ. The "amyloid hypothesis" states that Aβ is not just the main constituent of plaques but also causes neuronal toxicity (Fig. 3.1). There are numerous genetic and biochemical avenues of research that support this premise, and this topic is reviewed in detail elsewhere (Selkoe and Schenk 2003). Key findings in support of this theory initiated from the observation that the Aβ peptide is the main constituent of AD plaques. Aβ peptide is derived from the amyloid precursor protein (APP) by sequential protease action of a β-secretase and a γ-secretase (Mills and Reiner 1999; Goldgaber et al. 1987). The genetics of the rare cases of early onset autosomal dominant AD support the Aβ hypothesis. Genetic analysis of certain families has uncovered mutations in the APP gene itself (Citron et al. 1992). The familial AD mutations were found to cluster in or around the sites of cleavage activity and to promote a greater $A\beta_{42}$ to $A\beta_{40}$ ratio, where $A\beta_{42}$ is more prone to oligomerization and fibrillization than $A\beta_{40}$ (Hardy and Selkoe 2002). Rare AD inducing mutations within the APP gene did not

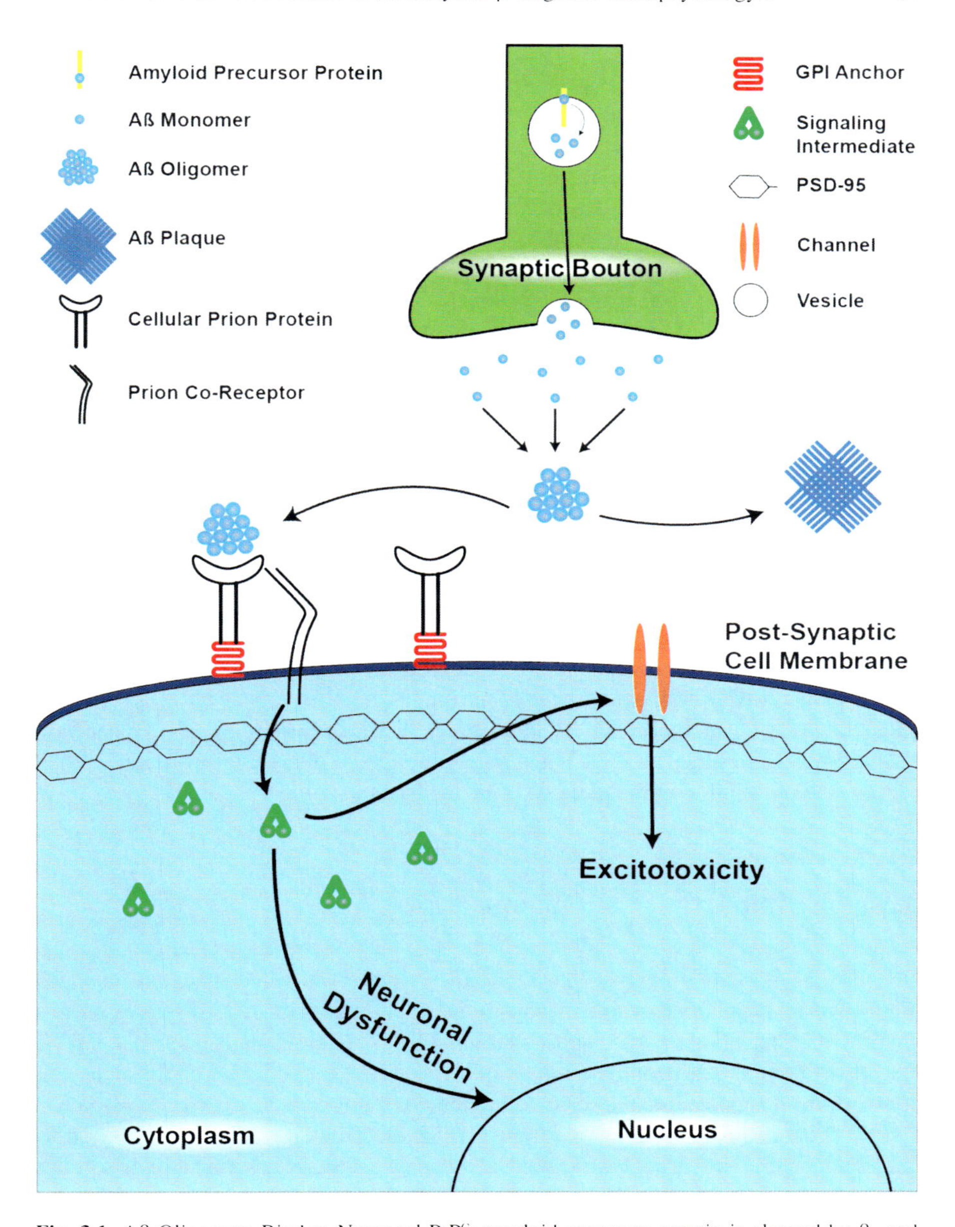

Fig. 3.1 Aβ Oligomers Bind to Neuronal PrP^{C}: amyloid precursor protein is cleaved by β- and γ-secretases within the presynaptic neuron to form 38–43 amino acid amyloid beta (Aβ) monomers. These monomers are then released into the synaptic cleft where they can oligomerize to form soluble Aβ oligomers. Alternatively, the monomers can continue to polymerize and form larger insoluble Aβ plaques. PrP^{C} has a high affinity for soluble Aβ oligomers while having limited affinity for both the monomers and the plaques. PrP^{C} on the postsynaptic neuron avidly binds Aβ oligomers and with the help of an unknown coreceptor initiates an intracellular cascade of events leading to neuronal dysfunction and excitotoxicity

affect Aβ processing directly, but increased rate of self-aggregation, leading indirectly to higher levels of Aβ plaques and fibrils (Wisniewski et al. 1991). Other cases of dominantly inherited early onset AD are caused by mutations in Presenilin-1 or 2, which are components of the γ-secretase. These AD mutations alter the enzymatic specificity of APP cleavage, leading to the same increase in Aβ42/Aβ40 ratio. Importantly, transfer of a human APPswe mutant transgene to mice was shown to recapitulate some aspects of human AD, including Aβ plaque and progressive memory deficits (Chapman et al. 1999).

It is also noteworthy that the APP gene is located on chromosome 21, and the Down's syndrome of trisomy 21 includes dementia and Aβ plaque deposition similar to non-syndromic AD (Masters et al. 1985). Moreover, a rare patient with Down's syndrome who did not develop memory loss was located and she had only a partial trisomy possessing the standard complement of two copies of APP (Prasher et al. 1998). Apart from Mendelian inheritance of early onset AD, genetic factors contribute to risk of late onset AD. Isoforms of the ApoE strongly affect risk, and these have been shown to alter Aβ clearance and aggregation (Kim et al. 2009). Variation at another risk genetic risk locus, clusterin, may cooperate ApoE to modulate Aβ levels (DeMattos et al. 2004).

More recently, consortium-based biomarker studies of aging and impaired cognition have demonstrated that Aβ alterations detected by either PET imaging or by CSF sampling are the first markers of AD, and that individuals with mild cognitive impairments (MCI) and Aβ marker abnormality have a very high likelihood of advancing to AD (Jack et al. 2010; Petersen et al. 2010; Sperling et al. 2011; Shaw et al. 2009; Heister et al. 2011). Thus, both genetic and disease progression studies support the amyloid hypothesis of AD.

3.3 The Importance of Oligomeric Aβ

One of the arguments against the amyloid hypothesis has been that the level of memory impairment and brain atrophy found in patients with AD correlates poorly with the number of plaques found in the brain (Terry et al. 1991; Josephs et al. 2008; Katzman 1986). Additionally, when neurons are exposed to fibrillar Aβ, the concentrations necessary to induce cell death were not consistent with physiologic levels. There has been increasing interest in Aβ oligomers (Aβo) as the solution to this conundrum (Fig. 3.1) (Walsh and Selkoe 2007). Oligomers are smaller soluble peptide polymers of Aβ monomers ranging in size from dimers up to 100-mers (Gunther and Strittmatter 2009). Consistent with a role in human neurodegeneration, nanomolar concentrations of Aβo derived from the cortices of patients with AD have been shown to inhibit long-term potentiation (LTP), reduce dendritic spine density, and impair recall of learned behavior (Wang et al. 2002; Walsh et al. 2002; Shankar et al. 2008). In the same assays, monomeric and polymeric Aβ had limited to no impact. Synthetically produced oligomers, also referred to as Aβ-derived diffusible ligands (ADDL), have been shown to induce memory dysfunction in the

AD mouse model independent of the presence of Aβ plaques (Lesne et al. 2006). Antibodies developed against the N-terminus of ADDL have been shown to block memory impairment.

While gathering evidence supports a crucial role for oligomeric Aβ, this begs the mechanistic question of how Aβo mediates its synaptotoxic and neurotoxic effects. Knowing that the effects of oligomeric Aβ are rapid, specific, and reversible all point to the existence of a high-affinity receptor. The existence of such a receptor would bring together many disparate facets within the field. Antagonizing this receptor would also represent a novel strategy for intervening in the progression of AD.

3.4 PrP^C Is the Binding Site for Oligomeric Aβ

It has been recently shown that cellular Prion Protein (PrP^C) acts as a high affinity binding site for Aβo (Laurén et al. 2009; Balducci et al. 2010). PrP^C has also been shown to transmit the synaptotoxic effect of Aβo (Laurén et al. 2009; Freir et al. 2011a; Barry et al. 2011; Chung et al. 2010). The original identification of PrP^C as an Aβo binding site utilized biotin-conjugated ADDLs (Laurén et al. 2009). Tagged ligand was then exposed to COS-7 cells that were expressing cDNA from an unbiased genome-wide adult mouse brain library in order to determine what gene products, if any, could bind ADDLs. COS-7 cells were chosen for this screening procedure because they bind less than 5% of the level of ADDL that endogenous receptors on hippocampal neurons bind. From within the 225,000 clones, there were only two positive hits, which both encoded a full-length version of PrP^C. The apparent dissociation constant for these clones was identical to that of hippocampal neurons, with nM affinity for ADDL. Depending on how the dissociation constant was calculated, it was found to be somewhere between 0.4 nM and 92 nM. PrP^C showed high selectivity for oligomerized Aβ versus monomeric Aβ, with a Kd difference of two orders of magnitude. Strong binding and specificity was also evident when PrP^C-Fc fragments are immobilized on resin and are exposed to ADDL.

A second library of 352 clones expressing transmembrane proteins was screened individually to identify hits with weaker interactions (Laurén et al. 2009). This produced a few hits; nonetheless the lowest dissociation constant found for any of these hits was 660 nM and there was minimal selectivity for oligomers over monomers. Previous papers had reported a possible interaction between monomeric Aβ and the receptor for advanced glycation products (Yan et al. 1996) or the α7 nicotinic acetylcholine receptor (Wang et al. 2000), but even with this lower stringency, direct Aβo binding did not indicate significant affinity.

E18 neurons have minimal affinity for Aβo immediately upon plating in vitro; however, the affinity for Aβo dramatically increases over a 15–20-day period that is contemporaneous with an equivalent increase in PrP^C expression levels in these cells (Laurén et al. 2009). There is broad colocalization of the immunoreactivity of bound Aβo and PrP^C. Neurons from *PRNP*$^{-/-}$ mice, which are PrP^C null, showed a

50% reduction in binding. Taken together, these data indicate that PrP^C contributes considerably to oligomeric Aβ binding, although there are likely other players or redundancy within the system.

3.5 Aβ Oligomers Bind to the Unstructured Central Domain of PrP^C

The specific domain of PrP^C that acts as the high affinity binding partner for Aβo was established by several methods. Mutant forms of PrP^C with different domains deleted were expressed on COS-7 cells to gauge the contributions of each domain to overall binding of ADDLs (Laurén et al. 2009). Removing the octapeptide repeat domain or the hydrophobic domain did not decrease the binding capacity, while cells expressing solely the globular domain were unable to bind oligomers. However, removing the unstructured central region, amino acids 95–110, dramatically lowered binding capability by 80%. In a DELFIA assay, human PrP^C fragments of amino acids 91–231 exhibited identical binding to ADDLs compared to that of full-length PrP^C, while fragments of amino acids 119–231 displayed almost no interaction (Freir et al. 2011a). This further emphasized the essential role that the amino acids 95–110 have for binding oligomers. Interestingly, the unstructured central domain has been implicated in contributing to neurodegeneration in mice (Baumann et al. 2007). Surface plasmon resonance studies demonstrated Aβo binding to both the 95–110 region and the extreme amino terminus 23–27, but not other regions of PrP^C (Chen et al. 2010).

The 6D11 antibody has as its epitope the amino acids 95–110 of the PrP^C protein. Preincubating the PrP^C expressing cells with 6D11 antibody effectively blocked the cells from interacting with oligomers (Laurén et al. 2009). The antibodies 8 G8 and ICSM-35 which both have epitopes that overlap with the epitope of 6D11 showed similar reduction in binding in a standard dose–response fashion (Laurén et al. 2009; Freir et al. 2011a). Anti-PrP^C antibodies that did not bind to this integral area did not impact binding, with one exception. Antibodies directed against the helix-1 domain appeared to lower affinity for ADDLs by up to 60%, which is surprising as this domain is quite far from the putative primary binding region (Freir et al. 2011a). It is possible that the antibodies at this region block a conformational shift within the PrP^C molecule that normally allows a stabilization of the binding of the oligomer, or it could potentially be a secondary binding site which could be consistent with the repetitive structure of Aβ.

Finally, although both Aβ and the octapeptide repeat domain of PrP^C are capable of binding copper ions with high affinity, the chelation of copper ions does not appear to contribute to their interaction. There was no change in binding affinity between COS-7 cells expressing PrP^C in copper-free F12 medium or in F12 medium with 1 mM of copper sulfate added (Laurén et al. 2009). The addition of up to 10 mM of EDTA, which would sequester any copper ions away from PrP^C and Aβ, had no impact on binding in hippocampal neuronal cultures (Freir et al. 2011a).

3.6 Aβ Oligomers Inhibit LTP Through PrP^{C}

LTP is a persistent increase in synaptic strength represented primarily by an increase in excitatory postsynaptic potentials (EPSP) that can last for hours in response to a high frequency train of electrical stimuli. It is believed to be a form of synaptic plasticity that likely forms the cellular and molecular basis for learning. Hippocampal LTP has been shown as necessary to form persistent spatial memories (Morris et al. 1986). In particular, Schaffer collateral LTP has been shown to be strongly inhibited by nanomolar concentrations of Aβo (Wang et al. 2002; Walsh et al. 2002). This makes LTP an excellent method to probe whether PrP^{C} participates in the pathogenicity of Aβo.

Hippocampal slices of brain from wild-type and *PRNP*$^{-/-}$ mice on a C57Bl6 background were stimulated to induce LTP in the presence of 2 nM Aβo in vitro (Laurén et al. 2009). Wild-type brain slices only had a 20% augmentation of the slope of the EPSP, a significant reduction in what would normally be expected. In contrast, the slope of the EPSP for the treated knockout brain slices had an 80% augmentation, which is identical to the EPSP of untreated brain slice. In order to rule out that congenital loss of PrP^{C} could induce some compensatory effects that could explain the unaffected LTP of the knockout brain slices in the face of Aβo challenge, wild-type brain slices were incubated with the 6D11 antibody followed by exposure to Aβo. These 6D11 antibody pretreated slices were completely protected from the expected loss of EPSP from Aβo.

The Malinow group also exposed hippocampal neurons of *PRNP*$^{+/+}$ or *PRNP*$^{-/-}$ genotypes to an Aβ42 preparation and monitored LTP (Kessels et al. 2010). In contrast to the findings described above, neither genotype had any augmentation of EPSP, even briefly, after LTP induction. These findings are also distinct from several previous studies of Aβo activity in wild-type neurons (Wang et al. 2002; Walsh et al. 2002), in which the peptide failed to abrogate short-term induction, but caused a diminution of long-term maintenance. The PrP-negative study (Kessels et al. 2010) also reported baseline inhibition by Aβ prior to induction. These two findings suggest that a general cytotoxic response was elicited by this incompletely characterized Aβ preparation. This led Collinge's group to demonstrate that a biochemically well-characterized Aβo preparation inhibited LTP in a PrP^{C}-dependent fashion (Freir et al. 2011b), replicating the original observation (Laurén et al. 2009).

PRNP$^{-/-}$ mice were crossed with APP-PS^{+} mice, which express human mutant forms of APP and PSen-1, to further evaluate the in vivo effects of Aβ on LTP (Calella et al. 2010). The Aguzzi group showed a deficiency in the augmentation of LTP at 4 months of age, regardless of PrP^{C} expression. Of note, this mouse model of AD is known for rapidly producing Aβ amyloid at an early age. The rapid production of Aβ might overwhelm PrP^{C} binding and bind to secondary receptors leading to irreversible damage. Intriguingly, APP-PS^{+} mice overexpressing an anchorless version of PrP^{C} were protected from LTP impairment. The secreted PrP^{C} likely bound to the soluble Aβ oligomers and protected the hippocampal neurons. This finding supports the hypothesis that PrP^{C} is the high affinity binding partner for relevant Aβo species.

Although synthetically produced Aβo is a potent synaptic toxin, it may not be identical to naturally occurring Aβo found within the brains of patients with AD. Importantly, water-soluble extracts derived from the brains of patients with AD have similar synaptotoxic effects to that of synthetic Aβo. Such AD extracts were preincubated with hippocampal slices from wild-type and *PRNP*$^{-/-}$ mice, followed by high frequency stimulation (HFS) (Freir et al. 2011a). The wild-type slices exposed to AD extracts had impaired LTP, while the knockout slices were resistant to LTP impairment. Furthermore, water-soluble extract from a non-demented individual was incubated on wild-type and *PRNP*$^{-/-}$ brain slices followed by stimulation, but had no impact on LTP for either genotype. Pretreating wild-type slices with an anti-PrP^{C} antibody directed against the unstructured central domain was also found to be protective against the loss of LTP from AD brain-derived extract. Therefore, PrP^{C} likely is necessary for human disease-derived Aβo to exert their plasticity-impairing effects.

To further evaluate the essential role of PrP^{C} for Aβo action in vivo, Wistar rats had one of their lateral ventricles cannulated. Through this cannula, water-soluble extract derived from the brains of patients with AD was infused. There was no change in baseline potentiation prior to induction, but there was a significant depression of LTP (Barry et al. 2011). A cohort of rats was infused with anti-PrP^{C} antibodies, D13 and ICSM-18, prior to receiving the brain-derived extract and HFS. These pretreated rats were fully protected from Aβo-induced loss of plasticity and LTP (Freir et al. 2011a; Barry et al. 2011). This strongly shows that the requirement of PrP^{C} for Aβo binding is relevant to AD.

3.7 PrP^{C} Is Necessary for Memory Impairment In Vivo

Until recently, the joint impact of Aβo and PrP^{C} on the performance of an in vivo learning and memory task had been unknown. Age-dependent memory loss is among the cardinal features of AD and can be tested in mice with a Morris water maze. In such a task, mice are placed in a large tank of water with a platform that is hidden from their view. Mice, being naturally averse to water, undertake a coordinated search strategy to find an exit. Over the course of repeated trials, the mice eventually learn the location of the hidden platform and escape quickly. Mice with spatial memory deficits take a significantly longer time in locating the platform to escape. This task is especially appropriate for better understanding AD as it has been shown that successfully completing the task relies on having a functioning hippocampus (Redish and Touretzky 1998).

A Morris water maze swim task was performed with wild-type mice, *PRNP*$^{-/-}$ mice, APPswe/Psen1 ΔE9 mice (an AD transgenic model), and APPswe/Psen1 ΔE9 *PRNP*$^{-/-}$ mice at 3 months and at 12 months (Gimbel et al. 2010). At 3 months, there was no apparent difference between any of the groups. At 12 months, the APPswe/Psen1 ΔE9 mice demonstrated significant impaired latencies to escape, while the APPswe/Psen1 ΔE9 mice lacking PrP^{C} had much faster latencies to escape, and were equivalent to the wild-type mice. To test retention of the learned location, the hidden platform was removed. At 12 months, APPswe/Psen1 ΔE9 mice crossed over the area where the platform had been significantly fewer times than the APPswe/Psen1 ΔE9 mice lacking PrP^C. The AD mice without PrP^C crossed the target area as many times as the wild-type group. The mice were also trained to avoid entering a darkened chamber by administration of an aversive shock. The APPswe/Psen1 ΔE9 mice did not remember this passive avoidance training and quickly went into the darkened chamber. In contrast, the PrP^C knockout APPswe/Psen1 ΔE9 mice demonstrated better learning by more prolonged avoidance of the darkened chamber (Gimbel et al. 2010). The levels of APP and Aβ were the same independent of genotype. These results are consistent with PrP^C being crucial for transgenic AD memory impairment.

Further support for the role of PrP^C in AD-related memory impairment comes from a study showing that short-term treatment with the 6D11 could reverse memory impairment in the APP/PS1 transgenic AD model (Chung et al. 2010). Transgenic mice received 10 high dose injections of the 6D11 antibody into their peritoneum over the course of 2 weeks. High doses were utilized so that a non-negligible amount of antibody would be able to successfully cross the blood–brain barrier. The mice were tested with a radial arm maze, and the number of errors that were made while completing the maze was counted. The number of errors that the treated APP/PS1 mice made was significantly fewer than that of the untreated APP/PS1 mice, and was not different from the error rate of wild-type mice. Again, treatment had no impact on amyloid burden, making a simple antagonism of the receptor the most likely mechanism for memory improvement.

Normally, when presented with a noveland familiar object, mice spend more time exploring the novel object compared to the familiar object. This forms the basis for the novel object recognition test, in which a memory-impaired mouse will not remember which object is novel and will show no preference for either object. $PRNP^{+/+}$ and $PRNP^{-/-}$ had a 100 μM solution of synthetic Aβo infused into their ventricles prior to testing over several days (Balducci et al. 2010). The pharmacokinetics of Aβo in this experiment are complicated because the starting dose is high, but the half-life of Aβ in the brain is very short, on the order of 1 h (Cirrito et al. 2003). The Aβ-injected $PRNP^{+/+}$ mice showed no preference for either object, consistent with memory impairment during some segments of the time. The Aβ-injected $PRNP^{-/-}$ mice did not show a preference for the novel object, but exhibited a preference for the familiar object. The authors interpreted these results to imply that PrP^C was not essential for Aβo-induced memory impairment. However, a preference for the familiar object rather than the novel object by the injected PrP knockout mice suggests intact memory, but altered novelty seeking. For transgenic AD mice, novel object recognition is less consistently impaired than is spatial memory (Chen et al. 2000).

Complicating the analysis further, hAPPJ20 mice, another transgenic AD model, had no preference for either object in the novel object recognition test with or without PrP^C (Cisse et al. 2011a). The hAPPJ20 mice also performed worse than the wild-type mice in a Morris water maze task, independent of PrP^C status. In fact, the

PRNP$^{-/-}$ hAPPJ20 mice did slightly worse than any other group in latency to escape and in the number of crosses over the platform area when the platform was removed. It has been previously shown however that the hAPPJ20 mice develop deficits at an early age that are not progressive (Harris et al. 2010). It can be hypothesized that PrP^{C} is necessary for the age-dependent loss of spatial memory seen in transgenic AD-like progression, but that juvenile-onset, age-independent impairment in hAPPJ20 mice occurs through a PrP^{C}-independent mechanism, perhaps involving EphB2 (Cisse et al. 2011b).

3.8 Neuronal Degeneration and Dysfunction Are Dependent upon PrP^{C}

Neurodegeneration is classically seen in AD, but most AD mouse models show limited neurodegeneration even in the face of significant amyloid burden. There have been reports however of monoamine neuronal degeneration in the AD model (Liu et al. 2008). Brains slices from APPswe/Psen1 ΔE9 show signs of axonal degeneration as evidenced by having fewer serotonin axons in the cerebral cortex than wild-type mice. The APPswe/Psen1 ΔE9/*PRNP*$^{-/-}$ brain have indistinguishable levels of serotonin-positive axons compared to wild-type mice, consistent with PrP^{C} being required for this form of AD transgene-induced degeneration (Gimbel et al. 2010).

Synaptophysin is a presynaptic marker and its level can be used to assay synaptic health. A loss of synapses is documented in AD, and APPswe/Psen1 ΔE9 mice show a decrease in levels of synaptophysin in the cortex (Gimbel et al. 2010). APPswe/Psen1 ΔE9 mice lacking PrP^{C} had similar levels of synaptophysin to that of wild-type mice (Gimbel et al. 2010). The postsynaptic marker PSD-95 was also preserved in APPswe/Psen1 ΔE9 PrP^{C} null mice (Gimbel et al. 2010). Excitingly, acute treatment with 6D11 anti-PrP antibody raises synaptophysin levels in the hippocampus of APPswe/Psen1 ΔE9 mice (Chung et al. 2010).

Transgenic AD mice have reduced survival with sudden unexplained deaths. It has been hypothesized that the sudden death may be mediated by hyperexcitability or status epilepticus (Minkeviciene et al. 2009). Over the course of 1 year, 40% of the APPswe/Psen1 ΔE9 mice died, while less than 4% of the APPswe/Psen1 ΔE9 *PRNP*$^{-/-}$ mice died (Gimbel et al. 2010). Wild-type mice experienced a less than 4% death rate as well. For this AD strain, PrP^{C} is essential for the early death phenotype.

Related to the sudden death phenotype, epileptiform discharges have been examined in hAPPJ20 mice with and without PrP^{C}. Knocking out PrP^{C} in this mouse strain slightly increased epileptiform spikes to about 15 per hour, although there were no convulsive seizures (Cisse et al. 2011a). Importantly, historical standards for hAPPJ20 have reported 100–1,000 spikes per hour (Roberson et al. 2011; Palop et al. 2007). Due to variability, single spikes may not be a robust phenotype. Consistent with the possible increase in spike discharges, the same group reported an increase in sudden death from the age of 30 days to 270 days for the hAPPJ20 mice without PrP^C compared to those with PrP^C (Cisse et al. 2011a). If the deaths during the first 30 days of life are included, the difference between the groups is nil. Either way, PrP^C does not appear to improve mortality in the hAPPJ20 mice, in contrast to the APPswe/Psen1 ΔE9 mice. This highlights the need for more research into the difference between the strains to explain the relevant factors. These studies simultaneously emphasize the importance and difficulty of modeling AD behavior in laboratory animals.

3.9 Human *PRNP* Genetics in AD

The possibility of an association between PrP^C genetic variation and AD has been considered in several studies. Most studies have focused on a common coding region variant, the presence of Met vs. Val at codon 129 (rs1799990). In particular, four studies found that the minor Val allele is underrepresented in the AD population (Gacia et al. 2006; Riemenschneider et al. 2004; Golanska et al. 2004; Dermaut et al. 2003). These studies also observed that M/V heterozygous state is less common among AD cases, suggesting that the homozygous state at residue 129 is a risk for AD. The interaction of the residue 129 status with age of onset and with ApoE genotype has not been consistent across these studies. A meta-analysis of published studies is available at AlzGene, and suggests limited, if any, association of the Val allele with AD http://www.alzforum.org/res/com/gen/alzgene/. In a genome-wide SNP study, Roses and colleagues confirmed a role of ApoE and identified new candidate risk loci for late onset Alzheimer's disease (LOAD) (Li et al. 2008). As part of that genome-wide study, a focused analysis of some 25 previously reported LOAD risk genes was completed and only *PRNP* achieved statistical significance in this large-scale genomic study (Li et al. 2008). The strongest association was with an intronic SNP of the *PRNP* gene. Altogether, the contribution of common genetic variants at the *PRNP* locus to AD does not appear to be strong. The potential presence of rare *PRNP* variants having a large effect for AD risk has not yet been explored.

3.10 Conclusion

A range of molecular, proteomic, electrophysiology, and behavioral data supports the hypothesis that PrP^C binding mediates a significant fraction of Aβo-specific pathophysiology in AD models. Additional work is required to understand the relative role of PrP^C in various mouse AD models, to elucidate coreceptors that function with PrP^C to mediate toxic effects, and to characterize the downstream signal transducers of PrP^C activation by amyloid oligomers (Fig. 3.1). Nonetheless, PrP^C remains an enticing target for pharmaceutical blockade, since deleting or antagonizing PrP^C function does not have substantial adverse effects in mice. Targeting PrP^C constitutes a unique strategy for rational disease-modifying AD therapy.

References

(1997) Consensus recommendations for the postmortem diagnosis of Alzheimer's disease. The National Institute on Aging, and Reagan Institute Working Group on Diagnostic Criteria for the Neuropathological Assessment of Alzheimer's Disease. Neurobiol Aging 18(4 Suppl):S1–S2

Alzheimer A et al (1995) An english translation of Alzheimer's 1907 paper, "Uber eine eigenartige Erkankung der Hirnrinde". Clin Anat 8(6):429–431

Balducci C et al (2010) Synthetic amyloid- oligomers impair long-term memory independently of cellular prion protein. Proc Natl Acad Sci USA 107(5):2295–2300

Barry AE et al (2011) Alzheimer's disease brain-derived amyloid-mediated inhibition of ltp in vivo is prevented by immunotargeting cellular prion protein. J Neurosci 31(20):7259–7263

Baumann F et al (2007) Lethal recessive myelin toxicity of prion protein lacking its central domain. EMBO J 26(2):538–547

Brookmeyer R et al (2007) Forecasting the global burden of Alzheimer's disease. Alzheimers Dement 3(3):186–191

Calella AM et al (2010) Prion protein and Abeta-related synaptic toxicity impairment. EMBO Mol Med 2(8):306–314

Chapman PF et al (1999) Impaired synaptic plasticity and learning in aged amyloid precursor protein transgenic mice. Nat Neurosci 2(3):271–276

Chen G et al (2000) A learning deficit related to age and beta-amyloid plaques in a mouse model of Alzheimer's disease. Nature 408(6815):975–979

Chen S, Yadav SP, Surewicz WK (2010) Interaction between human prion protein and amyloid-{beta} (A{beta}) oligomers: role of N-terminal residues. J Biol Chem 285(34):26377–26383

Chung E et al (2010) Anti-PrPC monoclonal antibody infusion as a novel treatment for cognitive deficits in an Alzheimer's disease model mouse. BMC Neurosci 11(1):130

Cirrito JR et al (2003) In vivo assessment of brain interstitial fluid with microdialysis reveals plaque-associated changes in amyloid-beta metabolism and half-life. J Neurosci 23(26):8844–8853

Cisse M et al (2011a) Ablation of cellular prion protein does not ameliorate abnormal neural network activity or cognitive dysfunction in the J20 line of human amyloid precursor protein transgenic mice. J Neurosci 31(29):10427–10431

Cisse M et al (2011b) Reversing EphB2 depletion rescues cognitive functions in Alzheimer model. Nature 469(7328):47–52

Citron M et al (1992) Mutation of the beta-amyloid precursor protein in familial Alzheimer's disease increases beta-protein production. Nature 360(6405):672–674

Cummings JL (2004) Alzheimer's disease. N Engl J Med 351(1):56–67

DeMattos RB et al (2004) ApoE and clusterin cooperatively suppress Abeta levels and deposition: evidence that ApoE regulates extracellular abeta metabolism in vivo. Neuron 41(2):193–202

Dermaut B et al (2003) PRNP Val129 homozygosity increases risk for early-onset Alzheimer's disease. Ann Neurol 53(3):409–412

Freir DB et al (2011) Interaction between prion protein and toxic amyloid β assemblies can be therapeutically targeted at multiple sites. Nat Commun 2:336

Gacia M et al (2006) Prion protein gene M129 allele is a risk factor for Alzheimer's disease. J Neural Transm 113(11):1747–1751

Gimbel DA et al (2010) Memory impairment in transgenic alzheimer mice requires cellular prion protein. J Neurosci 30(18):6367–6374

Glenner GG, Wong CW (1984) Alzheimer's disease and Down's syndrome: sharing of a unique cerebrovascular amyloid fibril protein. Biochem Biophys Res Commun 122(3):1131–1135

Golanska E et al (2004) Polymorphisms within the prion (PrP) and prion-like protein (Doppel) genes in AD. Neurology 62(2):313–315

Goldgaber D et al (1987) Characterization and chromosomal localization of a cDNA encoding brain amyloid of Alzheimer's disease. Science 235(4791):877–880

Gunther EC, Strittmatter SM (2009) β-amyloid oligomers and cellular prion protein in Alzheimer's disease. J Mol Med 88(4):331–338
Hardy J, Selkoe DJ (2002) The amyloid hypothesis of Alzheimer's disease: progress and problems on the road to therapeutics. Science 297(5580):353–356
Harris JA et al (2010) Many neuronal and behavioral impairments in transgenic mouse models of Alzheimer's disease are independent of caspase cleavage of the amyloid precursor protein. J Neurosci 30(1):372–381
Heister D et al (2011) Predicting MCI outcome with clinically available MRI and CSF biomarkers. Neurology 77(17):1619–1628
Helzner EP et al (2008) Survival in Alzheimer disease: a multiethnic, population-based study of incident cases. Neurology 71(19):1489–1495
Jack CR Jr et al (2010) Hypothetical model of dynamic biomarkers of the Alzheimer's pathological cascade. Lancet Neurol 9(1):119–128
Josephs KA et al (2008) Beta-amyloid burden is not associated with rates of brain atrophy. Ann Neurol 63(2):204–212
Katzman R (1986) Alzheimer's disease. N Engl J Med 314(15):964–973
Kessels HW et al (2010) The prion protein as a receptor for amyloid-beta. Nature 466(7308):E3–E4, discussion E4–E5
Kim J, Basak JM, Holtzman DM (2009) The role of apolipoprotein E in Alzheimer's disease. Neuron 63(3):287–303
Kosik KS, Joachim CL, Selkoe DJ (1986) Microtubule-associated protein tau (tau) is a major antigenic component of paired helical filaments in Alzheimer disease. Proc Natl Acad Sci USA 83(11):4044–4048
Laurén J et al (2009) Cellular prion protein mediates impairment of synaptic plasticity by amyloid-β oligomers. Nature 457(7233):1128–1132
Lesne S et al (2006) A specific amyloid-beta protein assembly in the brain impairs memory. Nature 440(7082):352–357
Li H et al (2008) Candidate single-nucleotide polymorphisms from a genomewide association study of Alzheimer disease. Arch Neurol 65(1):45–53
Liu Y et al (2008) Amyloid pathology is associated with progressive monoaminergic neurodegeneration in a transgenic mouse model of Alzheimer's disease. J Neurosci 28(51):13805–13814
Masters CL et al (1985) Amyloid plaque core protein in Alzheimer disease and Down syndrome. Proc Natl Acad Sci USA 82(12):4245–4249
McKhann GM et al (2011) The diagnosis of dementia due to Alzheimer's disease: recommendations from the National Institute on aging-Alzheimer's Association workgroups on diagnostic guidelines for Alzheimer's disease. Alzheimers Dement 7(3):263–269
Meek PD, McKeithan K, Schumock GT (1998) Economic considerations in Alzheimer's disease. Pharmacotherapy 18(2 Pt 2):68–73, discussion 79–82
Mills J, Reiner PB (1999) Regulation of amyloid precursor protein cleavage. J Neurochem 72(2):443–460
Minkeviciene R et al (2009) Amyloid beta-induced neuronal hyperexcitability triggers progressive epilepsy. J Neurosci 29(11):3453–3462
Morris RG et al (1986) Selective impairment of learning and blockade of long-term potentiation by an N-methyl-D-aspartate receptor antagonist, AP5. Nature 319(6056):774–776
Palop JJ et al (2007) Aberrant excitatory neuronal activity and compensatory remodeling of inhibitory hippocampal circuits in mouse models of Alzheimer's disease. Neuron 55(5):697–711
Petersen RC et al (2010) Alzheimer's Disease Neuroimaging Initiative (ADNI): clinical characterization. Neurology 74(3):201–209
Prasher VP et al (1998) Molecular mapping of Alzheimer-type dementia in Down's syndrome. Ann Neurol 43(3):380–383
Prince M et al (2009) Alzheimer's disease international world alzheimer report 2009. In: International AsD (ed) pp 1–96
Redish AD, Touretzky DS (1998) The role of the hippocampus in solving the Morris water maze. Neural Comput 10(1):73–111

Riemenschneider M et al (2004) Prion protein codon 129 polymorphism and risk of Alzheimer disease. Neurology 63(2):364–366
Roberson ED et al (2011) Amyloid-beta/Fyn-induced synaptic, network, and cognitive impairments depend on tau levels in multiple mouse models of Alzheimer's disease. J Neurosci 31(2):700–711
Selkoe DJ, Schenk D (2003) Alzheimer's disease: molecular understanding predicts amyloid-based therapeutics. Annu Rev Pharmacol Toxicol 43:545–584
Shankar GM et al (2008) Amyloid-beta protein dimers isolated directly from Alzheimer's brains impair synaptic plasticity and memory. Nat Med 14(8):837–842
Shaw LM et al (2009) Cerebrospinal fluid biomarker signature in Alzheimer's disease neuroimaging initiative subjects. Ann Neurol 65(4):403–413
Sperling RA et al (2011) Toward defining the preclinical stages of Alzheimer's disease: recommendations from the National Institute on Aging-Alzheimer's Association workgroups on diagnostic guidelines for Alzheimer's disease. Alzheimers Dement 7(3):280–292
Terry RD et al (1991) Physical basis of cognitive alterations in Alzheimer's disease: synapse loss is the major correlate of cognitive impairment. Ann Neurol 30(4):572–580
Thies W, Bleiler L (2011) 2011 Alzheimer's disease facts and figures. Alzheimers Dement 7(2):208–244
Walsh DM, Selkoe DJ (2007) A? Oligomers ? a decade of discovery. J Neurochem 101(5): 1172–1184
Walsh DM et al (2002) Naturally secreted oligomers of amyloid beta protein potently inhibit hippocampal long-term potentiation in vivo. Nature 416(6880):535–539
Wang HY et al (2000) Beta-Amyloid(1–42) binds to alpha7 nicotinic acetylcholine receptor with high affinity. Implications for Alzheimer's disease pathology. J Biol Chem 275(8):5626–5632
Wang HW et al (2002) Soluble oligomers of beta amyloid (1–42) inhibit long-term potentiation but not long-term depression in rat dentate gyrus. Brain Res 924(2):133–140
Wisniewski T, Ghiso J, Frangione B (1991) Peptides homologous to the amyloid protein of Alzheimer's disease containing a glutamine for glutamic acid substitution have accelerated amyloid fibril formation. Biochem Biophys Res Commun 179(3):1247–1254
Yan SD et al (1996) RAGE and amyloid-beta peptide neurotoxicity in Alzheimer's disease. Nature 382(6593):685–691

Chapter 4
Cellular Prion Protein and Cancers

Wei Xin, Man-sun Sy, and Chaoyang Li

Abstract Prion was first identified as the infectious agent of prion disease, since then the biological functions of PrP have been extensively studied. One of the functions of this glycosylphosphatidylinositol (GPI)-anchored protein is to act as an apoptotic regulator. Studies have shown that prion protein (PrP) is upregulated in some cancers including gastric, breast, and colorectal cancers. In these cancers, PrP has been postulated to regulate apoptosis through various pathways. However, the most recent data showed that in human pancreatic cancer and melanoma, PrP might play a different role. In these cancers, the upregulated PrP exist as a Pro-PrP instead of a mature, glycosylated, and GPI-anchored PrP. The Pro-PrP does not have the GPI anchor as it retains its GPI anchor peptide signal sequence (GPI-PSS). The GPI-anchor peptide signal sequence is normally removed in the endoplasmic reticulum prior to the addition of the GPI anchor. The GPI-PSS of PrP has a motif, which binds filament A (FLNA), a cytolinker protein. Binding of pro-PrP to FLNA disrupts the normal function of FLNA, which then facilitates the adhesion, migration, and invasion of the tumor cells. Most importantly, the upregulation of PrP is a marker of poorer prognosis in pancreatic cancer.

Keywords Prion • PrP • Cancer • Pancreas • Melanoma • Filamin A • Pro-PrP • Review

Disclaimer The authors have no financial interests to claim.

W. Xin, M.D., Ph.D. (✉) • M. Sy, Ph.D.
Department of Pathology, Case Western Reserve University,
11090 Euclid Ave, Cleveland, OH 44106, USA
e-mail: wxx10@case.edu

C. Li, Ph.D.
Wuhan Institute of Virology, Chinese Academy of Science, Wuhan, China

W.-Q. Zou and P. Gambetti (eds.), *Prions and Diseases: Volume 1, Physiology and Pathophysiology*, DOI 10.1007/978-1-4614-5305-5_4,

4.1 Introduction of Prion

Transmissible spongiform encephalopathies (TSE) or prion diseases are a group of fatal neurodegenerative disorders that affect both humans and animals. In humans, TSE include Creutzfeldt–Jakob disease (CJD), fatal insomnia (FI), and Gerstmann–Sträussler–Scheinker disease (GSS). In animals, TSE include scrapie in sheep and goat, bovine spongiform encephalopathy (BSE) (known as Mad Cow Disease), chronic wasting disease (CWD) in elk and deer, transmissible mink encephalopathy (TME), and transmissible spongiform encephalopathy of domestic and captive zoo animals (Bolton et al. 1982; Diener et al. 1982; Prusiner 1982, 1991).

Even though TSE has been known to be transmissible since early 1930s, the etiological agent remained elusive for decades. The infectious agent was too small to accommodate nucleic acid, resistant to agents that destroy nucleic acid but susceptible to agents that obliterate proteins. Griffith was the first to propose that the pathogen for TSE was a protein (Griffith 1967). Griffith proposed three mechanisms by which this might happen: a protein that turns on its own transcription; an altered form of a protein that catalyzes the conversion of the normal form into the same altered form through formation of an oligomer-like a crystal seed; and an antibody that stimulates its own production. However, it was Prusiner and his colleagues who made the fundamental discovery that led to current understanding of TSE. Prusiner and colleagues identified and sequenced the pathogen, which was subsequently found to be an abnormal form of a highly conserved normal protein in mammals. They named this agent as proteinaceous infectious particle, prion. Since then prion diseases have been used synonymously with TSE. All three forms of prion diseases: the infectious, the inherited, and the sporadic forms are believed to share the same pathogenic mechanism that is based on the conversion of the normal PrP into the pathogenic, scrapie PrP, PrP^{Sc} (Prusiner 1996).

The human prion gene, *PRNP*, is located on chromosome 20, at 20p13, with a three-exon structure. The third exon contains the entire open reading frame of the protein, which encodes PrP. PrP is a glycosylphosphatidylinositol (GPI)-anchored, highly conserved, and ubiquitously expressed glycoprotein (Kretzschmar et al. 1986; Harris 1999).

In human, the PrP is first synthesized as a pre-pro-PrP of 253 amino acids in the cytosol (Fig. 4.1). The first 22 amino acids at the N terminus contain the leader peptide sequence, while the last 22 amino acids at the C terminus encompass the GPI anchor peptide signal sequence (GPI-PSS). Both of these sequences are removed in the endoplasmic reticulum and thus are not present in the mature GPI-anchored PrP. Addition of a GPI anchor and two *N*-linked glycans co-translationally completes the synthesis of a mature GPI-anchored and glycosylated PrP.

The mature product of PrP contains 209 amino acids from residue 23–231, and can be divided into three major domains based on the structural motifs. The N-terminal domain includes the first 90 amino acids and is thought to be unstructured. This region also has a highly conserved motif of five repeating octapeptides. The central domain is located between amino acid 110 and 130. The C-terminal region

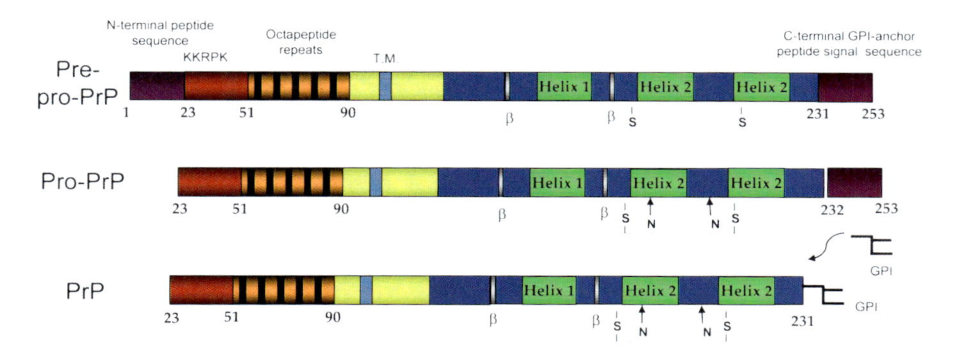

Fig. 4.1 The post-translational modification of the normal cellular PrP protein. Starting from Pre-pro-PrP, then Pro-PrP to final mature product of PrP

contains a well-defined, globular domain that has two potential *N*-linked glycosylation sites and a disulfide bridge (Donne et al. 1997; Prusiner et al. 1998; Safar and Prusiner 1998; Williamson et al. 1998). The protein backbone of the PrP has a molecular weight of approximately 23 kDa. However, with the addition of two *N*-linked glycans, and a GPI anchor, the final completed PrP has an approximately molecular weight of 34–39 kDa. Despite the fact that PrP is a relatively small protein, the synthesis, processing, and transit of PrP are complex, cell-context dependent, and not completely understood (Brown et al. 1997; Hope 1999; Hunter 1999; Kretzschmar 1999; Pergami et al. 1999).

In addition to cells in the CNS, PrP is expressed in many cell types, and many different PrP binding partners have been identified (Pergami et al. 1999). The octapeptide repeats at the N terminus of PrP contain four binding sites for divalent cations, such as Cu^{++} and Zn^{++}. Based on these findings it was proposed that PrP might function as a metal transporter (Viles et al. 1999; Wadsworth et al. 1999; Whittal et al. 2000). The N terminus of all mammalian PrP also contains a glycosaminoglycan (GAG)-binding motif. Binding of GAG has been speculated to be important in prion disease pathogenesis (Brown et al. 1997; Aguzzi 2000; Aguzzi et al. 2000; Aronoff-Spencer et al. 2000; Bonomo et al. 2000).

Like many other GPI anchored proteins, PrP is present in a microdomain on the cell surface commonly referred to as lipid rafts. Lipid rafts are special domain on the cell surface where signaling protein complexes are organized; thus PrP has been suggested to function as a signal transducing molecule (Harmey et al. 1995; Vey et al. 1996). PrP is detected in caveolae in association with caveollin-1. However, both neurons and lymphocyte express PrP but these cells lack caveolae (Harmey et al. 1995; Vey et al. 1996; Massimino et al. 2002; Prado et al. 2004). Therefore, the arrangement of PrP on the cell surface is likely to be cell type dependent.

In addition to binding metals and GAG, PrP also interacts with many other proteins, such as laminin receptor, *N*-CAM, lipids, heat shock proteins, chaperon protein, stress-inducible protein, and transcription factors (Martins and Brentani 2002; Martins et al. 2002). PrP also binds nucleic acid and lipids (Jaegly et al. 1998).

PrP has a putative nuclear localization signal and thus can function as a nuclear transport protein (Jaegly et al. 1998; Gu et al. 2003). PrP is expressed on murine bone marrow progenitor cells. $PrP^{-/-}$ stem cells are less efficient in engrafting irradiated host, suggesting that PrP is critical during hematopoietic development (Zhang et al. 2006). However, it should be noted that there is no obvious defect in hematopoietic development in *Prnp*$^{-/-}$ mice (Liu et al. 2001).

PrP has been reported to possess proapoptotic activity in primary murine neurons and in human HEK293 cells. PrP controls the functions of p53 at transcriptional and translational levels (Paitel et al. 2002; Kim et al. 2004a, b). On the other hand, other laboratories have reported that PrP has antiapoptotic activities in breast cancer cell lines (Roucou et al. 2005; Bounhar et al. 2006). PrP expressing neuronal cell lines are more resistant to apoptosis than PrP negative cell lines (Kuwahara et al. 1999). PrP transduces neruoprotective signals (Brown et al. 2002). PrP inhibits the functions of Bax and thus protects human neurons against Bax-mediated apoptosis in breast cancer cell lines (Bounhar et al. 2001, 2006). These differences may reflect the use of neurons from different species or the natures of the cell types studied.

4.2 PrP and Cancers

4.2.1 Breast Cancer

Since the discovery of PrP, most of the PrP studies have been focused on the role it plays in neurodegenerative disease. With the finding of PrP regulating apoptosis, more and more studies have been shifted on the possible role of this protein involving in cell survival and proliferation.

It has been reported that PrP was upregulated at both transcriptional and translational levels in TNF (tissue necrotizing factor)-resistant breast cancer cell lines compared to that of TNF-sensitive breast cancer MCF7 cell lines (Diarra-Mehrpour et al. 2004). These investigators identified many genes in PI3K/Akt pathways, which were involved in the TNF resistance. Also interestingly, these investigators showed that overexpression of PrP at both transcriptional and translational levels in TNF resistant cell lines compared to those of TNF-sensitive cell lines. By using recombinant adenoviruses, they could convert TNF-sensitive cells into TNF-resistant cells. Thus, PrP might induce cancer cells' resistance to TNF by involving the PI3K/Akt pathway. They also found that PrP might regulate P53 expression, and suggested that the resistant process might be related to the apoptotic cascades, involving P53.

Another study suggests that suppression of PrP expression may facilitate the activation of proapoptotic Bax by downregulation of Bcl-2 expression, and thus reduces the resistance to TRAIL induced apoptosis in breast cancer cells. These investigators studied the relationship between the resistance to the proapoptotic action of TNF-related apoptosis inducing ligand (TRAIL) and PrP function. They compared a TRAIL-sensitive MCF-7 human breast adenocarcinoma cell line with two TRAIL-resistant sublines: 2101 and MCF-7/ADR to Adriamycin, an apoptosis-inducing agent. It was found that the downregulation of PrP by small interfering RNA increased the sensitivity of Adriamycin- and TRAIL-resistant cells to TRAIL, but not to Epirubicin/Adriamycin. They also found that Bcl-2 expression was substantially decreased after PrP inhibition, but the levels of Bcl-X(L) and Mcl-1 were not affected and the downregulation of Bcl-2 expression was accompanied with Bax relocalization. Based on these findings, these investigators concluded that the inhibition of PrP expression promotes the activation of proapoptotic Bax by downregulation of Bcl-2 expression, thereby abolishing the resistance of breast cancer cells to TRAIL-induced apoptosis (Meslin et al. 2007b).

Expression of PrP was also associated with the resistance to adjuvant chemotherapy in patients with estrogen receptor (ER)-negative breast cancer (Meslin et al. 2007a). In this study, these investigators found that by immunohistochemical staining PrP was mainly expressed by myoepithelial cells in normal breast tissue. The tissue microarray analysis from 756 breast tumors showed that PrP was associated with ER-negative breast cancer subsets ($p<0.001$). The administration of the anthracycline-based adjuvant chemotherapy was not associated with a significant risk reduction for death in patients with ER-negative/PrP-positive disease, but it decreased the risk for death in patients with ER-negative/PrP-negative tumors. And they concluded that the ER-negative/PrP-negative phenotype is associated with an enhanced sensitivity to adjuvant chemotherapy (Meslin et al. 2007a; Mehrpour and Codogno 2010).

4.2.2 *Gastric Cancer*

Fan and his colleagues were the first group to report that PrP was overexpressed in some gastric carcinoma cell lines, and overexpression of PrP in gastric cancer cell lines was associated with the resistance to both P-glycoprotein (P-gp)-related and P-gp-nonrelated drugs. Inhibition of the PrP expression by antisense or RNAi partially reversed the multidrug resistance. PrP also suppresses adriamycin-induced apoptosis by altering the expression of Bcl-2 and Bax (Du et al. 2005). The inhibition of the PrP expression by RNAi in gastric cancer cell line could suppress ROS and slowed down apoptosis in these cells. These investigators proposed that PrP modulates the apoptotic pathway by functioning as an antiapoptotic protein through Bcl-2-dependent pathways (Liang et al. 2006).

Furthermore, by Immunohistochemical staining, gastric adenocarcinoma with increased PrP expression also correlated with the clinical staging. PrP was overexpressed in metastatic gastric cancers compared to nonmetastatic cancer. Expression of PrP promotes the adhesion, invasion, and in vivo metastasis of gastric cancer cell lines SGC7901 and MKN45 in xenograft models. Mechanistically, PrP appears to increase the promoter activity, and the expression of MMP1. It was suggested that the N-terminal region of PrP might promote the invasion and metastatic ability of the tumor cells partially through activation of MEK/ERK pathway, and consequently by transactivation of MMP11. They also reported that overexpression of PrP might promote the tumorigenesis and proliferation of gastric cancer cells partially through the activation of PI3K/Akt pathway and the activation of CyclinD1 to regulate the G1/S phase transition. It was reported that the octapeptide repeat region might play a role in promoting the proliferation of gastric cancer cells, as cancer cell proliferation with more octapeptide repeats has a more rapid proliferation rate (Pan et al. 2006; Liang et al. 2009).

4.2.3 Colorectal Cancers

By using expression microarray, a study showed that PrP was overexpressed in colorectal cancer. Along with other proteins, PrP had a significant difference in the expression levels between the right colonic and the rectal cancers. PrP expression constituted an independent prognostic factor of the 3-year survival in multivariate analysis (Antonacopoulou et al. 2008).

Another group utilized different antibodies against different PrP regions to investigate whether these antibodies could induce apoptosis and be utilized in the treatment of these cancers. They found that different antibodies against PrP had varying degrees of antiproliferative activity, and some antibodies were particularly potent and afforded >40% reduction in proliferation. In combination therapy experiments, antibodies to PrP could induce apoptosis and variably enhanced the antitumoral effect of irinotecan, 5-FU, cisplatin, and doxorubicin. In different colon cancer cell lines, antibody effectiveness correlated with tumor aggressiveness. The administration of PrP antibody in vivo nude mouse could inhibit human HCT 116 xenografts (McEwan et al. 2009).

4.2.4 Pancreatic Ductal Carcinoma

Pancreatic ductal carcinoma (PDAC) is one of most deadly solid cancer with a 5-year survival rate of about 6% (cancer statistics 2011, http://www.cancer.org/Research/CancerFactsFigures). Our group found that all human pancreatic ductal adenocarcinoma (PDAC) cell lines ($n=7$) have upregulated expression of PrP. On the other hand, in normal pancreas, only islet cells have detectable PrP; neither acinar cells nor ductal cells, which are thought to be the precursors of PDAC, have detectable PrP (Fig. 4.2) (Li et al. 2009, 2010; Sy et al. 2010). However, the PrP in pancreatic cancer cells is different from the normal forms, the PrP is neither glycosylated nor GPI-anchored, and it exists as pro-PrP retaining its GPI-PSS. Unexpectedly, in the PrP GPI-PSS there is a filamin A (FLNA)-binding motif, and thus binds FLNA. FLNA is an actin-binding protein that integrates cell mechanics and signaling (Stossel et al. 2001). FLNA links cell surface proteins, such as integrins and growth factor receptors, to the cytoskeleton. Binding of pro-PrP to FLNA disrupted cytoskeletal organization. Inhibition of PrP expression by shRNA in the PDAC cell lines altered the cytoskeleton and expression of multiple signaling proteins;

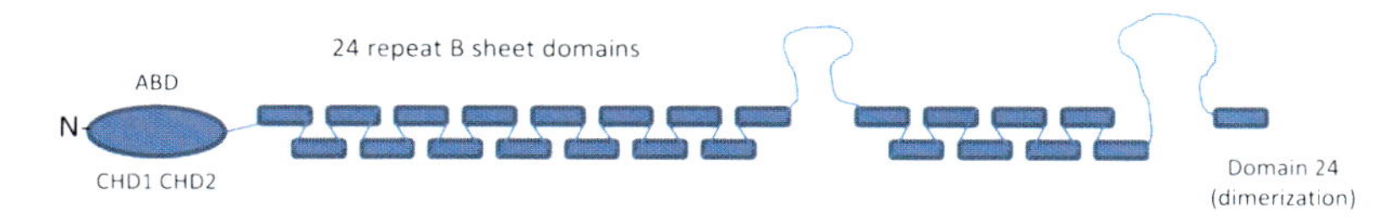

Fig. 4.2 The expression of PrP in benign pancreatic tissue, PanIN, and pancreatic ductal adenocarcinoma. (**a**) Normal pancreas. PrP expressed in normal islet cells not in pancreatic ductal cells. *Blue arrow*: pancreatic duct, *red arrow*: islet cells. (**b**) PanIN-2. The PrP expression not identified in PanIN-2. *Blue arrow*: neoplastic pancreatic ducts (**c**) and (**d**) pancreatic cancers. PrP are expressed in some pancreatic ductal adenocarcinoma (**c**) but not in others (**d**). *Blue arrow*, cancer glands

it also reduced cellular proliferation and invasiveness in vitro as well as tumor growth as xenografts in vivo.

A subgroup of human patients with pancreatic cancer was found to have tumors that expressed pro-PrP. Most importantly, PrP expression in tumors correlated with a marked decrease in patient survival. Therefore, the binding of pro-PrP to FLNA perturbs FLNA function, thus contributing to the aggressiveness of PDAC. Prevention of this interaction could provide an attractive target for therapeutic intervention in human PDAC.

4.3 Biological Function of GPI-PSS

The finding that PrP exists as Pro-PrP in PDAC is fascinating. Due to efficient processing and competent quality control system in normal cells, pro-PrP is undetectable in normal cells. What is the significance of the accumulation of pro-PrP in PDACs? The GPI modification pathway is complex and not completely understood (Ikezawa 2002; Maeda et al. 2006; Orlean and Menon 2007; Wiedman et al. 2007). More than 24 genes are involved in this process, and the biosynthesis of the GPI anchors and their attachment to proteins are complex, protein specific, and cell context dependent. The common core structure of the GPI anchor is synthesized in the endoplasmic reticulum in a stepwise mechanism. First catalyzed by a a1–6 GlcNAc transferase complex, which is composed of seven gene products: PIG-A, PIG-C, PIG-H, GPI-1, PIG-Y, PIG-P, and DPM2, it transfers *N*-acetyl-glucosamine (GlcNAc) from UDP-GlcNAc to phosphatidylinositol (PI) to form GlcNAc-PI. Second, the compound is de-*N*-acetylated by PIG-L to generate GlcN-PI. Then, three mannose residues are sequentially added. The last step is mediated by a transamidase. Subsequently, the GPI-PSS is cleaved and a completed GPI complex is then attached to the pro-protein. The cleavage site at the C-terminal of the GPI-PSS is known as the ω site, which is confined to amino acids glycine, serine, cysteine, alanine, aspartic acid, and asparagine. There is no other obvious motif in the GPI-PSS that signals the transamidase reaction.

The specific functions of GPI–PSS have been previously studied on human carcinoembryonic antigen (CEA). One study showed that ectopic expression of various members of the family of intercellular adhesion molecules in murine myoblasts either blocks or allows myogenic differentiation. CEA and CEACAM6 are GPI anchored, whereas CEACAM1 is transmembrane anchored. Overexpression of GPI-linked neural cell adhesion molecule (NCAM) accelerated the myogenic differentiation. After creating chimeric protein by exchanging C-terminal hydrophobic domains of CEA, CEACAM1, and NCAM, it was reported that the presence of the GPI-PSS from CEA in the chimeras was sufficient to convert both CEACAM1 and NCAM into differentiation-blocking proteins. Conversely, CEA could be converted into a neutral protein by exchanging its GPI anchor for the TM anchor of CEACAM1. These results suggest that significant functional information resides in the processed extreme C terminus of CEA and in the GPI anchor that it determines (Screaton et al. 2000).

Another study from the same group also showed that exchanging the GPI-PSS of NCAM for the GPI-PSS of CEA generates a mature protein that has a NCAM external domain, but CEA-like tumorigenic activity. Based on these findings, it is postulated that the GPI-PSS possesses a functional biological information that specifies the addition of a particular GPI anchor that, ultimately, determines the final function of the mature protein (Nicholson and Stanners 2007).

CDC91L1 is the gene encoding CDC91L1 also called phosphatidylinositol glycan class U (PIG-U) a transamidase complex unit in the GPI anchoring pathway. The germline mutation of translocation 20q11 in bladder cancer causes the CDC91L1 protein to overexpress, which could malignantly transform NIH3T3 cells in vitro and in vivo. Over expression of CDC91L1 also resulted in upregulation of the urokinase receptor (uPAR), a GPI-anchored protein, and in turn increased STAT-3 phosphorylation in bladder cancer cells. CDC91L1 could function as an oncogene in bladder cancer, and implicate the GPI anchoring system as a potential oncogenic pathway (Guo et al. 2004).

Evidence also showed that two other GPI transamidase complexes were involved in human breast cancer: PIG class T (PIG-T) and GPI anchor attachment 1 (GPAA1). The overexpression of PIG-T and GPAA1 transformed NIH3T3 cells in vitro and increased tumorigenicity and invasion of these cells in vivo. Suppression of PIG-T expression in breast cancer cell lines led to inhibition of anchorage-independent growth. In addition, PIG-T and GPAA1 expression levels could positively correlate with paxillin phosphorylation in invasive breast cancer cell lines. Furthermore, suppression of PIG-T and GPAA1 expression led to a decrease in paxillin phosphorylation with a concomitant decrease in invasion ability. These data suggest that the GPI transamidase complex can function as oncogenes (Wu et al. 2006).

In head squamous cell carcinomas, in addition to PIG-U, other proteins in the same family, such as GAA1, PIG-T, were also found to be significantly upregulated at transcriptional and translational levels, which further suggests the GPI anchor process involved in tumorigenesis (Jiang et al. 2007).

PrP in Pancreatic cancer cell lines is neither glycosylated nor GPI anchored; it exists as pro-PrP retaining its GPI-PSS (Fig. 4.1). This deficiency is not caused by a general defect of the GPI anchor process in the PDAC cell lines, as the two other GPI-anchored proteins, CD55 and flotillin 1, remain GPI anchored in the PDAC cell lines. Despite lacking a GPI anchor, the pro-PrP is present on the PDAC cell surface, using the GPI-PSS as a transmembrane domain, as the model proposed before (Waneck et al. 1988). While our immunoblotting results with multiple anti-PrP monoclonal antibodies (mAbs) suggest that pro-PrP is the only detectable PrP in the PDAC cell lines; however, we cannot completely rule out the possibility that a very small amount of normal, GPI-anchored PrP is also present in these cells.

The underlying reason that the GPI-PSS of PrP is not cleaved in the PDAC cell lines has not been elucidated so far. On the genetic level, we did not find any mutation in the coding region of the *PRNP* after sequencing all six PDAC cell lines. It is interesting to note that the GPI-PSS of PrP is intrinsically inefficient compared with other GPI-anchored proteins (Chen et al. 2001). Thus, a slight defect in the GPI anchor assembly machinery in PDAC may have a more dramatic effect on PrP than other GPI-anchored proteins with a more efficient GPI-PSS, such as CD55, which is GPI-anchored in the PDAC cell lines. A defect in lipid metabolism, which limits the availability of the GPI anchor precursor, can also impact the modification of PrP. In addition to defects in GPI anchor modification and lipid metabolism, defects in the quality-control system in the endoplasmic reticulum or in the removal of the unprocessed pro-PrP, presumably by the proteasomal degradation machinery, may also contribute to the accumulation of pro-PrP.

In the PDAC cell lines, PrP is also not glycosylated. Though the presence of the *N*-linked glycans on PrP is not required for GPI anchor modification (Cancellotti et al. 2005; Wiseman et al. 2005), the presence of a GPI anchor has been reported to influence the glycosylation of Thy-1, a GPI-anchored protein (Devasahayam et al. 1999). Thus, failure to remove the GPI-PSS may modulate PrP glycosylation. The lack of *N*-linked glycans may then alter the metabolism or transit of pro-PrP, contributing to its accumulation in the PDAC cell lines.

4.4 Filamin A and PrP Binding

Filamin A (*FLNA*) gene is located on chromosome Xq28 (Stossel et al. 2001). FLNA has a molecular mass of 280 kDa. After binding actin filaments, FLNA promotes high-angle branching of actin filaments to maintain a cytoskeletal network responsible for cell-shape maintenance and migration. In males, FLNA deficiency caused by a null mutation is embryonic lethal. In females, depending on the nature and location of the mutation, it causes several developmental syndromes involving neuronal, skeletal, and connective tissues (Feng and Walsh 2004).

Native FLNA is a homodimer and each subunit contains an N-terminal actin-binding domain (ABD) followed by the 24 long rod-like β-sheet, interrupted by two roughly 30-amino acid, flexible loops that are proposed to form hinge structures (Fig. 4.3). The C-terminal last domain 24 is responsible for the dimerization, forming a V-shaped flexible structure that is essential for function (Feng and Walsh 2004). FLNA interacts with numerous proteins, including proteins involving in signal-

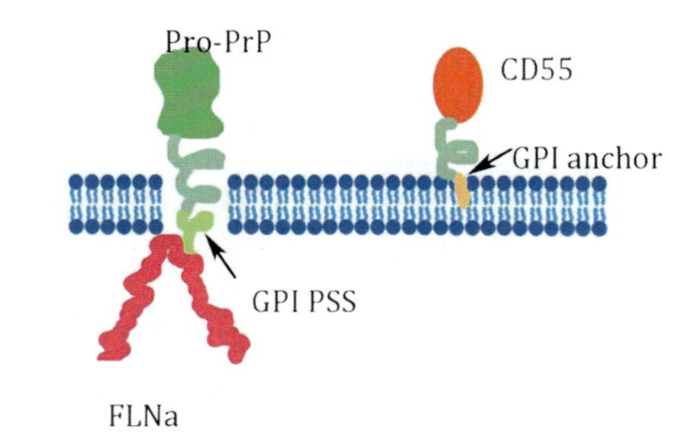

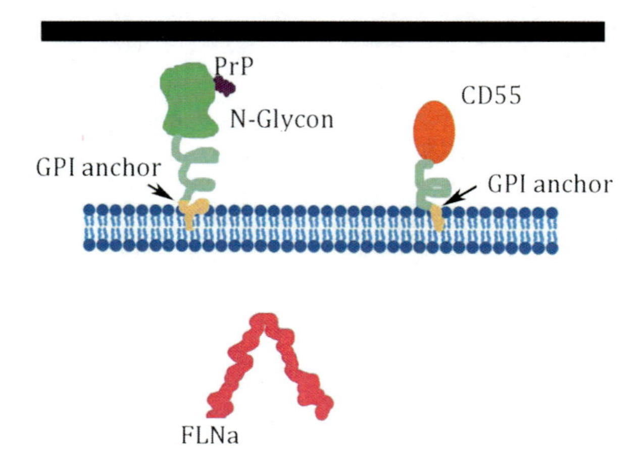

Fig. 4.3 The structure of Filamin A. CHD1 and CHD2 are two calponin homology domains. ABD is actin-binding domain

transducing pathway, adhesions, and growth factor receptors. Most of the proteins bind to domain 10 to domain 24 at C-terminal Ig-like domains of FLNA.

By coimmunoprecipiatation FLNA copurified with PrP and vice versa in pancreatic cell lines. Further in vitro studies show that FLNA only binds pro-PrP but not mature PrP, which lacks the GPI-PSS. In the PDAC cell lines, the binding of Pro-PrP and FLNA is stable, as PrP and FLNA colocalize in the cancer cells by immunofluorescent staining and observed in a confocal microscope (Li et al. 2009).

The presence of an FLNA-binding motif in the GPI-PSS appears to be specific for PrP. We examined 14 GPI-PSS from other normally GPI-anchored proteins and we found that only the GPI-PSS of PrP has the FLNA-binding motif. Therefore, even if some other normally GPI-anchored proteins also exist as pro-proteins, retaining their GPI-PSS, they are not expected to bind FLNA.

More recent studies using recombinant proteins in vitro reveal that pro-PrP has multiple binding sites at the C-terminal Ig-like domains of FLNA, including domains 10, 16, 17, 18, 20, 21, and 23. This finding is not unexpected because the Ig-like domains are highly conserved. However, whether all these binding sites are available for pro-PrP binding in native, dimeric, FLNA is not known. On the other hand, we found that the last five amino acids at the C-terminal end of the PrP GPI-PSS are critical for FLNA binding. Removal of these five amino acids completely eliminates its FLNA-binding capacity. The data suggest that the GPI-PSS of PrP is able to transverse the membrane bilayer, and binds FLNA.

Inhibition of PrP expression by PrP-specific shRNA in the PDAC cell lines did not affect the expression level of FLNA; however, it did alter the spatial distribution of FLNA (Li et al. 2009, 2010). Compared to control cells, in PrP downregulated cells, FLNA is more concentrated in the cytosol, away from the inner-membrane leaflet in the leading edges. Therefore, it appears that pro-PrP by binding to FLNA is able to pull FLNA closer to the inner membrane leaflet. As expected, downregulation of PrP also alters the organization of the actin filaments (Li et al. 2010). These morphological changes have significant behavior consequences, as PrP downregulated cell lines proliferate more slowly and are less invasive than control cells. Most importantly, the growth of the PrP downregulated tumor cells in nude mice was significantly diminished. Thus, the binding of pro-PrP to FLNA enables the PDAC cell lines to proliferate faster and more invasive. The binding of pro-PrP may physically remove FLNA from its normal environment and prevent its normal physiological function. Alternatively, binding of pro-PrP may compete for binding sites on FLNA that are normally occupied by its interaction partners.

At the molecular level, PrP downregulated cell lines have reduced levels of p-cofilin-1, a critical regulator of the actin filament polymerization. On the other hand, the levels of p-Rac1, a Rho-GTPase; p-ERK-1/2 and p-MEK-1, two serine/threonine kinases in the MAPK pathway; and p-Fyn, a *Src* family tyrosine kinase, are markedly increased in the PrP downregulated cells. Therefore, reducing the expression of PrP in the PDAC cell lines appears to have effects on multiple signal transduction pathways. As more than 40 proteins bind to FLNA, the aberrant binding of pro-PrP to FLNA will have rippling effects on the binding of FLNA to some of its binding partners, such as integrins, which are known to play critical roles in cellular adhesion, invasion and migration (Li et al. 2010).

4.5 Expression of pro-PrP Is a Marker of Poorer Prognosis in Pancreatic Cancer

Pancreatic cancer is the fourth leading cancer death in the USA, and it is responsible for more than 30,000 deaths a year in this country. Nearly 90% of pancreatic cancers are pancreatic ductal adenocarcinoma (PDAC). PDAC is still a lethal disease with a dismal overall median survival of 6 months for all stages and 6% of the 5-year survival rate (cancer statistics 2011, http://www.cancer.org/Research/CancerFactsFigures).

Progression of human PDAC correlates with a series of histological changes from a flat, normal columnar epithelium to a flat/papillary mucinous epithelium, with increasing complexity of cellular architecture and cytological atypia (Warshaw and Fernandez-del Castillo 1992; Hruban et al. 2001a, b). These precursor lesions are defined as pancreatic intraepithelial neoplasia (PanIN), which includes PanIN-1, -2, and -3, based on the cytological atypia and complex architecture, as well as accompanied with the increasing numbers of corresponding genetic mutations.

The molecular pancreatic carcinogenetic pathways are complex and not fully clarified; many genetic mutations been identified. The most common genetic lesions found in human PDAC are mutations in *KRAS*, *TP53*, *DPC4*, and *CDNK2A* (Hruban et al. 2001a, b). It is now generally accepted that the *KRAS* mutation is one of the earliest, and most important genetic lesion in the development of PDAC; the majority of PDAC cases have a mutation in codon 12 of *KRAS*, substituting a glycine with aspartate, valine, or arginine. However, many benign pancreatic lesions also have increased K-ras mutations.

In normal human pancreas, only islet cells demonstrated PrP immunoreactivity; neither acinar nor ductal epithelial cells stained for PrP. PrP was also undetectable in the duct cells in chronic pancreatitis, and PanIN-1 and -2. Approximately 13% PanIN-3 specimens showed weak PrP staining. However, among the 83 PDAC cases, 34 (41%) showed strong PrP staining. The PrP positive PDAC tumor cells also reacted strongly with the anti-GPI-PSS antiserum. Thus, as in the PDAC cell lines, PrP exists as pro-PrP in human PDAC lesions (Li et al. 2009).

Most importantly, the overexpression of pro-PrP is present only in a subset of pancreatic cancers associated with poorer clinical prognosis. PDAC patients with overexpression of PrP had a median survival time of 360 days, while those without PrP expressions had a median survival time of over 1,000 days. Furthermore, this association is independent of other clinical parameters, such as age, gender, size, or histological differentiation of the tumor. The PDAC tumors with PrP may have a growth advantage as in cell culture and, thus, are more aggressive.

Though there was a study reported that *PRNP* was upregulated in BxPC 3, Capan 1, and five other PDAC cell lines (Han et al. 2002). However, other gene profiling studies have not identified *PRNP* as a contributing factor in human PDAC (Aguirre et al. 2004; Holzmann et al. 2004; Bashyam et al. 2005). Whether other genetic mutations, especially *DPC4* and *TP53*, interact with PrP have not been fully studies. It has been reported that PrP could regulate TP53 in some cancer cells, and there was a potential TP53 binding site in the promoter region of *PRNP*As (Guillot-Sestier et al. 2009). Potentially PrP and P53 may act synergistically to modulate PDAC progression; the coexpression of these two molecules could deliver a much worse prognosis compared to that of either PrP or P53 alone.

4.6 PrP and Melanoma

Despite its importance in cellular responses, FLNA is dispensable for cell-autonomous survival. Some human melanoma cell lines, such as M2 and -3, do not express FLNA (Byers et al. 1991). Since FLNA is important in actin organization, cells lacking FLNA are devoid of actin fiber bundles and, thus, are impaired in their cellular migration in vitro. This deficiency is rescued by the transfection of a plasmid encoding FLNA into M2 cells (Cunningham et al. 1992).

More recently, we found that both M2 and A7 cells express pro-PrP. Similar to PDAC cell lines inhibition of PrP expression in A7 cells alters the spatial distribution of FLNA and reduces their spreading and migration. One of the best-characterized binding partners of FLNA is integrin β chain. Integrins are a family of cell adhesion molecules that are important in tumor cell growth, migration, invasion, and dissimulation (Stossel et al. 2001; Hynes 2002; Kim et al. 2011). Interestingly, in A7 cells, FLNA, PrP, and integrin do not exist as a stable trimeric complex. Instead, they exist as two independent, yet functionally linked, complexes; they are FLNA with PrP or FLNA with integrin beta1. Reducing PrP expression in A7 cells decreases the amount of integrin beta1 bound to FLNA. A PrP GPI-PSS synthetic peptide that crosses the cell membrane inhibits A7 cell spreading and migration. Thus, in A7 cells FLNA does not act alone; the binding of pro-PrP enhances association between FLNA and integrin beta1, which then promotes cell spreading and migration.

The underlying mechanisms of melanoma progression beginning from benign nevus, to aberrant growth of dysplastic cells, to radial growth phase, to vertical growth phase, and eventually to metastatic melanoma are complex and incompletely understood (Haass et al. 2005; Kuphal et al. 2005). Both FLNA and integrins have been implicated in melanoma progression. Human in situ melanoma cells growing along the dermal-epidermal junctions, as single cells, were largely FLNA negative, whereas tumor cells in nests and dermal components showed strong FLNA staining (Bouffard et al. 1994). It was postulated that FLNA might promote melanoma cell motility during tissue invasion from the epidermis to the dermis. With regard to integrin expression, it was reported that in situ melanoma stained either uniformly positive or uniformly negative for a2b1; the expression of this protein correlated with the later stages of melanoma progression (Duncan et al. 1996). With regard to the expression of PrP in normal human skin, only epithelial cells and sporadic mononuclear cells within the dermis demonstrated weak PrP immunoreactivity (Pammer et al. 1998).

We found that Pro-PrP is undetectable in normal melanocytes but is detected in melanoma in situ, and with significant upregulation of pro-PrP in invasive melanoma. The binding of pro-PrP to FLNA, therefore, also contributes to melanomagenesis. Immunostaining for pro-PrP, integrin, and FLNA in melanoma biopsies may provide new insights into the role these molecules play in human melanoma tumorigenesis.

4.7 Conclusion and Future Perspective

Multiple studies have shown that PrP are upregulated in some cancer types including breast, gastric, colorectal, and pancreatic cancers as well as melanoma. In breast, stomach, and colorectal cancers, the data suggest that PrP exert effects on drug resistance and invasiveness and protect the tumor cells by regulating apoptosis

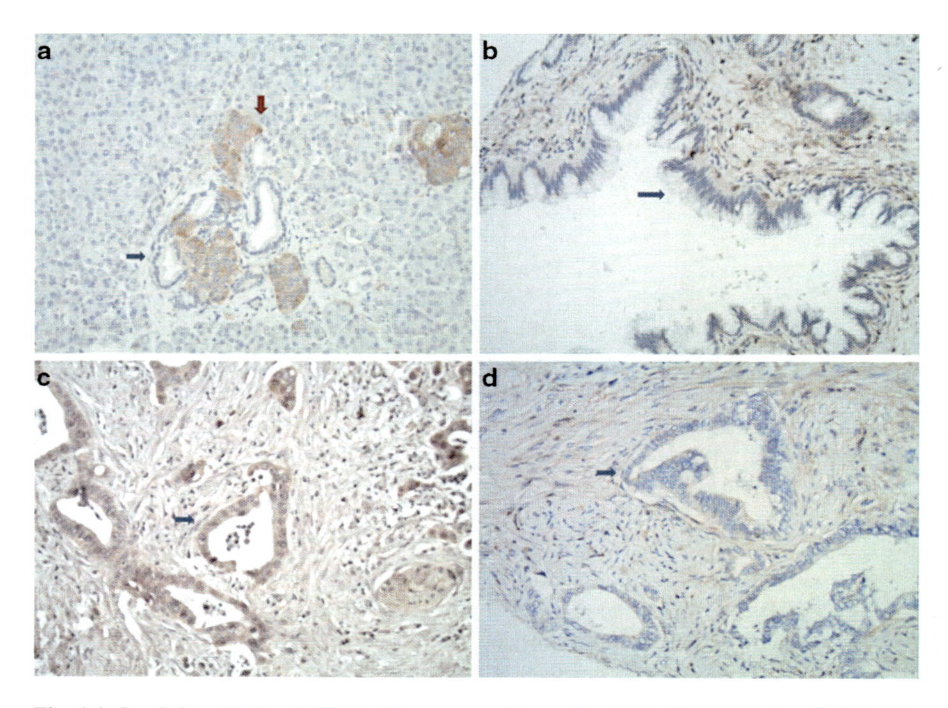

Fig. 4.4 Pro-PrP and PrP with FLNA. The *top panel* shows Pro-PrP in PDAC binds to FLNA. The GPI-PSS functions as the transmembrane domain. Other GPI anchor protein CD55 has the GPI anchor. The *bottom panel* shows normal PrP with GPI anchor and has no reaction with FLNA. Normal CD55 with GPI anchor

pathway. However, it should be noted that it is not clear whether in these tumor cell lines the PrP exists as a normal GPI-anchored PrP or pro-PrP as we have demonstrated in the PDAC cell lines as well as in melanoma cell lines.

In pancreas cancer and melanoma, the main form of PrP is pro-PrP but not normal mature PrP (Fig. 4.4). The pro-PrP is present on the cell surface as well as in the cytosol. Presumably, since pro-PrP and PrP have different biological functions, identifying the forms of PrP in other cancers will provide insights into the mechanisms by which expression of either PrP or pro-PrP modulates tumor cell biology. It should be noted that the expression of FLNA and pro-PrP does not to occur in all tumor types. For example, lung small cell carcinomas ($n=3$) express FLNA but not PrP; neuroblastomas ($n=3$) express neither FLNA nor PrP, and leukemia cell lines ($n=3$) express FLNA, but with a normal GPI anchored PrP.

Further study identifying the underlying mechanisms that cause the retention of the GPI-PSS on PrP in cancer cell lines will help us understand the cell biology of the GPI-anchor modification pathway and the roles they play in tumor biology. High levels of soluble PrP are detected in the culture supernatants of the PDAC cell lines. Therefore, soluble pro-PrP may be present in the circulation or body fluid of patients with PDAC. Detection of pro-PrP in fecal material or pancreatic ductal fluids may provide an early and noninvasive method for detecting PDAC. Finally, since pro-PrP is undetectable in normal cells, prevention of the interaction between pro-PrP and FLNA could provide a novel, specific target for therapeutic intervention in PDAC.

Acknowledgement Figure 4.4 was drawn by William Xin, and we appreciate his delicate art work.

References

Aguirre AJ, Brennan C et al (2004) High-resolution characterization of the pancreatic adenocarcinoma genome. Proc Natl Acad Sci USA 101(24):9067–9072

Aguzzi A (2000) Molecular pathology of prion diseases. Vox Sang 78(Suppl 2):25

Aguzzi A, Klein MA et al (2000) Prions: pathogenesis and reverse genetics. Ann NY Acad Sci 920:140–157

Antonacopoulou AG, Grivas PD et al (2008) POLR2F, ATP6V0A1 and PRNP expression in colorectal cancer: new molecules with prognostic significance? Anticancer Res 28(2B): 1221–1227

Aronoff-Spencer E, Burns CS et al (2000) Identification of the Cu^{2+} binding sites in the N-terminal domain of the prion protein by EPR and CD spectroscopy. Biochemistry 39(45):13760–13771

Bashyam MD, Bair R et al (2005) Array-based comparative genomic hybridization identifies localized DNA amplifications and homozygous deletions in pancreatic cancer. Neoplasia 7(6):556–562

Bolton DC, McKinley MP et al (1982) Identification of a protein that purifies with the scrapie prion. Science 218(4579):1309–1311

Bonomo RP, Imperllizzeri G et al (2000) Copper(II) binding modes in the prion octapeptide PHGGGWGQ: a spectroscopic and voltammetric study. Chemistry 6(22):4195–4202

Bouffard D, Duncan LM et al (1994) Actin-binding protein expression in benign and malignant melanocytic proliferations. Hum Pathol 25(7):709–714

Bounhar Y, Zhang Y et al (2001) Prion protein protects human neurons against Bax-mediated apoptosis. J Biol Chem 276(42):39145–39149

Bounhar Y, Mann KK et al (2006) Prion protein prevents Bax-mediated cell death in the absence of other Bcl-2 family members in *Saccharomyces cerevisiae*. FEMS Yeast Res 6(8): 1204–1212

Brown DR, Qin K et al (1997) The cellular prion protein binds copper in vivo. Nature 390(6661):684–687

Brown DR, Nicholas RS et al (2002) Lack of prion protein expression results in a neuronal phenotype sensitive to stress. J Neurosci Res 67(2):211–224

Byers HR, Etoh T et al (1991) Cell migration and actin organization in cultured human primary, recurrent cutaneous and metastatic melanoma. Time-lapse and image analysis. Am J Pathol 139(2):423–435

Cancellotti E, Wiseman F et al (2005) Altered glycosylated PrP proteins can have different neuronal trafficking in brain but do not acquire scrapie-like properties. J Biol Chem 280(52):42909–42918

Chen R, Knez JJ et al (2001) Comparative efficiencies of C-terminal signals of native glycophosphatidylinositol (GPI)-anchored proproteins in conferring GPI-anchoring. J Cell Biochem 84(1):68–83

Cunningham CC, Gorlin JB et al (1992) Actin-binding protein requirement for cortical stability and efficient locomotion. Science 255(5042):325–327

Devasahayam M, Catalino PD et al (1999) The glycan processing and site occupancy of recombinant Thy-1 is markedly affected by the presence of a glycosylphosphatidylinositol anchor. Glycobiology 9(12):1381–1387

Diarra-Mehrpour M, Arrabal S et al (2004) Prion protein prevents human breast carcinoma cell line from tumor necrosis factor alpha-induced cell death. Cancer Res 64(2):719–727

Diener TO, McKinley MP et al (1982) Viroids and prions. Proc Natl Acad Sci USA 79(17):5220–5224

Donne DG, Viles JH et al (1997) Structure of the recombinant full-length hamster prion protein PrP(29–231): the N terminus is highly flexible. Proc Natl Acad Sci USA 94(25):13452–13457

Du J, Pan Y et al (2005) Overexpression and significance of prion protein in gastric cancer and multidrug-resistant gastric carcinoma cell line SGC7901/ADR. Int J Cancer 113(2):213–220

Duncan LM, Bouffard D et al (1996) In situ distribution of integrin alpha 2 beta 1 and alpha-actinin in melanocytic proliferations. Mod Pathol 9(9):938–943

Feng Y, Walsh CA (2004) The many faces of filamin: a versatile molecular scaffold for cell motility and signalling. Nat Cell Biol 6(11):1034–1038

Griffith JS (1967) Self-replication and scrapie. Nature 215(5105):1043–1044

Gu Y, Hinnerwisch J et al (2003) Identification of cryptic nuclear localization signals in the prion protein. Neurobiol Dis 12(2):133–149

Guillot-Sestier MV, Sunyach C et al (2009) The alpha-secretase-derived N-terminal product of cellular prion, N1, displays neuroprotective function in vitro and in vivo. J Biol Chem 284(51):35973–35986

Guo Z, Linn JF et al (2004) CDC91L1 (PIG-U) is a newly discovered oncogene in human bladder cancer. Nat Med 10(4):374–381

Haass NK, Smalley KS et al (2005) Adhesion, migration and communication in melanocytes and melanoma. Pigment Cell Res 18(3):150–159

Han H, Bearss DJ et al (2002) Identification of differentially expressed genes in pancreatic cancer cells using cDNA microarray. Cancer Res 62(10):2890–2896

Harmey JH, Doyle D et al (1995) The cellular isoform of the prion protein, PrPc, is associated with caveolae in mouse neuroblastoma (N2a) cells. Biochem Biophys Res Commun 210(3): 753–759

Harris DA (1999) Cellular biology of prion diseases. Clin Microbiol Rev 12(3):429–444

Holzmann K, Kohlhammer H et al (2004) Genomic DNA-chip hybridization reveals a higher incidence of genomic amplifications in pancreatic cancer than conventional comparative genomic hybridization and leads to the identification of novel candidate genes. Cancer Res 64(13):4428–4433

Hope J (1999) Prions. Curr Biol 9(18):R673–R674

Hruban RH, Adsay NV et al (2001a) Pancreatic intraepithelial neoplasia: a new nomenclature and classification system for pancreatic duct lesions. Am J Surg Pathol 25(5):579–586

Hruban RH, Iacobuzio-Donahue C et al (2001b) Molecular pathology of pancreatic cancer. Cancer J 7(4):251–258

Hunter N (1999) Prion diseases and the central dogma of molecular biology. Trends Microbiol 7(7):265–266

Hynes RO (2002) Integrins: bidirectional, allosteric signaling machines. Cell 110(6):673–687

Ikezawa H (2002) Glycosylphosphatidylinositol (GPI)-anchored proteins. Biol Pharm Bull 25(4):409–417

Jaegly A, Mouthon F et al (1998) Search for a nuclear localization signal in the prion protein. Mol Cell Neurosci 11(3):127–133

Jiang WW, Zahurak M et al (2007) Alterations of GPI transamidase subunits in head and neck squamous carcinoma. Mol Cancer 6:74

Kim BH, Lee HG et al (2004a) The cellular prion protein (PrPC) prevents apoptotic neuronal cell death and mitochondrial dysfunction induced by serum deprivation. Brain Res Mol Brain Res 124(1):40–50

Kim CL, Karino A et al (2004b) Cell-surface retention of PrPC by anti-PrP antibody prevents protease-resistant PrP formation. J Gen Virol 85(Pt 11):3473–3482

Kim C, Ye F et al (2011) Regulation of integrin activation. Annu Rev Cell Dev Biol 27:321–45

Kretzschmar HA (1999) Molecular pathogenesis of prion diseases. Eur Arch Psychiatry Clin Neurosci 249(Suppl 3):56–63

Kretzschmar HA, Stowring LE et al (1986) Molecular cloning of a human prion protein cDNA. DNA 5(4):315–324

Kuphal S, Bauer R et al (2005) Integrin signaling in malignant melanoma. Cancer Metastasis Rev 24(2):195–222

Kuwahara C, Takeuchi AM et al (1999) Prions prevent neuronal cell-line death. Nature 400(6741):225–226

Li C, Yu S et al (2009) Binding of pro-prion to filamin A disrupts cytoskeleton and correlates with poor prognosis in pancreatic cancer. J Clin Invest 119(9):2725–2736

Li C, Xin W et al (2010) Binding of pro-prion to filamin A: by design or an unfortunate blunder. Oncogene 29(39):5329–5345

Liang J, Pan YL et al (2006) Overexpression of PrPC and its antiapoptosis function in gastric cancer. Tumour Biol 27(2):84–91

Liang J, Ge F et al (2009) Inhibition of PI3K/Akt partially leads to the inhibition of PrP(C)-induced drug resistance in gastric cancer cells. FEBS J 276(3):685–694

Liu T, Li R et al (2001) Normal cellular prion protein is preferentially expressed on subpopulations of murine hemopoietic cells. J Immunol 166(6):3733–3742

Maeda Y, Ashida H et al (2006) CHO glycosylation mutants: GPI anchor. Methods Enzymol 416:182–205

Martins VR, Brentani RR (2002) The biology of the cellular prion protein. Neurochem Int 41(5):353–355

Martins VR, Linden R et al (2002) Cellular prion protein: on the road for functions. FEBS Lett 512(1–3):25–28

Massimino ML, Griffoni C et al (2002) Involvement of caveolae and caveolae-like domains in signalling, cell survival and angiogenesis. Cell Signal 14(2):93–98

McEwan JF, Windsor ML et al (2009) Antibodies to prion protein inhibit human colon cancer cell growth. Tumour Biol 30(3):141–147

Mehrpour M, Codogno P (2010) Prion protein: from physiology to cancer biology. Cancer Lett 290(1):1–23

Meslin F, Conforti R et al (2007a) Efficacy of adjuvant chemotherapy according to Prion protein expression in patients with estrogen receptor-negative breast cancer. Ann Oncol 18(11): 1793–1798

Meslin F, Hamai A et al (2007b) Silencing of prion protein sensitizes breast adriamycin-resistant carcinoma cells to TRAIL-mediated cell death. Cancer Res 67(22):10910–10919

Nicholson TB, Stanners CP (2007) Identification of a novel functional specificity signal within the GPI anchor signal sequence of carcinoembryonic antigen. J Cell Biol 177(2):211–218

Orlean P, Menon AK (2007) Thematic review series: lipid posttranslational modifications. GPI anchoring of protein in yeast and mammalian cells, or: how we learned to stop worrying and love glycophospholipids. J Lipid Res 48(5):993–1011

Paitel E, Alves da Costa C et al (2002) Overexpression of PrPc triggers caspase 3 activation: potentiation by proteasome inhibitors and blockade by anti-PrP antibodies. J Neurochem 83(5): 1208–1214

Pammer J, Weninger W et al (1998) Human keratinocytes express cellular prion-related protein in vitro and during inflammatory skin diseases. Am J Pathol 153(5):1353–1358

Pan Y, Zhao L et al (2006) Cellular prion protein promotes invasion and metastasis of gastric cancer. FASEB J 20(11):1886–1888

Pergami P, Poloni TE et al (1999) Prions and prion diseases. Funct Neurol 14(4):241–252

Prado MA, Alves-Silva J et al (2004) PrPc on the road: trafficking of the cellular prion protein. J Neurochem 88(4):769–781

Prusiner SB (1982) Novel proteinaceous infectious particles cause scrapie. Science 216(4542): 136–144

Prusiner SB (1991) Molecular biology of prion diseases. Science 252(5012):1515–1522

Prusiner SB (1996) Molecular biology and pathogenesis of prion diseases. Trends Biochem Sci 21(12):482–487

Prusiner SB, Scott MR et al (1998) Prion protein biology. Cell 93(3):337–348

Roucou X, Giannopoulos PN et al (2005) Cellular prion protein inhibits proapoptotic Bax conformational change in human neurons and in breast carcinoma MCF-7 cells. Cell Death Differ 12(7):783–795

Safar J, Prusiner SB (1998) Molecular studies of prion diseases. Prog Brain Res 117:421–434

Screaton RA, DeMarte L et al (2000) The specificity for the differentiation blocking activity of carcinoembryonic antigen resides in its glycophosphatidyl-inositol anchor. J Cell Biol 150(3):613–626

Stossel TP, Condeelis J et al (2001) Filamins as integrators of cell mechanics and signalling. Nat Rev Mol Cell Biol 2(2):138–145

Sy MS, Li C et al (2010) The fatal attraction between pro-prion and filamin A: prion as a marker in human cancers. Biomark Med 4(3):453–464

Vey M, Pilkuhn S et al (1996) Subcellular colocalization of the cellular and scrapie prion proteins in caveolae-like membranous domains. Proc Natl Acad Sci USA 93(25):14945–14949

Viles JH, Cohen FE et al (1999) Copper binding to the prion protein: structural implications of four identical cooperative binding sites. Proc Natl Acad Sci USA 96(5):2042–2047

Wadsworth JD, Hill AF et al (1999) Strain-specific prion-protein conformation determined by metal ions. Nat Cell Biol 1(1):55–59

Waneck GL, Stein ME et al (1988) Conversion of a PI-anchored protein to an integral membrane protein by a single amino acid mutation. Science 241(4866):697–699

Warshaw AL, Fernandez-del Castillo C (1992) Pancreatic carcinoma. N Engl J Med 326(7): 455–465

Whittal RM, Ball HL et al (2000) Copper binding to octarepeat peptides of the prion protein monitored by mass spectrometry. Protein Sci 9(2):332–343

Wiedman JM, Fabre AL et al (2007) In vivo characterization of the GPI assembly defect in yeast mcd4-174 mutants and bypass of the Mcd4p-dependent step in mcd4Delta cells. FEMS Yeast Res 7(1):78–83

Williamson RA, Peretz D et al (1998) Mapping the prion protein using recombinant antibodies. J Virol 72(11):9413–9418

Wiseman F, Cancellotti E et al (2005) Glycosylation and misfolding of PrP. Biochem Soc Trans 33(Pt 5):1094–1095

Wu G, Guo Z et al (2006) Overexpression of glycosylphosphatidylinositol (GPI) transamidase subunits phosphatidylinositol glycan class T and/or GPI anchor attachment 1 induces tumorigenesis and contributes to invasion in human breast cancer. Cancer Res 66(20):9829–9836

Zhang CC, Steele AD et al (2006) Prion protein is expressed on long-term repopulating hematopoietic stem cells and is important for their self-renewal. Proc Natl Acad Sci USA 103(7):2184–2189

Chapter 5
Insoluble Cellular Prion Protein

Wen-Quan Zou

Abstract The detergent-soluble cellular prion protein (PrP^C) and its detergent-insoluble infectious isoform (PrP^{Sc}) are two major conformers of the prion protein. Soluble PrP^C has been the only isoform detected in the normal mammalian brain. In 2006, however, we identified an insoluble PrP^C conformer (termed $iPrP^C$) in uninfected human and animal brains. This article highlights the physiochemical properties of $iPrP^C$, a conformer distinct from PrP^C or PrP^{Sc}, and discusses its formation and probable pathophysiology.

Keywords Prion protein • Prion disease • Insoluble prion protein • Alzheimer disease • Variably protease-sensitive prionopathy • Dementia • Memory

5.1 Introduction

The cellular prion protein (PrP^C) is a universally expressed membrane protein present predominantly in the central nervous system (CNS). Deposition in the CNS of its pathologic isoform (PrP^{Sc}), derived from PrP^C via a conformational transition, is a molecular hallmark of prion diseases, a group of fatal transmissible neurodegenerative disorders in humans and animals. Although the physiologic functions of PrP^C are unclear, it has nevertheless been proposed that PrP^C has beneficial and deleterious effects on cognition (Collinge et al. 1994; Laurén et al. 2009; Linden et al. 2008; Westaway et al. 2011). Moreover, it has been well demonstrated that the coexistence of PrP^C and PrP^{Sc} is the prerequisite for the emergence of prion diseases (PrDs).

W.-Q. Zou, M.D., Ph.D. (✉)
Department of Pathology and Department of Neurology, National Prion Disease Pathology Surveillance Center, Case Western Reserve University,
2085 Adelbert Rd, Cleveland, OH 44106, USA
e-mail: wenquan.zou@case.edu

W.-Q. Zou and P. Gambetti (eds.), *Prions and Diseases: Volume 1, Physiology and Pathophysiology*, DOI 10.1007/978-1-4614-5305-5_5,

The two PrP conformers mainly studied so far are believed to be implicated in these diseases. PrP^C and PrP^{Sc} share the same primary sequence but have distinct secondary structures (Meyer et al. 1986; Caughey et al. 1991; Pan et al. 1993). PrP^C is monomeric, rich in α-helical structure, sensitive to proteinase K (PK) digestion, soluble in non-denaturing detergents, non-precipitable by anti-DNA antibodies or DNA-binding proteins, noninfectious, and present in both uninfected and scrapie-infected brains. PrP^{Sc}, on the other hand, is oligomeric or aggregate, rich in β-sheet structure, partially resistant to PK digestion, insoluble in detergents, precipitable by anti-DNA antibodies or DNA-binding proteins, infectious, and present only in infected brains. Soluble PrP^C has been the only conformer detected in the uninfected mammalian brain. In contrast, insoluble PrP^{Sc} exhibits chameleon-like conformations, which may underlie the distinct prion strains and phenotypes of PrDs identified in animals and humans (Bessen and Marsh 1992; Parchi et al. 1996; Caughey et al. 1998; Safar et al. 1998; Zou and Gambetti 2007; Collinge and Clarke 2007). Recent identification of insoluble cellular PrP ($iPrP^C$) in the uninfected human and animal brain raises two possibilities: that the PrP^C species in the brain may also exhibit chameleon-like conformations that are implicated in the beneficial or deleterious effects of PrP^C, and that these species may play a role in the pathogenesis of PrDs and other neurodegenerative disorders (Yuan et al. 2006; Zou 2010; Zou et al. 2011b).

5.2 Prion Protein Is Characterized by the Presence of an Intrinsically Chameleon-Like Conformation

Studies using recombinant PrP (rPrP) in vitro indicated that PrP possesses a highly variable conformation. In aqueous solutions, rPrP could be folded into pH-dependent α-helical conformations, a thermodynamically more stable β-sheet, and various stable or transient intermediates (Zhang et al. 1997). A stopped-flow kinetic study demonstrated that PrP folded by a three-state mechanism involving a monomeric intermediate (Apetri and Surewicz 2002). It was found that the population of this partially structured PrP intermediate increased in the presence of relatively low concentrations of urea and was more stable at acidic pH 4.8, compared to neutral pH 7.0. Moreover, this approach revealed that PrP mutations, linked with naturally occurring familial prion diseases, showed a pronounced stabilization of the folding intermediate (Apetri et al. 2004). This characteristic strongly suggested that these intermediates play a crucial role in PrP conversion and serve as direct precursors of the pathologic PrP^{Sc} isoform. The existence of a PrP folding intermediate was also indicated by hydrogen exchange experiments (Nicholson et al. 2002), and by studies using high pressure NMR and fluorescence spectroscopy (Kuwata et al. 2002; Martins et al. 2003). In addition to a β-oligomer and an amyloid fibril (Baskakov et al. 2001; Morillas et al. 2001; Lu and Chang 2002; Sokolowski et al. 2003; Baskakov et al. 2004), two additional polymeric transient intermediates were also identified during fibrillogenesis of rPrP in vitro (Baskakov et al. 2002).

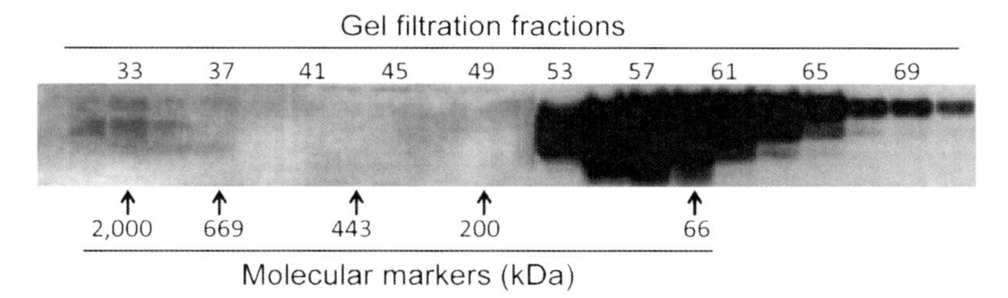

Fig. 5.1 Gel filtration of PrP from uninfected human brains. Gel filtration fractions of uninfected brain homogenates were subjected to SDS-PAGE and Western blotting with 3F4. Molecular mass (kDa) of various PrP species recovered in different fractions is indicated by an *arrow* and molecular mass markers used include dextran blue (2,000 kDa), thyroglobulin (669 kDa), apoferritin (443 kDa), β-amylase (200 kDa), and albumin (66 kDa). PrP was detected not only in fractions with molecular mass less than 66 kDa after fraction 59 but also in fractions with molecular mass greater than 66 kDa before fraction 59 including fraction 33 containing large PrP aggregates (2,000 kDa)

PrP^C in vivo is anchored to the cell membrane. Several experiments have indicated that the PrP conformation is affected by its local conditions. For example, the interaction of the anchorless recombinant PrP with lipids in a membrane-like environment resulted in a conformational transition (Wang et al. 2007; Re et al. 2008). Increasing the local concentration of membrane-anchored PrP^C seems to induce a conformational transition accompanied by oligomerization of PrP^C (Elfrink et al. 2008). Therefore, the tendency of PrP to form multiple nonnative β-sheet-rich isoforms in vitro, as demonstrated in biophysical studies on rPrP, may represent a unique intrinsic feature of this protein.

5.3 Insoluble Cellular Prion Protein Is Present in Normal Mammalian Brains

If the tendency of PrP to form multiple conformations in vitro represents a unique intrinsic feature of this protein, it is conceivable that other PrP conformers would be present in the normal brain in addition to the well-characterized PrP^C. To test for this possibility, we examined uninfected human and animal brains using a combination of biophysical and biochemical approaches to determine whether there are additional PrP conformers (Yuan et al. 2006). We identified a novel conformer which forms insoluble cellular PrP aggregates and protease-resistant PrP species in uninfected human brains (Yuan et al. 2006). Using gel filtration, we revealed that PrP in uninfected human brains is present not only in monomers but also in oligomers and large aggregates (Fig. 5.1). The new PrP conformer, which we termed insoluble cellular PrP ($iPrP^C$), accounts for approximately 5–25% of total PrP including full-length and N terminally truncated forms, and a portion of $iPrP^C$ is resistant to PK digestion even at 50 μg/ml (Yuan et al. 2006). Notably, the PK-resistant $iPrP^C$ has an immunoreactive behavior different from that of classic PrP^{Sc} detected in prion-infected brains; its affinity is much lower for 3F4 while higher for 1E4, compared to the

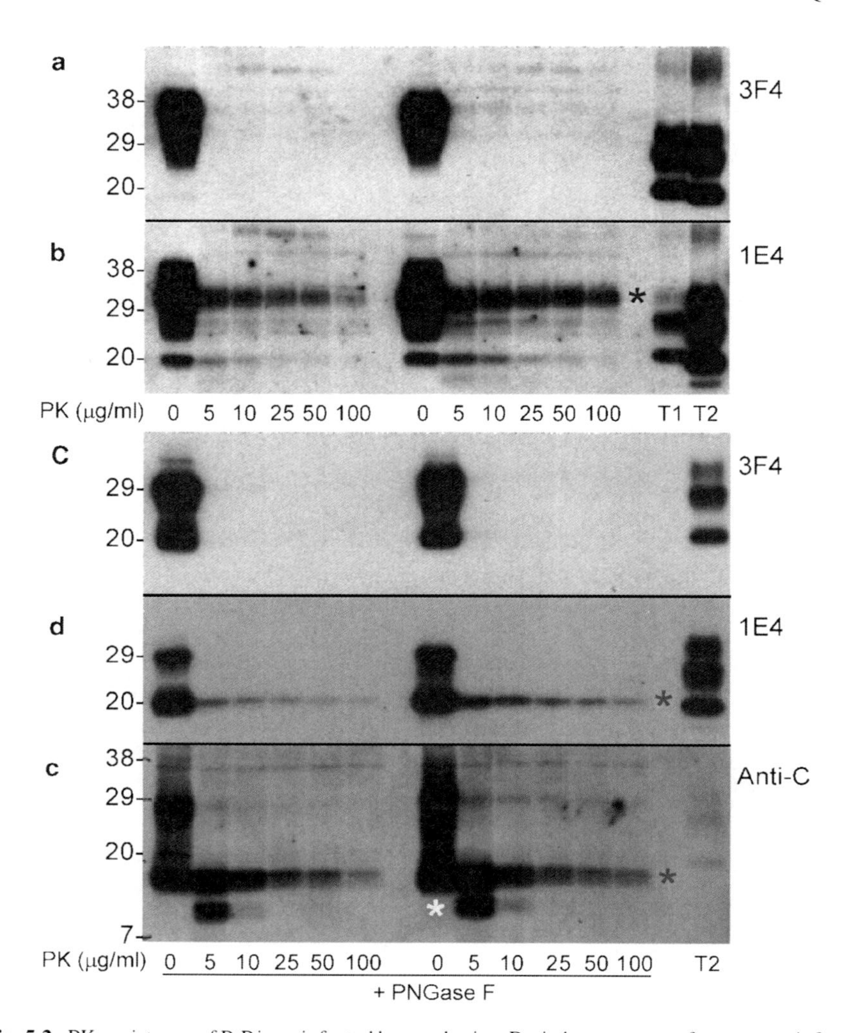

Fig. 5.2 PK-resistance of PrP in uninfected human brains. Brain homogenates from two uninfected human brains received at autopsy were treated with PK at 0, 5, 10, 25, 50, 100 μg/ml (*upper two panels A* and *B*) or PK plus PNGase F (*lower three panels C, D*, and *E*). The samples were subjected to SDS-PAGE and Western blotting with 3F4, 1E4, and Anti-C antibodies. No PK-resistant PrP was detectable with 3F4 antibody. In contrast, PK-resistant PrP was detected with 1E4 and Anti-C up to 100 μg/ml. With PK alone, three PrP bands migrating at 30-29 kDa, 27-26 kDa, and 21-20 kDa were detected, in which the *upper* band (~30-29 kDa, *dark asterisk*) was predominant while the intensity of the *middle* band was *lowest*, which is apparently different from those of PrP^{Sc} type 1 (T1) and type 2 (T2). After PNGase F treatment, only one band was detected with 1E4 and Anti-C migrating at ~20 kDa and ~18 kDa, respectively (PrP^{*20} and PrP^{*18}, *gray asterisk*). Interestingly, a band migrating at ~12–13 kDa was also detected with Anti-C at low PK concentration (5–10 μg/ml, *white asterisk*)

affinity of those antibodies for classic PrP^{Sc} (Yuan et al. 2006, 2008; Zou et al. 2010a, 2011a) (Fig. 5.2). The epitopes of the two antibodies 3F4 and 1E4 are adjacent and the C terminus of the 1E4 epitope between PrP97-105 is connected to the N terminus of the 3F4 epitope between PrP 106-112 (Yuan et al. 2008; Zou et al.

2010a). Antibody 3F4 is the most widely used antibody in the detection of human PrP^C and PrP^{Sc}, including PrP^{Sc} types 1 and 2 seen in sCJD and inherited CJD, and the internal PrP^{Sc} fragment PrP7-8 seen in GSS. In addition, the new conformer reveals high affinity for the gene 5 protein (g5p, a single-stranded DNA-binding protein) and sodium phosphotungstate (NaPTA), both of which also specifically bind to PrP^{Sc} but not to soluble PrP^C (Zou et al. 2004; Yuan et al. 2006; Safar et al. 1998; Wadsworth et al. 2001). To rule out the possibility that PrP aggregates detected in the uninfected human brain result from postmortem autolysis of autopsy tissues or from other neurodegenerative disorders, we also examined frozen uninfected human biopsy brain tissues or normal animal brain tissues from hamsters and cows. We discovered that the insoluble PrP^C was also detectable in these tissues, a finding which confirmed that $iPrP^C$ is a de novo generated PrP conformer (Yuan et al. 2006). Using gel filtration, we recently further demonstrated that not only soluble PrP^C monomers but also soluble PrP^C oligomers are present in the uninfected human brain (Xiao et al. 2012).

The presence of PrP conformers besides the typical PrP^C in uninfected brains was also implied in observations reported by other groups. Consistent with our findings, small amounts of PrP (less than 5% of total PrP^C) were also reported to be precipitated by NaPTA from uninfected human brains (Wadsworth et al. 2001). Moreover, by a differential SDS solubility assay, PrP^C species with either lower or higher solubility were differentiated in brain homogenates of noninfected humans, sheep, and cattle (Kuczius et al. 2009, 2011). Notably, a purified hamster brain PrP^C displayed an unexpectedly high β-sheet component under native conditions; this finding provided the first evidence that the full-length native PrP^C isolated from animal brains exhibited intrinsic conformational plasticity (Pergami et al. 1999). Moreover, mammalian brain PrP^C from six species was observed to be initially degraded to an intermediate fragment prior to complete proteolysis, suggesting an intrinsic partial PK-resistance (Buschmann et al. 1998). Interestingly, PrP aggregates have also been reported in pancreatic beta-cells of uninfected rats in response to hyperglycemia (Strom et al. 2007). In sum, the cumulative evidence shows that insoluble PrP^C is present in tissues and organs from uninfected animals and humans.

5.4 Spontaneous Formation of the Insoluble Cellular Prion Protein Has Been Modeled with Cultured Cells and May Result from PrP Cytosolic Accumulation

Lehmann and Harris (1996) modeled spontaneous formation of PrP^{Sc}-like insoluble PrP in cultured Chinese hamster ovary (CHO) cells expressing wild-type or mutant mouse PrP. Significant amounts of mutant PrP with point mutation at residue 199 (E199K) (~60%) or six octapeptide repeat insertion mutation between residues 51 and 90 (~90%) linked to inherited human prion disease were detergent; notably approximately 15% wild-type PrP^C was also detergent insoluble (Lehmann and Harris 1996). While approximately 5% mutant PrP was resistant to the digestion by PK at 3.3 μg/ml for 20 min, wild-type PrP was completely degraded. Because the two mutant PrP molecules but not wild-type PrP were tightly associated with the plasma membrane, it was hypothesized that acquisition of PrP^{Sc}-like properties results from an alternation in membrane topology or affinity (Lehmann and Harris 1996). Using the same models, the same group further identified a three-step endocytic pathway by which mutant PrP forms a PrP^{Sc}-like conformer: initially hydrophobic, then detergent insoluble, and finally partially PK resistant (Daude et al. 1997). Using human neuroblastoma cells, Singh et al. also revealed that PrP with Q217R mutation linked to GSS formed a PrP^{Sc}-like form (Singh et al. 1997).

In addition to above PrP mutations, the two N-linked glycosylation sites located at residue 181, Asn–Ile–Thr residues 181–183, and at residue 197, Asn–Phe–Thr residues 197–199 (Puckett et al. 1991) are believed to play a crucial role in the stabilization of prion protein conformation. The naturally occurring mutations at residue 183, Thr to Ala (PrP^{T183A}), or at residue 198, Phe to Ser (PrP^{F198S}), falling in the two consensus sites, are linked to two distinct familial prion diseases (Nitrini et al. 1997; Tagliavini et al. 1991). Elimination of either site, or of both by mutagenesis of hamster PrP in CV1 cells, induced intracellular accumulation of mutant proteins (Rogers et al. 1990). Lehmann and Harris observed that mouse PrP mutated at T182 alone, or at both T182 and T198 in CHO cells, failed to reach the cell surface but the PrP with T198 mutation did. Moreover, all three mutant PrP's acquired PrP^{Sc}-like physicochemical properties reminiscent of PrP^{Sc}; PrP^{Wt} did so only when synthesized in the presence of N-linked glycosylation inhibitor tunicamycin (Lehmann and Harris 1997). Using M17 cells expressing human PrP^{N181G} or PrP^{T183A}, Capellari et al. observed that PrP^{N181G}, but not PrP^{T183A}, reached the cell surface even though both mutations eliminated glycosylation at the first site (Capellari et al. 2000). This observation indicates that the Thr to Ala mutation itself, rather than the elimination of the first glycosylation site, altered the physical properties of the mutant protein (Capellari et al. 2000). Although the F198S mutation falls within the second glycosylation site, Asn–Phe–Thr residues 197–199, PrP^{F198S} slightly increased the efficiency of glycosylation at the first glycosylation site (N181), and greatly increased the efficiency of glycosylation at the second site (N197) in cultured cells (Zaidi et al. 2005).

To further investigate the formation of $iPrP^{C}$ and the effect of mutations on the formation of $iPrP^{C}$, we examined $iPrP^{C}$ in cultured M17 cells expressing human wild-type (PrP^{Wt}) and mutant PrP (Yuan et al. 2008; Zou et al. 2011a). We confirmed that the de novo generated iPrP was detectable not only in cells expressing mutant PrP (PrP^{T183A} or PrP^{F198S}) linked to naturally occurring genetic Creutzfeldt–Jakob disease and Gerstmann–Sträussler–Scheinker disease, respectively, but also in cells expressing wild-type PrP. Compared to cells expressing wild-type PrP, significantly increased amounts of iPrP forming PrP aggregates and PK-resistant PrP were found in cells expressing mutant PrP. Most of PrP^{T183A} was composed of oligomers and large aggregates; virtually no monomeric form was present. In PrP^{F198S}, however, monomeric species were still dominant despite an increase in the amounts of aggregates. The enhanced tendency of PrP^{T183A} to form aggregates may result from the intracellular accumulation of the mutant protein. The F198S mutation did not significantly diminish the ability of PrP^{F198S} to reach the cell surface (Zaidi et al. 2005), although the mutation may change the structure around the V14 epitope previously found to be localized between human PrP185-196 (Zou et al. 2011a; Moudjou et al. 2004; Rezaei et al. 2005). Therefore, the majority of the $iPrP^{C}$ associated with the T183A mutation may result from PrP intracellular accumulation, raising the possibility that $iPrP^{C}$ is derived predominantly from intracellular PrP species. Immunofluorescence microscopy of tagged PrP also indicated that PrP^{T183A} accumulates within the cell, whereas PrP^{F198S} was distributed both inside the cell and on the cell surface, consistent with previous observations (Zou et al. 2011a; Capellari et al. 2000; Zaidi et al. 2005).

In uninfected cultured cells, we also confirmed that the PK-resistant PrP exhibited higher affinity for 1E4 than for 3F4 that was initially observed in tissue samples (Zou et al. 2011a; Yuan et al. 2006, 2008). In Western blotting with cell lysates, virtually no PrP was detected by 1E4, and PrP was detectable only after PK treatment. However, PrP was stainable by 1E4 in fixed cultured cells treated with or without PK although the PrP signal was weaker in treated than in untreated cells (Zou et al. 2011a). It is worth noting that an antibody against human PrP95-110 (termed 8G8), that actually extends merely two more amino acids toward the N and C terminuses of the 1E4 epitope, respectively, stained PrP-expressing cells with a brilliant cytoplamic fluorescence (Krasemann et al. 1999). However, the number of positive cells was smaller than that of cells strained with antibodies against other PrP regions. Moreover, despite sharing a similar amino acid sequence within the corresponding region, only cattle, but not mouse and hamster PrP, was observed to react with 8G8 (Krasemann et al. 1999). In contrast to 3F4, 1E4 indeed seems to detect intracellular PrP in cultured cells (Zou et al. 2011a). Therefore, like 8G8, 1E4 may recognize a PrP species with a unique conformation in its epitope region.

In the absence of scrapie infection, aggregation of the cellular wild-type prion protein in cultured cells was first observed only when proteasome inhibitors were used (Yedidia et al. 2001). It was later reported that PrP^{Wt} accumulated in the cytoplasm of cultured cells under other conditions as well, such as in a reducing environment, or when expressing PrP without both N and C terminal signal peptides (Ma and Lindquist 2001, 2002; Drisaldi et al. 2003; Grenier et al. 2006). Cytosolic PrP forms aggregates that are insoluble in non-ionic detergents and partially resistant to PK (Ma and Lindquist 2001). Accumulated cytosolic PrP aggregates induced by ER stress and inhibition of proteasomal activity were recently observed to travel through the secretory pathway and reach the plasma membrane (Nunziante et al. 2011). Cytosolic PrP was observed not only in cultured cells but also in subpopulations of neurons in the hippocampus, neocortex, and thalamus in uninfected wild-type mice (Mironov et al. 2003). In addition, soluble PrP^{C} was observed to switch to insoluble PrP^{C} by treatment with acidic buffers in vitro (Zou and Cashman 2002).

The above observations with cell models may suggest that the formation of $iPrP^{C}$ or the aggregation of PrP^{C} is associated not only with mutations of the protein but also with altered cellular conditions that cause abnormal traffic and distribution of PrP in cells including reductive/oxidative stress.

5.5 Physiology and Pathophysiology of Insoluble PrP^C

5.5.1 Long-Term Memory Storage

The $iPrP^C$ with a conformation potentially different from soluble PrP^C may have a physiologic function. It has been hypothesized that prion-like conformational changes are indispensable for the maintenance of structural synaptic changes required for long-term memory (Si et al. 2003, 2010; Papassotiropoulos et al. 2005; Shorter and Lindquist 2005). Conceivably, the conversion of soluble PrP^C monomers into insoluble PrP oligomers or aggregates could be associated with long-term memory storage in the normal human brain (Zou et al. 2011c). The $iPrP^C$ molecule is able to bind to g5p, the single-stranded DNA-binding protein (Yuan et al. 2006, 2008). The possible binding of $iPrP^C$ to mRNA in vivo cannot be ruled out. Based on the observation that 24 h after a word-list learning task, carriers of either the polymorphism methionine/methionine (M/M) at residue 129 (129MM) or M/valine (V) (129 MV) genotype recalled 17% more information than did 129VV carriers (Papassotiropoulos et al. 2005). The PrP gene is believed to be genetically associated with human long-term memory performance. Therefore, the polymorphism at the residue 129 of PrP may participate in mediating human memory, in which the 129 M allele may have a beneficial effect on long-term memory. Interestingly, the impact of a putative PrP conformation rather than pathologic PrP^{Sc} on long-term memory in healthy humans was proposed to be related to physiologically occurring conformational changes (Tompa and Friedrich 1998; Papassotiropoulos et al. 2005).

5.5.2 Prion Disease

The in vivo pathway by which PrP^C forms PrP^{Sc} remains poorly understood. Two non-exclusive conversion models were proposed: *refolding* (Griffith 1967; Prusiner 1991) and *seeding* (Jarrett and Lansbury 1993). In the former, the exogenous PrP^{Sc} binds to the PrP^C species that has been partially unfolded and the PrP^{Sc}-bound PrP^C molecule undergoes a refolding process during which the nascent PrP^{Sc} is derived from this PrP^C species via a conformational transition. The latter proposes that a small amount of abnormal PrP^{Sc} or PrP^{Sc}-like form (PrP*) is present in the normal brain and is in reversible equilibrium with PrP^C. When several monomeric PrP* molecules form a highly ordered nucleus, PrP^C is converted to PrP^{Sc} polymers. Obviously, two key elements are required by the seeding model. One is the presence in the uninfected brain of a small amount of endogenous PrP^{Sc} or PrP* and the second is the formation of PrP^{Sc}-derived oligomers. The seeding model, with the two elements, has been recapitulated in vitro using PrP from various fungal and mammalian sources (Ross et al. 2005; Castilla et al. 2005; Tanaka et al. 2005).

Because $iPrP^C$ possesses PrP^{Sc}-like physicochemical properties, it is possible that $iPrP^C$ represents endogenous PrP^{Sc}, an intermediate form (PrP*) between PrP^C and

PrP^{Sc}, or silent prion (Hall and Edskes 2004; Weissmann 2004; Yuan et al. 2006; Zou et al. 2011a). Based on the observation that the brains of bigenic mice are capable of clearing prions, it has been proposed that the normal brain contains low levels of PrP^{Sc} (Safar et al. 2005). Under normal circumstances, despite the presence of a small amount of PrP^{Sc}, the brain may maintain equilibrium between the formation and clearance of PrP^{Sc}. The amount of PrP^{Sc} may be too small to induce a neurodegenerative disorder, which presumably, remain in a silent state. However, prion diseases may be triggered when the levels of the silent prions are significantly increased due to infection, PrP mutation, or unknown causes. Using protein misfolding cyclic amplification (PMCA), Barria and coworkers generated a new infectious prion without adding exogenous PrP^{Sc} seeds (Barria et al. 2009). This study may raise two possibilities (1) PMCA replicated an intermediate PrP^{Sc} which was present in the brain homogenate; or (2) the silent prion was activated by the sonication–incubation cycles during PMCA.

As mentioned above, $iPrP^{C}$ possesses a unique immunoreactive behavior showing poor affinity for 3F4 and higher affinity for 1E4, different from other types of human PrP^{Sc} identified so far (Yuan et al. 2006, 2008; Zou et al. 2011a). The two antibodies have adjacent epitopes on PrP (Yuan et al. 2008; Zou et al. 2010b). Thus, the possibility cannot be ruled out that iPrP is a distinct PrP species with an altered conformation and that $iPrP^{C}$ may be a conformer which, when it increases, induces an atypical form of prion disease. Some previous observations with experimental animals may favor this hypothesis. A novel neurologic syndrome was reported in Tg mice overexpressing wild-type PrP and these mice exhibited degeneration of skeletal muscle, peripheral nerves, and the central nervous system (Westaway et al. 1994). The increased amounts of wild-type PrP^{C} might form aggregates that induce degeneration in those mice. Chiesa et al. indeed observed that homozygous Tg mice overexpressing wild-type PrP at approximately tenfold but not hemizygous mice overexpressing wild-type PrP at approximately fivefold developed a spontaneous neurodegenerative disorder manifesting tremor and paresis (Chiesa et al. 2008). Nevertheless, abnormal PrP deposits and enlarged synaptic terminals with a dramatic proliferation of membranous structures were found in both types of mice. It was also observed that the overexpressed PrP assembled into insoluble aggregates with mild PK resistance but acquired no infectivity (Chiesa et al. 2008). Misfolding and neurotoxicity of wild-type PrP in transgenic flies were observed to be sequence dependent: Hamster PrP formed large amounts of PrP aggregates with spongiform degeneration, whereas rabbit PrP formed only small amounts of PrP aggregates without spongiform degeneration (Fernandez-Funez et al. 2010). Moreover, the same study also found that although small amounts of PrP aggregates were similarly detected in young flies (day 1) expressing hamster PrP, spongiform degeneration was not evident. Therefore, the small amounts of PrP aggregates were unable to induce spongiform degeneration. Interestingly, spongiform degeneration was observed to occur in older flies (day 30) only when the concentrations of PrP aggregates increased.

The same unique immunoreactivity behavior with 1E4 has also been observed in a new PrP^{Sc} species we recently identified from variably protease-sensitive prionopathy (VPSPr), a novel human prion disease (Gambetti et al. 2008; Zou et al. 2010b). VPSPr exhibits an abnormal PrP species with peculiar glycosylation and enzymatic proteolysis (Zou et al. 2010b, 2011c). The 1E4-detected pathogenetic PK-resistant PrP^{Sc} with a ladder-like electrophoretic profile is the molecular hallmark of VPSPr. PrP^{Sc} from VPSPr exhibits not only the peculiar immunoreactivity behavior but also three PK-resistant core fragments, which is similar to $iPrP^{C}$ (Zou et al. 2010b, 2011c). These similarities may suggest that they share a common molecular metabolic pathway. Similar to sCJD, VPSPr affects patients regardless of their PrP genotypes defined by 129 MV polymorphism; however, the allelic prevalence is distinct in the two diseases (Zou et al. 2010b; Gambetti et al. 2011a). Notably, the amounts of PK-resistant PrP^{Sc} in VPSPr seem to be dependent on the polymorphism, a characteristic which has not been observed in sCJD. Preliminary data revealed no clinical phenotype during the normal life span of the transgenic mice expressing human PrP-129V at sixfold inoculated with brain homogenates from cases of VPSPr-129VV (Gambetti et al. 2011a), suggesting that the infectivity of PrP^{Sc} from VPSPr may be much lower than that of PrP^{Sc} from sCJD. Only 30% of the mice exhibited peculiar PrP plaques with a distinctive topography and minimal or no spongiform degeneration, compared to the typical neuropathological changes found in 100% mice inoculated with the classical sCJD control. Some of these mice inoculated with VPSPr also had the PK-resistant PrP^{Sc} whose profile exhibited the ladder-like electrophoresis detected by 1E4. Therefore, it is possible that VPSPr characterized by the deposition in the brain of $iPrP^{C}$-like PrP^{Sc} represents a prion disease, distinct from classical prion diseases and bearing more resemblance to other neurodegenerative diseases such as Alzheimer disease and tauopathies (Gambetti et al. 2011b). Because of the similarities between $iPrP^{C}$ and PrP^{Sc} from VPSPr, the possibility that VPSPr results from an increase in the amount of $iPrP^{C}$ cannot be excluded.

5.5.3 *Alzheimer Disease*

The insoluble PrP^{C} has been recently demonstrated to be the main species that interacts with Aβ in AD (Zou et al. 2011b). Moreover, using a peptide membrane array involving 13-mer human PrP peptides, and two Aβ peptides (Aβ42 and Aβ40), we identified 17 Aβ binding regions distributed on N terminal, internal, and C terminal PrP domains. Two distinct types of Aβ-binding sites were differentiated: one specifically binds to Aβ42 and the other binds to both Aβ42 and Aβ40. Notably, Aβ42-specific binding sites are localized predominantly in the octapeptide repeat region, whereas sites that bind both Aβ40 and Aβ42 are mainly in the extreme N terminal and C terminal domains of PrP (Zou et al. 2011b). Our study is consistent with other observations. PrP deposits often histologically accompany Aβ-positive plaques in AD brains (Esiri et al. 2000; Ferrer et al. 2001; Kovacs et al. 2002). In addition, Freir et al. also observed that interaction between PrP and toxic Aβ assemblies can be therapeutically targeted at multiple sites (Freir et al. 2011), indicating that their binding sites are not limited only to the internal domain. Remarkably, Kudo et al. showed more recently that not only anti-PrP antibodies but also PrP^{C} peptides identified in our previous study (Zou et al. 2011b) rescued Aβ oligomer-induced neurotoxiciy (Kudo et al. 2012).

Although the exact biological relevance of the interaction between iPrP^{C} and Aβ remains unclear, aggregation of one protein was observed to facilitate aggregation of the other (Morales et al. 2010). Moreover, synergistic interactions between other amyloidogenic proteins associated with neurodegeneration have also been reported to promote each other's fibrillization, amyloid deposition, and formation of filamentous inclusions in transgenic mice (Schwarze-Eicker et al. 2005). An increase in the efficiency of Aβ42 aggregation in vitro was dependent on PrP^{Sc} dosage (Morales et al. 2010). Moreover, insoluble PrP^{Sc} aggregates also seemed to facilitate Aβ42 aggregation in vivo; AD mice developed a strikingly higher load of cerebral amyloid plaques that appeared much faster in prion infected than in uninfected mice (Morales et al. 2010). Our finding that Aβ42 binds to iPrP may suggest that iPrP (the PrP^{Sc}-like forms in uninfected human brains) facilitates the fibrillization of Aβ42 in AD. Similarly, the possibility should be considered that a significant increase in the total number of Aβ plaques observed in bigenic mice overexpressing PrP (Schwarze-Eicker et al. 2005) might result from an increase in the formation of iPrP. Since the less toxic insoluble Aβ42 aggregates constitute the end products of highly toxic soluble Aβ42 oligomers, it is conceivable that formation of the large aggregates facilitated by iPrP^{C} may reduce the amount of Aβ42 oligomers. The decrease in the levels of toxic Aβ42 oligomers would then attenuate the cognitive impairment induced by Aβ42 oligomers in AD. If this is the case, iPrP^{C} may play a protective role in AD. Given that iPrP^{C} interacts with insoluble Aβ42, whereas soluble PrP^{C} binds soluble Aβ42 in vivo (Zou et al. 2011b), it is possible that distinct PrP conformers binding to different Aβ42 species thereby function either as receptors for soluble Aβ42 oligomers or as modulators of insoluble Aβ42 deposition. It would be interesting to test this hypothesis by intracerebrally injecting anti-PrP antibodies against either soluble or insoluble PrP species in AD animal models. This experiment would establish that the multiple conformers of PrP^{C} are coupled with its beneficial and deleterious effects.

5.6 Conclusions

Like PrP^{Sc} whose chameleon-like conformations are believed to link to transmissible and non-transmissible prion diseases with highly heterogeneous phenotypes (Zou 2007; Zou and Gambetti 2007), the chameleon-like conformations of PrP^{C} may be linked to its beneficial and deleterious effects (Zou et al. 2011c). Demonstration of the presence of insoluble PrP in normal mammalian brains and its potential association with AD and atypical prion disease may open a new avenue in the exploration of prion formation and in the physiology and pathophysiology of prion protein.

Acknowledgements The author wants to thank Dr. Xiangzhu Xiao and Jue Yuan for the technical support in Western blotting analysis for the two figures. This work was supported by the National Institutes of Health R01NS062787, the University Center on Aging and Health with the support of the McGregor Foundation and the President's Discretionary Fund (Case Western Reserve University), the Alliance BioSecure, and the CJD Foundation.

References

Apetri AC, Surewicz WK (2002) Kinetic intermediate in the folding of human prion protein. J Biol Chem 277:44589–44592

Apetri AC, Surewicz K, Surewicz WK (2004) The effect of disease-associated mutations on the folding pathway of human prion protein. J Biol Chem 279:18008–18014

Barria MA, Mukherjee A, Gonzalez-Romero D, Morales R, Soto C (2009) De novo generation of infectious prions in vitro produces a new disease phenotype. PLoS Pathog 5:e1000421

Baskakov IV, Legname G, Prusiner SB, Cohen FE (2001) Folding of prion protein to its native alpha-helical conformation is under kinetic control. J Biol Chem 276:19687–19690

Baskakov IV, Legname G, Baldwin MA, Prusiner SB, Cohen FE (2002) Pathway complexity of prion protein assembly into amyloid. J Biol Chem 277:21140–21148

Baskakov IV, Legname G, Gryczynski Z, Prusiner SB (2004) The peculiar nature of unfolding of the human prion protein. Protein Sci 13:586–595

Bessen RA, Marsh RF (1992) Identification of two biologically distinct strains of transmissible mink encephalopathy in hamsters. J Gen Virol 73:329–334

Buschmann A, Kuczius T, Bodemer W, Groschup MH (1998) Cellular prion proteins of mammalian species display an intrinsic partial proteinase K resistance. Biochem Biophys Res Commun 253:693–702

Capellari S, Zaidi SI, Long AC, Kwon EE, Petersen RB (2000) The Thr183Ala mutation, not the loss of the first glycosylation site, alters the physical properties of the prion protein. J Alzheimers Dis 2:27–35

Castilla J, Saá P, Hetz C, Soto C (2005) In vitro generation of infectious scrapie prions. Cell 121:195–206

Caughey BW, Dong A, Bhat KS, Ernst D, Hayes SF, Caughey WS (1991) Secondary structure analysis of the scrapie-associated protein PrP 27-30 in water by infrared spectroscopy. Biochemistry 30:7672–7680

Caughey B, Raymond GJ, Bessen RA (1998) Strain-dependent differences in beta-sheet conformations of abnormal prion protein. J Biol Chem 273:32230–32235

Chiesa R, Piccardo P, Biasini E, Ghetti B, Harris DA (2008) Aggregated, wild-type prion protein causes neurological dysfunction and synaptic abnormalities. J Neurosci 28:13258–13267

Collinge J, Clarke AR (2007) A general model of prion strains and their pathogenicity. Science 318:930–936

Collinge J, Whittington MA, Sidle KC, Smith CJ, Palmer MS, Clarke AR et al (1994) Prion protein is necessary for normal synaptic function. Nature 370:295–297

Daude N, Lehmann S, Harris DA (1997) Identification of intermediate steps in the conversion of a mutant prion protein to a scrapie-like form in cultured cells. J Biol Chem 272(17): 11604–11612, Erratum in J Biol Chem 2000;275:1520

Drisaldi B, Stewart RS, Adles C, Stewart LR, Quaglio E, Biasini E, Fioriti L, Chiesa R, Harris DA (2003) Mutant PrP is delayed in its exit from the endoplasmic reticulum, but neither wild-type nor mutant PrP undergoes retrotranslocation prior to proteasomal degradation. J Biol Chem 278:21732–21743

Elfrink K, Ollesch J, Stöhr J, Willbold D, Riesner D, Gerwert K (2008) Structural changes of membrane-anchored native PrP(C). Proc Natl Acad Sci USA 105:10815–10819

Esiri MM, Carter J, Ironside JW (2000) Prion protein immunoreactivity in brain samples from an unselected autopsy population: findings in 200 consecutive cases. Neuropathol Appl Neurobiol 26:273–284

Fernandez-Funez P, Zhang Y, Casas-Tinto S, Xiao X, Zou WQ, Rincon-Limas DE (2010) Sequence-dependent prion protein misfolding and neurotoxicity. J Biol Chem 285:36897–36908

Ferrer I, Blanco R, Carmona M, Puig B, Ribera R, Rey MJ et al (2001) Prion protein expression in senile plaques in Alzheimer's disease. Acta Neuropathol 101:49–56

Freir DB, Nicoll AJ, Klyubin I, Panico S, Mc Donald JM, Risse E, Asante EA, Farrow MA, Sessions RB, Saibil HR, Clarke AR, Rowan MJ, Walsh DM, Collinge J (2011) Interaction between prion protein and toxic amyloid β assemblies can be therapeutically targeted at multiple sites. Nat Commun 2:336–340

Gambetti P, Dong Z, Yuan J, Xiao X, Zheng M, Alshekhlee A et al (2008) A novel human disease with abnormal prion protein sensitive to protease. Ann Neurol 63:697–708

Gambetti P, Puoti G, Zou WQ (2011a) Variably protease-sensitive prionopathy: a novel disease of the prion protein. J Mol Neurosci 45:422–424

Gambetti P, Zou WQ, Torres JM, Soto C, Notari S, Espinosa JC et al (2011b) Variably protease-sensitive prionopathy: transmissibility and PMCA studies. Prion 5:14

Grenier C, Bissonnette C, Volkov L, Roucou X (2006) Molecular morphology and toxicity of cytoplasmic prion protein aggregates in neuronal and non-neuronal cells. J Neurochem 97:1456–1466

Griffith JS (1967) Self-replication and scrapie. Nature 215:1043–1044

Hall D, Edskes H (2004) Silent prions lying in wait: a two-hit model of prion/amyloid formation and infection. J Mol Biol 336:775–786

Jarrett JT, Lansbury PT Jr (1993) Seeding "one-dimensional crystallization" of amyloid: a pathogenic mechanism in Alzheimer's disease and scrapie? Cell 73:1055–1058

Kovacs GG, Zerbi P, Voigtländer T, Strohschneider M, Trabattoni G, Hainfellner JA et al (2002) The prion protein in human neurodegenerative disorders. Neurosci Lett 329:269–272

Krasemann S, Jürgens T, Bodemer W (1999) Generation of monoclonal antibodies against prion proteins with an unconventional nucleic acid-based immunization strategy. J Biotechnol 73:119–129

Kuczius T, Karch H, Groschup MH (2009) Differential solubility of prions is associated in manifold phenotypes. Mol Cell Neurosci 42:226–233

Kuczius T, Wohlers J, Karch H, Groschup MH (2011) Subtyping of human cellular prion proteins and their differential solubility. Exp Neurol 227:188–194

Kudo W, Lee HP, Zou WQ, Wang X, Perry G, Zhu X, Smith MA, Petersen RB, Lee HG (2012) Cellular prion protein is essential oligomeric amyloid-β induced neuronal cell death. Hum Mol Genet 21(5):1138–1144

Kuwata K, Li H, Yamada H, Legname G, Prusiner SB, Akasaka K et al (2002) Locally disordered conformer of the hamster prion protein: a crucial intermediate to PrPSc? Biochemistry 41:12277–12283

Laurén J, Gimbel DA, Nygaard HB, Gilbert JW, Strittmatter SM (2009) Cellular prion protein mediates impairment of synaptic plasticity by amyloid-beta oligomers. Nature 457:1128–1132

Lehmann S, Harris DA (1996) Two mutant prion proteins expressed in cultured cells acquire biochemical properties reminiscent of the scrapie isoform. Proc Natl Acad Sci USA 93:5610–5614

Lehmann S, Harris DA (1997) Blockade of glycosylation promotes acquisition of scrapie-like properties by the prion protein in cultured cells. J Biol Chem 272:21479–21487

Linden R, Martins VR, Prado MA, Cammarota M, Izquierdo I, Brentani RR (2008) Physiology of the prion protein. Physiol Rev 88:673–728

Lu BY, Chang JY (2002) Isolation and characterization of a polymerized prion protein. Biochem J 364:81–87

Ma J, Lindquist S (2001) Wild-type PrP and a mutant associated with prion disease are subject to retrograde transport and proteasome degradation. Proc Natl Acad Sci USA 98:14955–14960

Ma J, Lindquist S (2002) Conversion of PrP to a self-perpetuating PrPSc-like conformation in the cytosol. Science 298:1785–1788

Martins SM, Chapeaurouge A, Ferreira ST (2003) Folding intermediates of the prion protein stabilized by hydrostatic pressure and low temperature. J Biol Chem 278:50449–50455

Meyer RK, McKinley MP, Bowman KA, Braunfeld MB, Barry RA, Prusiner SB (1986) Separation and properties of cellular and scrapie prion proteins. Proc Natl Acad Sci USA 83:2310–2314

Mironov A Jr, Latawiec D, Wille H, Bouzamondo-Bernstein E, Legname G, Williamson RA, Burton D, DeArmond SJ, Prusiner SB, Peters PJ (2003) Cytosolic prion protein in neurons. J Neurosci 23:7183–7193

Morales R, Estrada LD, Diaz-Espinoza R, Morales-Scheihing D, Jara MC, Castilla J et al (2010) Molecular cross talk between misfolded proteins in animal models of Alzheimer's and Prion diseases. J Neurosci 30:4528–4535

Morillas M, Vanik DL, Surewicz WK (2001) On the mechanism of alpha-helix to beta-sheet transition in the recombinant prion protein. Biochemistry 40:6982–6987

Moudjou M, Treguer E, Rezaei H, Sabuncu E, Neuendorf E, Groschup MH, Grosclaude J, Laude H (2004) Glycan-controlled epitopes of prion protein include a major determinant of susceptibility to sheep scrapie. J Virol 78:11449

Nicholson EM, Mo H, Prusiner SB, Cohen FE, Marqusee S (2002) Differences between the prion protein and its homolog Doppel: a partially structured state with implications for scrapie formation. J Mol Biol 316:807–815

Nitrini R, Rosemberg S, Passos-Bueno MR, da Silva LS, Iughetti P, Papadopoulos M, Carrilho PM, Caramelli P, Albrecht S, Zatz M, LeBlanc A (1997) Familial spongiform encephalopathy associated with a novel prion protein gene mutation. Ann Neurol 42:138–146

Nunziante M, Ackermann K, Dietrich K, Wolf H, Gädtke L, Gilch S, Vorberg I, Groschup M, Schätzl HM (2011) Proteasomal dysfunction and endoplasmic reticulum stress enhance trafficking of prion protein aggregates through the secretory pathway and increase accumulation of pathologic prion protein. J Biol Chem 286:33942–33953

Pan KM, Baldwin M, Nguyen J, Gasset M, Serban A, Groth D et al (1993) Conversion of alpha-helices into beta-sheets features in the formation of the scrapie prion proteins. Proc Natl Acad Sci USA 90:10962–10966

Papassotiropoulos A, Wollmer MA, Aguzzi A, Hock C, Nitsch RM, de Quervain DJ (2005) The prion gene is associated with human long-term memory. Hum Mol Genet 14:2241–2246

Parchi P, Castellani R, Capellari S, Ghetti B, Young K, Chen SG et al (1996) Molecular basis of phenotypic variability in sporadic Creutzfeldt–Jakob disease. Ann Neurol 39:767–778

Pergami P, Poloni TE, Corato M, Camisa B, Ceroni M (1999) Prions and prion diseases. Funct Neurol 14:241–252

Prusiner SB (1991) Molecular biology of prion diseases. Science 252:1515–1522

Puckett C, Concannon P, Casey C, Hood L (1991) Genomic structure of the human prion protein gene. Am J Hum Genet 49:320–329

Re F, Sesana S, Barbiroli A, Bonomi F, Cazzaniga E, Lonati E, Bulbarelli A, Masserini M (2008) Prion protein structure is affected by pH-dependent interaction with membranes: a study in a model system. FEBS Lett 582:215–220

Rezaei H, Eghiaian F, Perez J, Doublet B, Choiset Y, Haertle T, Grosclaude J (2005) Sequential generation of two structurally distinct ovine prion protein soluble oligomers displaying different biochemical reactivities. J Mol Biol 347:665–679

Rogers M, Taraboulos A, Scott M, Groth D, Prusiner SB (1990) Intracellular accumulation of the cellular prion protein after mutagenesis of its Asn-linked glycosylation sites. Glycobiology 1:101–109

Ross ED, Minton A, Wickner RB (2005) Prion domains: sequences, structures and interactions. Nat Cell Biol 7:1039–1044

Safar J, Wille H, Itri V, Groth D, Serban H, Torchia M et al (1998) Eight prion strains have PrP(Sc) molecules with different conformations. Nat Med 4:1157–1165

Safar JG, DeArmond SJ, Kociuba K, Deering C, Didorenko S, Bouzamondo-Bernstein E et al (2005) Prion clearance in bigenic mice. J Gen Virol 86:2913–2923

Schwarze-Eicker K, Keyvani K, Görtz N, Westaway D, Sachser N, Paulus W (2005) Prion protein (PrPc) promotes beta-amyloid plaque formation. Neurobiol Aging 26:1177–1182

Shorter J, Lindquist S (2005) Prions as adaptive conduits of memory and inheritance. Nat Rev Genet 6:435–450

Si K, Lindquist S, Kandel ER (2003) A neuronal isoform of the aplysia CPEB has prion-like properties. Cell 115:879–891

Si K, Choi YB, White-Grindley E, Majumdar A, Kandel ER (2010) Aplysia CPEB can form prion-like multimers in sensory neurons that contribute to long-term facilitation. Cell 140:421–435

Singh N, Zanusso G, Chen SG, Fujioka H, Richardson S, Gambetti P, Petersen RB (1997) Prion protein aggregation reverted by low temperature in transfected cells carrying a prion protein gene mutation. J Biol Chem 272:28461–28470

Sokolowski F, Modler AJ, Masuch R, Zirwer D, Baier M, Lutsch G et al (2003) Formation of critical oligomers is a key event during conformational transition of recombinant syrian hamster prion protein. J Biol Chem 278:40481–40492

Strom A, Wang GS, Reimer R, Finegood DT, Scott FW (2007) Pronounced cytosolic aggregation of cellular prion protein in pancreatic beta-cells in response to hyperglycemia. Lab Invest 87:139–149

Tagliavini F, Prelli F, Ghiso J, Bugiani O, Serban D, Prusiner SB, Farlow MR, Ghetti B, Frangione B (1991) Amyloid protein of Gerstmann-Sträussler-Scheinker disease (Indiana kindred) is an 11 kd fragment of prion protein with an N-terminal glycine at codon 58. EMBO J 10:513–519

Tanaka M, Chien P, Yonekura K, Weissman JS (2005) Mechanism of cross-species prion transmission: an infectious conformation compatible with two highly divergent yeast prion proteins. Cell 121:49–62

Tompa P, Friedrich P (1998) Prion proteins as memory molecules: an hypothesis. Neuroscience 86:1037–1043

Wadsworth JD, Joiner S, Hill AF, Campbell TA, Desbruslais M, Luthert PJ, Collinge J (2001) Tissue distribution of protease resistant prion protein in variant Creutzfeldt–Jakob disease using a highly sensitive immunoblotting assay. Lancet 358:171–180

Wang F, Yang F, Hu Y, Wang X, Wang X, Jin C, Ma J (2007) Lipid interaction converts prion protein to a PrPSc-like proteinase K-resistant conformation under physiological conditions. Biochemistry 46:7045–7053

Weissmann C (2004) The state of the prion. Nat Rev Microbiol 2:861–871

Westaway D, DeArmond SJ, Cayetano-Canlas J, Groth D, Foster D, Yang SL et al (1994) Degeneration of skeletal muscle, peripheral nerves, and the central nervous system in transgenic mice overexpressing wild-type prion proteins. Cell 76:117–129

Westaway D, Alier K, Vergote D, MacTavish D, Mercer R, Fu W et al (2011) Prion proteins and the Alzheimer disease Aβ amyloid cascade. Prion 5:1–2

Xiao X, Yuan J, Zou WQ (2012) Isolation of soluble and insoluble PrP oligomers in the normal human brain. J Vis Exp 68:e3788, DOI: 10.3791/3788

Yedidia Y, Horonchik L, Tzaban S, Yanai A, Taraboulos A (2001) Proteasomes and ubiquitin are involved in the turnover of the wild-type prion protein. EMBO J 20:5383–5391

Yuan J, Xiao X, McGeehan J, Dong Z, Cali I, Fujioka H et al (2006) Insoluble aggregates and protease-resistant conformers of prion protein in uninfected human brains. J Biol Chem 281:34848–34858

Yuan J, Dong Z, Guo JP, McGeehan J, Xiao X, Wang J et al (2008) Accessibility of a critical prion protein region involved in strain recognition and its implications for the early detection of prions. Cell Mol Life Sci 65:631–643

Zaidi SI, Richardson SL, Capellari S, Song L, Smith MA, Ghetti B, Sy MS, Gambetti P, Petersen RB (2005) Characterization of the F198S prion protein mutation: enhanced glycosylation and defective refolding. J Alzheimers Dis 7:159–171, discussion 173–180

Zhang H, Stockel J, Mehlhorn I, Groth D, Baldwin MA, Prusiner SB et al (1997) Physical studies of conformational plasticity in a recombinant prion protein. Biochemistry 36:3543–3553

Zou WQ (2007) Transmissible spongiform encephalopathy and beyond (E-letter). Science. http://www.sciencemag.org/content/308/5727/1420.long/reply#sci_el_10316. Accessed 20 Sep 2007

Zou WQ (2010) Chameleon-like prion protein and human cognition. Curr Top Biochem Res 12:1–8

Zou WQ, Cashman NR (2002) Acidic pH and detergents enhance in vitro conversion of human brain PrPC to a PrPSc-like form. J Biol Chem 277:43942–43947

Zou WQ, Gambetti P (2007) Prion: the chameleon protein. Cell Mol Life Sci 64:3266–3270

Zou WQ, Zheng J, Gray DM, Gambetti P, Chen SG (2004) Antibody to DNA detects scrapie but not normal prion protein. Proc Natl Acad Sci USA 101:1380–1385

Zou WQ, Langeveld J, Xiao X, Chen S, McGeer PL, Yuan J et al (2010a) PrP conformational transitions alter species preference of a PrP-specific antibody. J Biol Chem 285:13874–13884

Zou WQ, Puoti G, Xiao X, Yuan J, Qing L, Cali I et al (2010b) Variably protease-sensitive prionopathy: a new sporadic disease of the prion protein. Ann Neurol 68:162–172

Zou RS, Fujioka H, Guo JP, Xiao X, Shimoji M, Kong C, Chen C, Tasnadi M, Voma C, Yuan J, Moudjou M, Laude H, Petersen RB, Zou WQ (2011a) Characterization of spontaneously generated prion-like conformers in cultured cells. Aging 3:968–984

Zou WQ, Xiao X, Yuan J, Puoti G, Fujioka H, Wang X et al (2011b) Amyloid-{beta}42 interacts mainly with insoluble prion protein in the Alzheimer brain. J Biol Chem 286:15095–15105

Zou WQ, Zhou X, Yuan J, Xiao X (2011c) Insoluble cellular prion protein and its association with prion and Alzheimer diseases. Prion 5:172–178

Chapter 6
Protein Misfolding Cyclic Amplification

Fabio Moda, Sandra Pritzkow, and Claudio Soto

Abstract Prion diseases are caused by a conformational conversion of the cellular prion protein (PrP^C) to a pathological conformer (PrP^{Sc}). The "prion-only" hypothesis suggests that PrP^{Sc} is the infectious agent that propagates the disease acting as a template for the conversion of PrP^C. In 2001, we developed a novel in vitro technique, called Protein misfolding cyclic amplification (PMCA), which mimics this pathological process in an accelerated way. Thereby, minimal amount of PrP^{Sc} can be amplified to several millions fold, providing an important tool for diagnosis and investigation of prion biology, and the molecular mechanism of prion conversion. PMCA also offers a great platform for the study and amplification of the protein misfolding process associated with other neurodegenerative disorders, such as Alzheimer's and Parkinson's diseases.

Keywords Prion diseases • Transmissible spongiform encephalopathies • Protein misfolding cyclic amplification • PMCA • Prion transmission • Prion decontamination procedures

6.1 PMCA: A Great Tool to Study Prion Biology

Prion diseases, or transmissible spongiform encephalopathies (TSEs), are a group of fatal disorders that affect both humans and animals. Prions are the proteinaceous infectious agent that is responsible for TSEs. Prions replicate through a nucleation-dependent process which is characterized by a long and silent incubation period followed by a rapid clinical phase. Thereby, a minute quantity of the pathological

F. Moda, Ph.D. • S. Pritzkow, Ph.D. • C. Soto, Ph.D. (✉)
Department of Neurology, Mitchell Center for Alzheimer's disease and Related Brain Disorders, University of Texas Houston Medical School, Houston, TX, USA
e-mail: fabio.moda@uth.tmc.edu; sandra.pritzkow@uth.tmc.edu

W.-Q. Zou and P. Gambetti (eds.), *Prions and Diseases: Volume 1, Physiology and Pathophysiology*, DOI 10.1007/978-1-4614-5305-5_6,

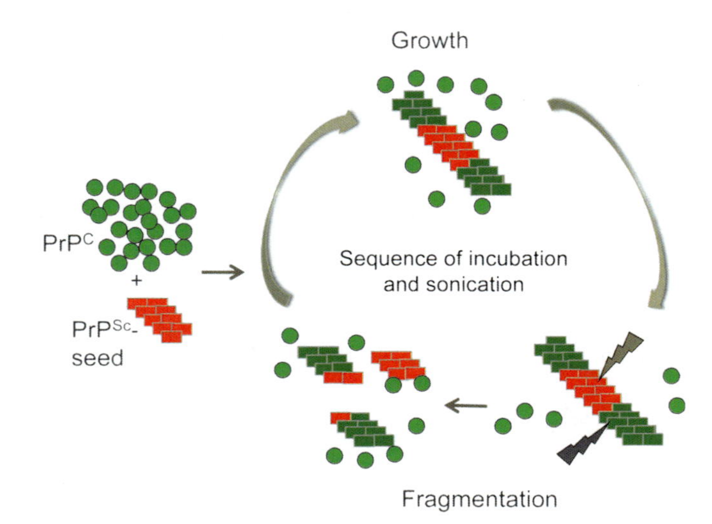

Fig. 6.1 Schematic representation of the PMCA principle. PMCA offers the chance to amplify minute quantities of PrP^{Sc} to a detectable level. In a cyclic manner consisting of two phases (incubation and sonication), PrP^{Sc} seeds from a sample are amplified at expenses of an excess of PrP^{C}. During the incubation phase, polymers of PrP^{Sc} grow by incorporation of PrP^{C}. In the following sonication phase, the large polymers are fragmented to generate multiple smaller PrP^{Sc} seeds for further prion replication

prion protein (PrP^{Sc}) works as a template and induces the conformational conversion of the cellular prion protein (PrP^{C}) to the pathogenic isoform (Prusiner 1998).

In 2001, we described an efficient technique to reproduce prion replication in the test tube in an accelerated manner, which is called protein misfolding cyclic amplification (PMCA) (Saborio et al. 2001). PMCA consists of cycles of incubation and sonication of a sample containing small amounts of PrP^{Sc} in the presence of an excess of PrP^{C}. During the incubation step, PrP^{Sc} aggregates grow through recruitment and conversion of PrP^{C} molecules. The following sonication phase is responsible for fragmenting these polymers to create new PrP^{Sc} seeds, which can induce further conversion of the cellular prion protein (Saborio et al. 2001; Soto et al. 2002). This method allows the exponential amplification of PrP^{Sc} in a PCR-like manner, and can begin the reaction with the equivalent to a single molecule of PrP^{Sc}, which after amplification can give rise to billions of PrP^{Sc} molecules (Saa et al. 2006a). The principle of PMCA is schematically illustrated in Fig. 6.1.

In following years, PMCA was improved through automation and the development of serial PMCA (sPMCA) (Fig. 6.2). Thereby, an aliquot of a PMCA sample, already subjected to many cycles of incubation and sonication, was diluted into fresh uninfected brain homogenate and subsequently exposed to further PMCA cycles. In this way, minute amounts of PrP^{Sc} can be detected through autocatalytic in vitro amplification, while the original inoculum is continuously diluted (Bieschke et al. 2004; Castilla et al. 2005a). An additional improvement was the addition of Teflon beads, which increase the efficiency and reproducibility of prion amplification (Gonzalez-Montalban et al. 2011).

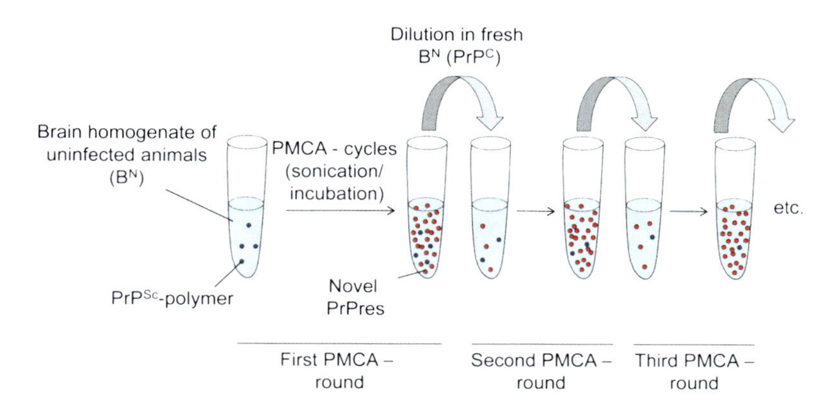

Fig. 6.2 Schematic design of serial PMCA. An aliquot of a PMCA sample, previously exposed to several PMCA cycles of incubation and sonication, is diluted in fresh brain homogenate and exposed to further PMCA cycles. Through sequences of serial PMCA rounds, the inoculum will be infinitely diluted and, in this way, prions can be maintained replicating indefinitively in vitro

Further experiments showed that the in vitro generated prions were fully infectious when injected into wild-type animals (Castilla et al. 2005a). They caused a similar disease with analog biochemical, biological, and structural properties observed in animals injected with brain derived PrP^{Sc} (Castilla et al. 2005a; 2008a; Weber et al. 2007). Studies of the components required to sustain PMCA amplification demonstrated the importance of cellular cofactors (e.g., nucleic acids and lipids) for efficient prion amplification (Deleault et al. 2003, 2007; Abid et al. 2010). Moreover, extensive PMCA cycling allows de novo formation of infectious prions mimicking the sporadic appearance of the disease (Deleault et al. 2007; Barria et al. 2009). In some of these cases, the prions produced through de novo creation in the test tube produced a new disease phenotype with unique clinical, neuropathological, and biochemical characteristics, never seen in nature (Barria et al. 2009).

An important development was the use of bacterially expressed recombinant prion protein (rPrP) as a substrate for PMCA (Atarashi et al. 2007). Wang and coworkers demonstrated that infectious prions can be generated from rPrP in the presence of synthetic lipids together with total RNA from normal mouse liver. When injected into wild-type mice, they caused a prion disease with similar incubation periods compared to naturally occurring prions (Wang et al. 2010). Recombinant PrP could also be labeled to perform structural studies of the prion protein.

The ability of PMCA to mimic the process of prion conversion in vitro provides great opportunities to analyze many aspects of prion biology, including (1) the biochemical mechanism of prion conversion and replication, (2) the species barrier and prion strain phenomena, (3) the potential role of cellular cofactors in PrP^C to PrP^{Sc} conversion, (4) the sensitive detection of prions for an early diagnosis of patients silently incubating the disease, (5) the evaluation of methods to remove and decontaminate prions, (6) the identification of prions in biological and environmental samples, and (7) the discovery and development of novel drugs to halt the prion conversion process.

6.2 PMCA Applications to Understand the Mechanism of Prion Transmission, Species Barrier and Strain Phenomena

Interspecies prion transmission is a process not well understood and limited by the so-called "species barrier" that corresponds to the ability of prions coming from one species to infect only a limited number of other species (Hill and Collinge 2004). This phenomenon is manifested as an incomplete attack rate and prolongation of the time to develop the disease in animals injected with infectious material from another species. The molecular basis of this event is not clear but convincing evidences indicate that the sequence of PrP controls this process; however the degree of the species barrier cannot be measured only by comparing the sequence of the proteins (Moore et al. 2005). The best way to investigate the species barrier is by infectivity experiments using animal models of the disease. However, these studies are costly and time-consuming because it is necessary to wait for several months or even years until the animals develop the clinical symptoms. Furthermore, the assessment of the species barrier for prion transmission to humans is compromised by the use of transgenic animal models expressing human PrP^{C}. PMCA can provide an in vitro alternative for studying the species barrier by combining PrP^{Sc} and PrP^{C} from different sources in distinct quantities. In this way, it is possible to quantitatively evaluate the efficiency of the conversion. Several studies confirmed that PMCA exhibits species specificity that faithfully reflects the same transmission barrier observed in animals (Castilla et al. 2008b; Green et al. 2008; Meyerett et al. 2008).

Transmission of sCJD between humans occurred through neurosurgical procedures as a consequence of using inappropriate techniques to sterilize instruments or devices that had been in contact with the brain tissue of sCJD infected individuals. Treatment with human-derived pituitary growth hormones or cornea or dura mater transplants, derived from infected recipients, also efficiently transmitted the disease (Brown et al. 2000). Conversely to vCJD, numerous studies have shown no evidence of human-to-human transmission of sCJD through the transfusion of blood or plasma, or the administration of plasma-derived therapeutic products (Operalski and Mosley 1995). Prions can also be transmitted from animals to humans. Epidemiological evidence suggests that among the animal TSEs, only BSE has been transmitted to humans through the consumption of contaminated beef products, generating the variant form of CJD (Will et al. 1996). Another concern is CWD, a disorder affecting mule deer and elk (Sigurdson and Aguzzi 2006) with high incidence in North America. CWD is highly transmissible within deer and elk populations. The mechanism of transmission is not well understood, but evidence supports the possibility that the disease is spread through direct animal-to-animal contact or as a result of indirect exposure to prions in the environment (e.g., in contaminated food and water sources). Transmission of CWD to humans cannot be excluded at this moment and transmissibility studies have been performed in many species to predict the spreading of the disease (e.g., in consequence of the consumption of CWD infected meat) (Sigurdson and Aguzzi 2006). In a recent study, we showed that cervid PrP^{Sc} can induce the conversion of human PrP^{C}, but only after the CWD prion strain has been stabilized by successive passages in vitro or in vivo (Barria et al. 2011). Interestingly, the newly generated human PrP^{Sc} exhibits a distinct biochemical pattern that differs from any of the currently known forms of human PrP^{Sc}. These findings imply that CWD prions have the potential to infect humans, and that this ability depends on CWD strain adaptation.

An intriguing feature of prions that has been often used against the prion hypothesis is the existence of prion strains (Soto 2011). Nearly all TSEs are known to exhibit various strains characterized by different incubation periods, clinical features, and neuropathology (Morales et al. 2007). In traditional infectious diseases, different strains generally arise from mutations or polymorphisms in the genetic makeup of the infectious agent. To reconcile the infectious agent composed exclusively of a protein with the strain phenomenon, it has been proposed that PrPSc obtained from different prion strains has slightly different conformation or aggregation states that can faithfully replicate at the expense of the host PrPC (Bessen et al. 1995; Telling et al. 1996; Safar et al. 1998). Various reports have shown that PMCA allows the faithful replication of prion strains in many different species of prions, indicating that all the elements required for strain determination are enciphered in the folding of PrP^{Sc} (Castilla et al. 2008a; Jones et al. 2009; Shikiya and Bartz 2011).

6.3 PMCA Applications in Prion Detection and Diagnosis

Clinical diagnosis of definite CJD can only be made postmortem by histological analysis of spongiform changes and accumulation of PrP^{Sc} in the brain (Soto 2004). Since presymptomatic detection of sCJD or variant CJD (vCJD) in living people is currently not possible, it is important to develop an objective and sensitive test which has the potential to identify infected individuals at presymptomatic stages of the disease.

To date, PrP^{Sc} represents the main component of the infectious agent and is the only disease-specific marker for CJD (Prusiner 1998; Brown et al. 2001; Soto 2004). It is abundant in the brain at late stage of the disease, while minute amount are present in peripheral tissues and biological fluids, such as lymphoid organs, cerebrospinal fluid (CSF), urine, and blood (Aguzzi 2000; Brown et al. 2001; Wadsworth et al. 2001; Gonzalez-Romero et al. 2008). The latter two fluids could be the best candidates for routine noninvasive diagnostic tests, but there is no validated method to detect PrP^{Sc} in these biological fluids (Soto 2004). In this regard, we and others recently reported that PMCA enables detection of PrP^{Sc} in samples of blood and/or urine from prion-infected hamsters, mice, sheep, and cervids (Castilla et al. 2005b; Gonzalez-Romero et al. 2008; Thorne and Terry 2008; Haley et al. 2009; Tattum et al. 2010). We also showed that PrP^{Sc} can be detected during the presymptomatic phase of the disease in blood (Saa et al. 2006b). These results are extremely important since it has been demonstrated that vCJD transmission occurred in patients after blood transfusion (Llewelyn, et al. 2004). Undetectable levels of PrP^{Sc} could be present in the blood of individuals silently incubating vCJD who may never develop clinical symptoms but remaining asymptomatic carriers able to transmit the disease to other individuals (Bishop et al. 2006). This could be a big problem for public health, especially for individuals who routinely rely on the blood supply and blood therapies.

In contrast, it is completely unknown if patients with sCJD have PrP^{Sc} circulating in blood and urine. Considering that most urine proteins originate from blood, it is likely that during disease progression, PrP^{Sc} is released from brain or peripheral organs into the blood at low concentrations, which is then excreted into the urine. Additionally, using a modified PMCA procedure, detection of PrP^{Sc} in CSF of humans affected by sCJD has been recently reported (Atarashi et al. 2011).

These findings suggest that PMCA enables efficient, specific, and rapid detection of prions in a variety of samples, offering a high promise for developing a noninvasive early diagnosis of prion diseases. Serial PMCA have been also applied for the diagnosis of different forms of animal prion diseases, including scrapie in hamster, mice and sheep, BSE in cattle, and CWD in cervids. In particular, PMCA allowed detecting PrP^{Sc} in the brain of presymptomatic hamsters, enabling a clear identification of infected animals as early as two weeks after inoculation (Soto et al. 2005). We demonstrated as well the presence of PrP^{Sc} in an experimentally infected cow 32 months postinoculation, that did not show clinical signs and was negative by standard western blot analysis (Soto et al. 2005).

Early diagnosis is very important for improving therapeutic perspectives, as treatment should start in an early stage, before the appearance of clinical signs and the occurrence of irreversible brain alterations. In addition, it should be possible to screen blood banks, reduce the iatrogenic transmission of the disease, and identify populations at risk (Soto 2004).

6.4 PMCA Applications in Development of Drugs and Prion Decontamination Procedures

One of the best targets for TSE therapy is the inhibition and reversal of PrP^C to PrP^{Sc} conversion. In drug development, it is crucial to have a relevant and robust in vitro assay to screen compounds for activity before testing them in more time-consuming and expensive in vivo assays. PMCA represents a convenient biochemical tool to identify and evaluate the activity of drug candidates for TSE treatment, because it mimics in vitro the central pathogenic process of the disease. Inhibitors and promoters could be tested quickly in different contexts using even human and bovine prions, for which no prion-permissive culture cells have been generated. Also the simplicity of the method and the relatively rapid outcome are important features for this type of studies. Moreover, the fact that PMCA can be applied to prion conversion in different species provides the opportunity to validate the use in humans of drugs that have been evaluated in experimental animal models of the disease.

In a similar way, the efficacy of devices and procedures to remove infectious prions from biological or environmental samples can be investigated in a rapid and efficient way using PMCA. The fact that PMCA enables to detect quantities of prions several order of magnitude smaller than infectivity bioassay makes PMCA more effective in studying prion removal procedures. Particularly useful for this type of application is the recent development of the quantitative PMCA technology which in addition to detect prions also permits to estimate the concentration of PrP^{Sc} present in the sample (Chen et al. 2010). Various recent articles have been published using PMCA to evaluate prion inactivation and removal from biological and environmental samples using diverse procedures (Morales et al. 2008; Pritzkow et al. 2011; Saunders et al. 2011; Ding et al. 2012).

6.5 Expanding PMCA Beyond Prion Diseases

As prion diseases, most of the neurodegenerative disorders like Alzheimer's disease, Parkinson's disease, amyotrophic lateral sclerosis, and Huntington's disease are thought to be caused by the brain accumulation of misfolded protein aggregates (Soto 2003). Protein misfolding and aggregation in other neurodegenerative diseases also follows a seeding-nucleation model involving the formation of similar intermediates and end products as in TSEs (Soto et al. 2006). Indeed, acceleration of protein aggregation by the addition of seeds has been convincingly reported in vitro for several proteins implicated in diverse diseases (Krebs et al. 2004). These theoretical considerations suggest that protein misfolding processes have the inherent ability to be transmissible (Soto et al. 2006). Strikingly, a series of recent and exciting reports, using cellular and/or animal models, have provided evidence suggesting that the transmission of protein misfolding by a prion-like mechanism might be at the heart of the most common neurodegenerative diseases (Meyer-Luehmann et al. 2006; Clavaguera et al. 2009; Ren et al. 2009; Frost et al. 2009; Munch et al. 2011; Morales et al. 2011; Mougenot et al. 2012). The similarities between TSEs and other neurodegenerative diseases in terms of their molecular mechanisms suggest that PMCA might be adapted to amplify the abnormal folding of these proteins as well. In very recent studies, we have been able to optimize PMCA for the detection of minute quantities of amyloid-beta misfolded oligomers in biological fluids of patients affected by Alzheimer's disease (Salvadores-Bersezio et al., manuscript submitted).

6.6 Concluding Remarks

PMCA was first published in a Nature article in 2001 (Saborio et al. 2001) and is today widely considered as a major breakthrough in science and technology. PMCA enabled for the first time to cyclically amplify the folding and biochemical properties of a protein in a manner conceptually analogous to the amplification of DNA by PCR. PMCA has enabled the generation infectious prions in vitro providing the strongest proof in favor of the prion hypothesis and has permitted to detect for the first time infectious prions in blood, offering a great possibility for early diagnosis. Over the past 5 years, PMCA has become widely used and invaluable technique to study the diverse aspects of prions. The PMCA technology has been used by several groups to understand the molecular mechanism of prion replication, the cellular factors involved in prion propagation, the intriguing phenomena of prion strains and species barriers, to detect PrPSc in tissues and biological fluids, and to screen for inhibitors against prion replication. The impact of PMCA is not only restricted to replication of prions, because it represents a platform technology to amplify the process of protein misfolding of the many proteins in which this mechanism occurs.

References

Abid K, Morales R, Soto C (2010) Cellular factors implicated in prion replication. FEBS Lett 584:2409–2414

Aguzzi A (2000) Prion diseases, blood and the immune system: concerns and reality. Haematologica 85:3–10

Atarashi R et al (2007) Ultrasensitive detection of scrapie prion protein using seeded conversion of recombinant prion protein. Nat Methods 4:645–650

Atarashi R et al (2011) Ultrasensitive human prion detection in cerebrospinal fluid by real-time quaking induced conversion. Nat Med 17:175–178

Barria MA, Mukherjee A, Gonzalez-Romero D, Morales R, Soto C (2009) De novo generation of infectious prions in vitro produces a new disease phenotype. PLoS Pathog 5(5):e1000421

Barria MA, Telling GC, Gambetti P, Mastrianni JA, Soto C (2011) Generation of a New Form of Human PrP^{Sc} in Vitro by Interspecies Transmission from Cervid Prions. J Biol Chem 286:7490–7495

Bessen RA et al (1995) Non-genetic propagation of strain-specific properties of scrapie prion protein. Nature 375:698–700

Bieschke J et al (2004) Autocatalytic self-propagation of misfolded prion protein. Proc Natl Acad Sci USA 101:12207–12211

Bishop MT et al (2006) Predicting susceptibility and incubation time of human-to-human transmission of vCJD. Lancet Neurol 5:393–398

Brown P et al (2000) Iatrogenic Creutzfeldt-Jakob disease at the millennium. Neurology 55:1075–1081

Brown P, Cervenakova L, Diringer H (2001) Blood infectivity and the prospects for a diagnostic screening test in Creutzfeldt-Jakob disease. J Lab Clin Med 137:5–13

Castilla J, Saá P, Hetz C, Soto C (2005a) In vitro generation of infectious scrapie prions. Cell 121:195–206

Castilla J, Saa P, Soto C (2005b) Detection of prions in blood. Nat Med 11:982–985

Castilla J et al (2008a) Cell-free propagation of prion strains. EMBO J 27:2557–2566

Castilla J, Gonzalez-Romero D, Saá P, Morales R, De Castro J, Soto C (2008b) Crossing the species barrier by PrPSc replication in vitro generates unique infectious prion. Cell 134:575–768

Chen B, Morales R, Barria MA, Soto C (2010) Estimating prion concentration in fluids and tissues by quantitative PMCA. Nat Methods 7:519–520

Clavaguera F et al (2009) Transmission and spreading of tauopathy in transgenic mouse brain. Nat Cell Biol 11:909–913

Deleault NR, Lucassen RW, Supattapone S (2003) RNA molecules stimulate prion protein conversion. Nature 425:717–720

Deleault NR, Harris BT, Rees JR, Supattapone S (2007) Formation of native prions from minimal components in vitro. Proc Natl Acad Sci USA 104:9741–9746
Ding N et al (2012) Inactivation of template-directed misfolding of infectious prion protein by ozone. Appl Environ Microbiol 78(3):613–620
Frost B, Jacks RL, Diamond MI (2009) Propagation of tau misfolding from the outside to the inside of a cell. J Biol Chem 284:12845–12852
Gonzalez-Montalban N et al (2011) Highly efficient protein misfolding cyclic amplification. PLoS Pathog 7(2):e1001277
Gonzalez-Romero D, Barria MA, Leon P, Morales R, Soto C (2008) Detection of infectious prions in urine. FEBS Lett 582:3161–3166
Green KM et al (2008) Accelerated high fidelity prion amplification within and across prion species barriers. PLoS Pathog 4:e1000139
Haley NJ, Seelig DM, Zabel MD, Telling GC, Hoover EA (2009) Detection of CWD prions in urine and saliva of deer by transgenic mouse bioassay. PLoS One 4:e4848
Hill AF, Collinge J (2004) Prion strains and species barriers. Contrib Microbiol 11:33–49
Jones M et al (2009) Human platelets as a substrate source for the in vitro amplification of the abnormal prion protein (PrP) associated with variant Creutzfeldt-Jakob disease. Transfusion 49:376–384
Krebs MR, Morozova-Roche LA, Daniel K, Robinson CV, Dobson CM (2004) Observation of sequence specificity in the seeding of protein amyloid fibrils. Protein Sci 13:1933–1938
Llewelyn CA et al (2004) Possible transmission of variant Creutzfeldt-Jakob disease by blood transfusion. Lancet 363:417–421
Meyer-Luehmann M et al (2006) Exogenous induction of cerebral beta-amyloidogenesis is governed by agent and host. Science 313:1781–1784
Meyerett C et al (2008) In vitro strain adaptation of CWD prions by serial protein misfolding cyclic amplification. Virology 382:267–276
Moore RA, Vorberg I, Priola SA (2005) Species barrier in prion disease-brief review. Arch Virol Suppl 19:187–202
Morales R, Abid K, Soto C (2007) The prion strain phenomenon: molecular basis and unprecedented features. Biochim Biophys Acta 1772:681–691
Morales R et al (2008) Reduction of prion infectivity in packed red blood cells. Biochem Biophys Res Commun 377:373–378
Morales R, Duran-Aniotz C, Castilla J, Estrada LD, Soto C (2011) De novo induction of amyloid-β deposition in vivo. Mol Psychiatry. doi:10.1038/mp. 2011.120
Mougenot AL et al (2012) Prion-like acceleration of a synucleinopathy in a transgenic mouse model. Neurobiol Aging 33(9):2225–2228
Munch C, O'Brien J, Bertolotti A (2011) Prion-like propagation of mutant superoxide dismutase-1 misfolding in neuronal cells. Proc Natl Acad Sci USA 108:3548–3553
Operalski EA, Mosley JW (1995) Pooled plasma derivatives and Creutzfeldt-Jakob disease. Lancet 346:1224
Pritzkow S et al (2011) Quantitative detection and biological propagation of scrapie seeding activity in vitro facilitate use of prions as model pathogens for disinfection. PLoS One 6:e20384
Prusiner SB (1998) Prions. Proc Natl Acad Sci USA 95:13363–13383
Ren PH et al (2009) Cytoplasmic penetration and persistent infection of mammalian cells by polyglutamine aggregates. Nat Cell Biol 11:219–225
Saa P, Castilla J, Soto C (2006a) Ultra-efficient replication of infectious prions by automated protein misfolding cyclic amplification. J Biol Chem 281:35245–35252
Saa P, Castilla J, Soto C (2006b) Presymptomatic detection of prions in blood. Science 313:92–94
Saborio GP, Permanne B, Soto C (2001) Sensitive detection of pathological prion protein by cyclic amplification of protein misfolding. Nature 411:810–813
Safar J et al (1998) Eight prion strains have PrP(Sc) molecules with different conformations. Nat Med 4:1157–1165
Saunders SE, Bartz JC, Vercauteren KC, Bartelt-Hunt SL (2011) An enzymatic treatment of soil-bound prions effectively inhibits replication. Appl Environ Microbiol 77:4313–4317

Shikiya RA, Bartz JC (2011) In vitro generation of high-titer prions. J Virol 85:13439–13442
Sigurdson CJ, Aguzzi A (2006) Cronic wasting disease. Biochim Biophys Acta 1772:610–618
Soto C, Saborio GP, Anderes L (2002) Cyclic amplification of protein misfolding: application to prion-related disorders and beyond. Trends Neurosci 25:390–394
Soto C (2003) Unfolding the role of protein misfolding in neurodegenerative diseases. Nat Rev Neurosci 4:49–60
Soto C (2004) Diagnosing prion diseases: needs, challenges and hopes. Nat Rev Microbiol 2:809–819
Soto C et al (2005) Pre-symptomatic detection of prions by cyclic amplification of protein misfolding. FEBS Lett 579:638–642
Soto C, Estrada L, Castilla J (2006) Amyloids, prions and the inherent infectious nature of misfolded protein aggregates. Trends Biochem Sci 31:150–155
Soto C (2011) Prion hypothesis: The end of the controversy? Trends Biochem Sci 36:151–158
Tattum MH et al (2010) Discrimination between prion-infected and normal blood samples by protein misfolding cyclic amplification. Transfusion 50:996–1002
Telling GC et al (1996) Evidence for the conformation of the pathologic isoform of the prion protein enciphering and propagating prion diversity. Science 274:2079–2082
Thorne L, Terry LA (2008) In vitro amplification of PrP^{Sc} derived from the brain and blood of sheep infected with scrapie. J Gen Virol 39:3177–3184
Wadsworth JD et al (2001) Tissue distribution of protease resistant prion protein in variant Creutzfeldt-Jakob disease using a highly sensitive immunoblotting assay. Lancet 358:171–180
Wang F, Wang X, Yuan CG, Ma J (2010) Generating a prion with bacterially expressed recombinant prion protein. Science 327:1132–1135
Weber P et al (2007) Generation of genuine prion infectivity by serial PMCA. Vet Microbiol 123:346–357
Will RG et al (1996) A new variant of Creutzfeldt Jakob disease in the UK. Lancet 347:921–925

Chapter 7
Cofactor Involvement in Prion Propagation

Surachai Supattapone and Michael B. Miller

Abstract Pure amyloid proteins are responsible for the transmissible properties of yeast prions (Tanaka et al., Nature 428(6980):323–328, 2004; Tanaka et al., Cell 121(1):49–62, 2005; King and Diaz-Avalos, Nature 428(6980):319–323, 2004). However, it is currently unknown whether the infectious properties of mammalian prions can also be explained by a "protein-only" mechanism in which a host-encoded protein, PrP^{C}, undergoes a conformational change into an infectious conformer, PrP^{Sc}. Recent studies have shown that non-proteinaceous cofactors are necessary for the formation of PrP^{Sc} and mammalian prion infectivity in vitro. Reconstitution studies suggest that different prion variants may preferentially propagate with specific classes of cofactor molecules. The pathogenic roles played by putative prion cofactors remain to be elucidated.

Keywords Prion • Cofactor • RNA • Lipid • Protein-only hypothesis • Strains • Neurotropism • Polyanion

7.1 The "Protein-Only" Hypothesis

Prions are the infectious agents of fatal neurodegenerative diseases affecting humans and other animals, such as Creutzfeldt–Jakob disease (CJD), kuru, and scrapie (Prusiner 1998); and "prion-like" mechanisms have recently been implicated in the pathogenesis of other disorders, such as Alzheimer's and Parkinson's diseases (Brundin et al. 2010; Cushman et al. 2010; Kim and Holtzman 2010; Lee et al. 2010). Despite decades of investigation, the essential composition of mammalian prions and their mechanism of formation remain unknown (Supattapone 2010).

S. Supattapone, M.D., Ph.D., D.Phil. (✉) • M.B. Miller, Ph.D.
Department of Biochemistry, Dartmouth Medical School,
7200 Vail Building, Hanover, NH 03755, USA
e-mail: Supattapone@dartmouth.edu1

W.-Q. Zou and P. Gambetti (eds.), *Prions and Diseases: Volume 1, Physiology and Pathophysiology*, DOI 10.1007/978-1-4614-5305-5_7,

Experiments showing that scrapie and kuru were transmissible led early investigators to search for a causative pathogen for these diseases (Cuillé and Chelle 1939; Gajdusek et al. 1966). The accepted paradigm for identifying and proving that a pathogen causes a infectious disease, proposed by Robert Koch (1893), required isolation of the diseased organism in pure culture and demonstrating its subsequent ability to cause the disease. As such, efforts were made to characterize the pathogen, in order to facilitate isolation. Aided by the advance of adapting the scrapie agent to mice (Chandler 1961), it was demonstrated that scrapie infectivity could pass through filters with pores as small as 43 nm (Hunter 1969), indicating that the agent was not a bacterium, since the smallest known bacteria measure approximately 300 nm (Robertson et al. 1975). During this time, all infectious agents smaller than bacteria were thought to be viruses, intracellular parasites with a nucleic acid genome of deoxyribonucleic acid (DNA) or ribonucleic acid (RNA) surrounded by a protein capsid and, in some, a lipid envelope. As a result, the infectious agent of scrapie and other spongiform encephalopathies, which display a very long incubation period (Mead et al. 2009; Prusiner 1997), were originally described as slow viruses (Sigurdsson 1954; Gajdusek 1967).

A significant step forward in understanding the nature of the infectious pathogen came from experiments performed by Tikvah Alper in 1967. She found that scrapie infectivity was resistant to high doses of ultraviolet (UV) irradiation (Alper et al. 1967). UV irradiation, known to abolish infectivity of viruses, is thought to inactivate gene-coding nucleic acids by inducing dimerization of pyrimidine nucleotides (Barnhart et al. 1976). These experiments indicated that the scrapie agent lacked a nucleic acid genome, suggesting that a novel class of agent may be responsible.

Griffith proposed three possible molecular mechanisms that could accommodate the experimental observations, including a hypothesis that the scrapie agent may contain only one essential component, a protein (Griffith 1967). Under this "protein-only" hypothesis, this protein would bear a certain conformation and replicate by changing the conformation of a host cell protein. If this hypothesis is correct, then the natural occurrence of multiple prion strains with distinct PrP^{Sc} conformations violates the most fundamental principle of protein folding, originally proposed by Anfinsen, that primary sequence determines tertiary structure (Anfinsen et al. 1961). In 1982, Prusiner and colleagues successfully isolated and characterized infectious prions biochemically (Prusiner 1982). This landmark achievement confirmed that prions are indeed unorthodox infectious agents, identified PrP^{Sc} as a critical component of infectious prions, and greatly facilitated subsequent efforts to determine the molecular basis of prion infectivity.

7.2 Components of Purified Native Prions

Pulsed-field flow fractionation analysis of purified prion preparations has indicated that the most infectious prion particles are 17–27 nm in size (Silveira et al. 2005). Protease-resistant PrP^{Sc} has been the most consistent and principal substance identified in biochemically purified prion infectivity (Bolton et al. 1982; Prusiner et al. 1984). Still, many efforts have searched for other components in the infectious particles. Studies from the Manuelidis laboratory have identified 25 nm virus-like particles and various nucleic acids in prion-infected brains (Manuelidis et al. 2007; Manuelidis 2011), suggested as the "likely cause" of prion diseases. Various other studies have reported no specific nucleic acids copurifying with prion infectivity (Hunter et al. 1976; Meyer et al. 1991) or only molecules of variable sequence (Safar et al. 2005). The sum of these findings, put together with UV resistance (Alper et al. 1967) and the successful propagation of prion infectivity in cell-free systems (Castilla et al. 2005), suggests that prions do not contain gene-coding nucleic acids. Thus, neither a virus nor a viroid (RNA lacking protein coat) is likely to be the agent causing prion disease. There has also been a report of copurifying polysaccharides distinct from the *N*-linked PrP glycans (Appel et al. 1999). On the basis of disinfection studies with organic solvents and heat, another hypothesis suggests that prions may contain a lipid component in addition to PrP (Gale 2006). Currently, it remains unclear whether native prions contain any essential components other than PrP^{Sc}.

7.3 Prion Replication in Cell-Free Conditions

While purified prion preparations, animals of various *Prnp* sequences, and prion-infected cultured cells have been excellent tools for learning about prion behavior, the development of several in vitro PrP^{Sc} formation techniques has been particularly helpful for studying the composition and propagation mechanism of prions. In a significant advance, Caughey and colleagues carried out the first cell-free conversion of PrP^{C} into PrP^{Sc} (Kocisko et al. 1994). In this method, a stoichiometric excess of infectious PrP^{Sc} is mixed with radiolabeled PrP^{C} molecules, and newly formed, radioactive PrP^{Sc} is detected by its acquisition of protease resistance. Using this technique, it was demonstrated that the distinct PrP^{Sc} biochemical characteristics of prion strains (Bessen and Marsh 1994) were maintained during PrP^{Sc} propagation in vitro (Bessen et al. 1995), providing evidence that another biologic characteristic of prions could be observed under cell-free conditions. However, a large excess of PrP^{Sc} was required to convert a small amount of PrP^{C}, precluding measurements of the infectivity of in vitro-generated PrP^{Sc} molecules.

Subsequently, Soto and colleagues reported a more efficient method for propagating prions in vitro, protein misfolding cyclic amplification (PMCA) (Saborio et al. 2001). Using alternating steps of incubation and sonication, PMCA facilitated robust PrP^{Sc} amplification in the context of homogenized brain tissue. PMCA was subsequently adapted into a serial format, where the newly generated PrP^{Sc} molecules were used to seed fresh brain homogenate containing unconverted PrP^{C} substrate. Using many serial amplifications in this manner, the input prion infectivity was diluted to undetectable and mathematically negligible levels, and reactions containing newly generated PrP^{Sc} were shown to contain prion infectivity by bioassay (Castilla et al. 2005). Serial PMCA (sPMCA) has also been used to show that specific clinical and neuropathological properties of prion strains may be propagated in a cell-free environment (Castilla et al. 2008; Green et al. 2008), building on the finding of strain-specific PrP^{Sc} pattern propagation in vitro (Bessen et al. 1995).

An alternative method for native PrP^{Sc} formation in vitro employs high frequency shaking of brain homogenates instead of sonication (Lucassen et al. 2003). Like PMCA, this non-sonication method amplifies PrP^{Sc} levels several fold over the input seed, suggesting that PrP^{Sc} amplification is primarily dependent upon the presence of cofactors in normal brain homogenate rather than sonication. Indeed, subsequent enzyme treatment and reconstitution studies showed that amplification of hamster PrP^{Sc} in this system is dependent upon the endogenous RNA present within the brain homogenate (Deleault et al. 2003).

7.4 Formation of Infectious Prions from Minimal Components: Requirement of Non-PrP Cofactor

The "protein-only" hypothesis provides a simple explanation for the infectivity of mammalian prions despite their lack of replicating nucleic acids. One prediction of this hypothesis is that, since PrP^{Sc} molecules in infectious prions are thermodynamically more stable than PrP^{C} molecules, it should be possible to produce infectious PrPSc molecules in vitro by refolding pure recombinant PrP (recPrP) substrate (Cohen 1999). However, attempts to form infectious prions from purified PrP alone have not yielded products that are consistently infectious to wild-type animals. Based on the observation of amyloid fibrils containing PrP in the brains of infected animals (Merz et al. 1987) and potential parallels to self-propagating fungal protein conformations (Wickner et al. 1995; Balbirnie et al. 2001), PrP amyloid fibrils were prepared in vitro from bacterially expressed recombinant PrP (Baskakov et al. 2002). When injected into mice expressing 16-fold greater PrP than endogenous levels, a transmissible neurologic disease resulted after 380–600 days (Legname et al. 2004). However, uninoculated 16X PrP control animals are prone to neurologic dysfunction after approximately 600 days (Colby et al. 2010), suggesting that the injected amyloid fibrils may have accelerated a preexisting disease, similar to the transmission experiments of GSS from mice (Hsiao et al. 1994). Furthermore, this PrP amyloid did not consistently transmit disease to wild-type mice (Colby et al. 2010). A subsequent study also found that PrP amyloid fibrils failed to transmit prion disease to wild-type animals, though fibrils annealed by high-temperature with brain homogenate could trigger infectious PrP^{Sc} formation (Makarava et al. 2010). In another study, PrP fibrillar aggregates formed by PMCA without adding cofactors showed minimal and inconsistent infectivity in animals (Kim et al. 2010).

Preparations formed from purified PrP alone have not reproducibly shown prion infectivity in wild-type animals. However, PrP^{Sc} generated from purified PrP^{C} mixed with cofactors is infectious to wild-type animals (Deleault et al. 2007; Wang et al. 2010). Both of these recipes included purified PrP^{C} (or recombinant α-helical PrP), a polynuceleotide polyanion, and a lipid, suggesting that non-PrP components may be necessary to form *bona fide* infectious prions.

7.5 The Protein X Hypothesis

Specific mutant MoPrP molecules can act in a dominant negative manner to prevent the propagation of human prions with HuPrP molecules in transgenic mice (Telling et al. 1995). A potential explanation for this dominant negative effect is that mutant MoPrP^C molecules bind and sequester a cofactor that is necessary for prion propagation. Such a cofactor was hypothesized to be a protein, Protein X (Telling et al. 1995). Subsequent investigation identified four C terminal PrP residues which, when mutated, are capable of exhibiting dominant negative inhibition of prion propagation in cultured cells (Kaneko et al. 1997; Perrier et al. 2002). It was postulated that these residues form a discontinuous epitope that interacts with Protein X. However, in a polymerization reaction of purified recombinant PrP, one such mutant PrP reduced polymerization of wild-type PrP (Lee et al. 2007). Furthermore, the dominant negative effect can be observed with prions propagating in vitro in with purified PrP^C substrate and accessory non-protein cofactors (Geoghegan et al. 2009), indicating that Protein X is not responsible for the dominant negative effect. Thus, it is not likely that non-PrP proteins serve as cofactors in prion formation.

7.6 Non-Proteinaceous Prion Cofactors

Many different molecules have been proposed to participate in prion propagation. Sulfated glycosaminoglycans (GAGs), such as heparan sulfate proteoglycan (HSPG), can stimulate formation of protease-resistant PrP^{Sc} (Wong et al. 2001) and may play a role in PrP^{Sc} formation in cells (Ben-Zaken et al. 2003; Taylor et al. 2009). Copper ions can induce PrP^C to form a protease-resistant state (Quaglio et al. 2001; Kuczius et al. 2004), but copper also inhibits PrP^{Sc} propagation in vitro (Orem et al. 2006) and in cultured cells (Hijazi et al. 2003). Plasminogen (Mays and Ryou 2010) and the laminin receptor (Leucht et al. 2003) have also been proposed to participate in prion propagation. PrP also interacts with nucleic acids (Grossman et al. 2003; Cordeiro and Silva 2005; Adler et al. 2003) and lipid membrane vesicles (Morillas et al. 1999; Gabizon et al. 1987).

Specific evidence of a role for RNA in prion propagation came from the observation that transformation of PrP^C into PrP^{Sc} in vitro in brain homogenates is reduced after RNase digestion and increased after RNA supplementation (Deleault et al. 2003). Subsequently, the PrP^C substrate was purified, and various preparations were tested for their ability to reconstitute PrP^{Sc} amplification (Deleault et al. 2005). PrP-null mouse brain homogenate control and RNA from various sources enabled amplification. Interestingly, various homopolymeric nucleic acids also stimulated PrP^{Sc} amplification, suggesting that the mechanism did not rely on information-coding nucleic acids but instead on polyanionic molecules. Other such polyanions, like HSPG, stimulated conversion to some degree, but less than nucleic acid polyanions (Deleault et al. 2005). Using PMCA, further studies found that polyanions must be at least 40 nucleotides in length to act as PrP^{Sc} propagation cofactors (Geoghegan et al. 2007). Furthermore, during PrP^{Sc} propagation, polyanion cofactors are incorporated into a complex with PrP (Geoghegan et al. 2007). This suggests that the polyanions may act as a structural component of infectious prions.

Not only do polyanion cofactors permit PrP^{Sc} amplification in vitro, but propagation in this minimal component reaction system proceeds indefinitely, and robust in vivo prion infectivity is likewise propagated (Deleault et al. 2007). Thus, infectious prions can be made from a defined mixture of minimal components: prion seed, PrP^{C} substrate, polyanion cofactor, and stoichiometric lipids co-purifying with PrP^{C}. From calculations of the maximal prion seed dilution that could be detected after amplification, these authors estimated that infectious prions could contain as few as 7 PrP^{Sc} monomers. Also, using this recipe but omitting the PrP^{Sc} seed, infectious prions were formed de novo (Deleault et al. 2007), suggesting a potential mechanism for the genesis of sporadic prion diseases such as CJD.

Interaction with anionic phospholipids, such as phosphatidylglycerol (POPG) and phosphatidylserine (POPS), induces a conformational change in recombinant PrP in which β-sheet content increases (Kazlauskaite et al. 2003; Wang et al. 2007). Under some conditions, POPG induces a portion of PrP molecules to adopt a small protease-resistant C terminal core, reminiscent of PrP^{Sc}. When the POPG–PrP complex was subjected to PMCA supplemented with RNA polyanion cofactors (purified from liver), the reactions generated PrP^{Sc} infectious to wild-type mice (Wang et al. 2010). This first demonstration of prion infectivity generated from bacterially expressed recombinant PrP further implicated cofactors in the prion propagation mechanism.

Prions of different species may display distinct cofactor requirements for propagation. While RNA polyanion cofactors support propagation of hamster PrP^{Sc}, they do not support mouse PrP^{Sc} propagation under the same conditions. Other cofactor molecules, present in PrP-null mouse brain homogenate and resistant to protease and nuclease digestion, appear required for mouse PrP^{Sc} propagation (Deleault et al. 2010).

7.7 Potential Roles of Cofactors in Prion Formation and Encoding Infectivity

PrP in vitro conversion studies and biological infectivity assays have shown a clear role for non-PrP cofactors in prion propagation (Legname et al. 2004; Makarava et al. 2010; Deleault et al. 2007; Wang et al. 2007, 2010). The function of such cofactors is not known. They could either act as an integral component of the infectious prion or as a catalyst for PrP conformational change (Fig. 7.1). Polyanions may be incorporated into a complex with PrP^{Sc} during propagation in vitro (Geoghegan

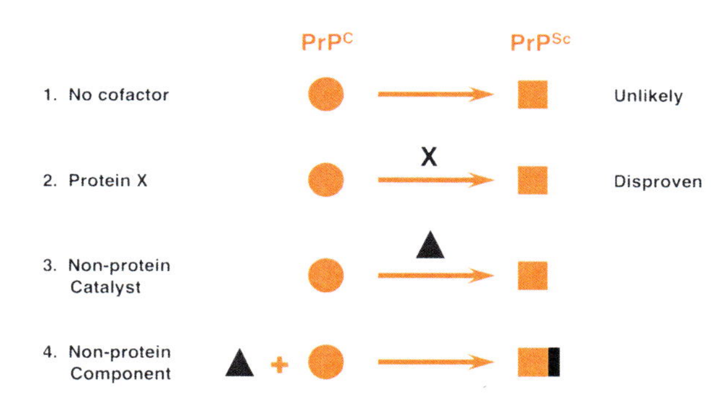

Fig. 7.1 Potential mechanisms of prion formation. A non-proteinaceous cofactor (*triangle*) is likely to assist the conversion of PrP^{C} (*circle*) into PrP^{Sc} (*rectangle*), either as a catalyst or an essential component

et al. 2007), possible evidence that they are an integral component. However, photofragmentation of incorporated photolabile nucleic acid polyanion cofactor molecules down to pentanucleotide units did not reduce prion infectivity (Piro et al. 2011), suggesting that cofactor function may be more catalytic in nature. Put another way, while polyanions >40 nucleotides in length are required for such propagation (Geoghegan et al. 2007), fragmentation to 5 base oligonucleotides permits retention of formed prion infectivity. This finding does not provide definitive proof for the "protein only" hypothesis since copurified lipids and short oligonucleotides remain present after photodegradation, but it places significant constraints on the possible mechanism by which cofactors facilitate prion formation in vitro.

If cofactors function as an integral component of infectious prions, their contribution could be structural or informational. The resistance of prions to UV irradiation (Alper et al. 1967) and lack of requirement for gene-coding sequence of nucleic acid cofactors (Deleault et al. 2005) argues against such a classical genetic informational role, though such a function could be more subtle. For example, different types of cofactors could support PrP^{Sc} structures in distinct conformations, whereby the cofactor would serve both a structural and informational role. As structure or information, cofactors could also play a role in modulating interactions between PrP^{Sc} and host PrP^{C} molecules, where PrP^{C} polybasic domains appear to provide PrP^{Sc}-binding sites (Miller et al. 2011).

Questions about potential information that cofactors may convey in prions lead to the issue of whether they are universal or specific. The same cofactor molecule could be universally required for the propagation of all prions, or distinct cofactors could participate in propagation of different strains or species of prions (Fig. 7.2). Reconstitution studies suggest that certain PrP^{Sc} molecules propagate best with certain cofactors (Deleault et al. 2010). However, this in vitro finding does not preclude that a single molecule in brain tissue acts in vivo as a universal cofactor for prion propagation.

One of the most important challenges to the "protein-only" hypothesis is the existence of multiple prion "strains." Strains are defined as natural isolates of infectious prions characterized by distinctive clinical and neuropathological features,

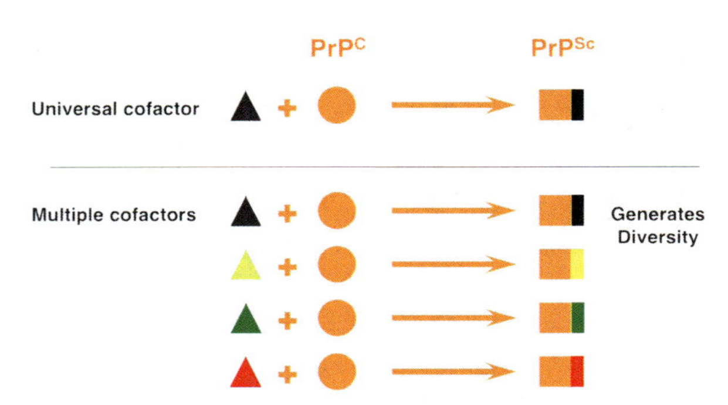

Fig. 7.2 How many cofactors? It is not currently known whether a single, universal cofactor can facilitate the formation of multiple prion species and strains or, alternatively, whether different cofactors are preferentially used by different prions to generate diversity

which are faithfully recapitulated upon serial passage within the same animal species (Bruce 1993; Carlson 1996). Because prions lack a nucleic acid genome, the mechanism of prion strain variation cannot involve gene mutation (Li et al. 2009).

Studies with yeast models and recombinant mammalian PrP show that pure proteins can adopt multiple, self-propagating conformations (Tanaka et al. 2004, 2005; King and Diaz-Avalos 2004; Jones and Surewicz 2005; Makarava and Baskakov 2008). However, it is difficult to explain the selective neurotropism of native mammalian prion strains on the basis of differential PrP polypeptide folding alone (DeArmond et al. 1997; Mahal et al. 2007).

Some investigators have speculated that strain-dependent differences in PrP^{Sc} glycosylation might encipher the selective neurotropism of prion strains since PrP^{C} glycosylation patterns vary in different regions of the brain (Vorberg and Priola 2002; Beringue et al. 2003; Cancellotti et al. 2005; Khalili-Shirazi et al. 2005; Tuzi et al. 2008). However, this hypothesis was refuted by as study showing that unglycosylated PrP^{Sc} molecules successfully transmit the strain-specific neurotropism of several mouse prion strains (Piro et al. 2009).

Another possible explanation for cellular tropism is that perhaps only a subset of cell types contains the specific cofactor(s) needed to propagate a particular prion strain (Supattapone 2010). In this setting, each prion strain might require a unique set of endogenous cofactors to propagate, i.e., a "cofactor variation" hypothesis of strain diversity (Fig. 7.2). The existence of multiple classes of cofactors for prion propagation in vitro is consistent with this hypothesis (Deleault et al. 2010).

7.8 Additional Roles and Applications for Prion Cofactors

Beyond participating in the propagation mechanism of infectious prions, cofactors could also play a role in the mechanism of neurotoxicity. For example, prion infection could deplete or modulate the normal activity of an essential endogenous cofactor molecule. Such a scenario would be compatible with the observation that symptomatic prion disease occurs a long time after maximal infectious titers accumulate in the brains of infected animals, and that the interval period to symptomatic disease is inversely proportional to PrP expression level (Sandberg et al. 2011).

Cofactors may also be required for the pathogenesis of other neurodegenerative diseases involving protein misfolding. For instance, although inoculation brain homogenates containing ABeta plaques can stimulate the spread of similar plaques in the brains of recipient transgenic mice, inoculation of pure synthetic ABeta amyloid into the same recipient mice fails to induce plaque formation (Meyer-Luehmann et al. 2006). One possible explanation for this discrepancy is that additional cofactors are required for amyloid plaques to mature into a form that can propagate in the brain.

It is possible to envision a number of practical applications for prion cofactors. For instance, they could be used to produce large quantities of infectious prions for biophysical studies. Prion cofactors also represent new potential therapeutic targets, and antagonists that block their interaction with PrP might prove to be useful drugs for treating clinical prion disease. In addition, cofactors could be used in the area of prion diagnostics, either by facilitating the amplification of prions in vitro, or by serving as a biomarker of prion disease in histological studies.

References

Adler V, Zeiler B, Kryukov V, Kascsak R, Rubenstein R, Grossman A (2003) Small, highly structured RNAs participate in the conversion of human recombinant PrP(Sen) to PrP(Res) in vitro. J Mol Biol 332(1):47–57

Alper T, Cramp WA, Haig DA, Clarke MC (1967) Does the agent of scrapie replicate without nucleic acid? Nature 214(90):764–766

Anfinsen CB, Haber E, Sela M, White FH Jr (1961) The kinetics of formation of native ribonuclease during oxidation of the reduced polypeptide chain. Proc Natl Acad Sci USA 47:1309–1314

Appel TR, Dumpitak C, Matthiesen U, Riesner D (1999) Prion rods contain an inert polysaccharide scaffold. Biol Chem 380(11):1295–1306

Balbirnie M, Grothe R, Eisenberg DS (2001) An amyloid-forming peptide from the yeast prion Sup35 reveals a dehydrated beta-sheet structure for amyloid. Proc Natl Acad Sci USA 98(5):2375–2380

Barnhart BJ, Cox SH, Jett JH (1976) Prophage induction and inactivation by UV light. J Virol 18(3):950–955

Baskakov IV, Legname G, Baldwin MA, Prusiner SB, Cohen FE (2002) Pathway complexity of prion protein assembly into amyloid. J Biol Chem 277(24):21140–21148

Ben-Zaken O, Tzaban S, Tal Y, Horonchik L, Esko JD, Vlodavsky I, Taraboulos A (2003) Cellular heparan sulfate participates in the metabolism of prions. J Biol Chem 278(41):40041–40049

Beringue V, Mallinson G, Kaisar M, Tayebi M, Sattar Z, Jackson G, Anstee D, Collinge J, Hawke S (2003) Regional heterogeneity of cellular prion protein isoforms in the mouse brain. Brain 126(Pt 9):2065–2073

Bessen RA, Marsh RF (1994) Distinct PrP properties suggest the molecular basis of strain variation in transmissible mink encephalopathy. J Virol 68(12):7859–7868

Bessen RA, Kocisko DA, Raymond GJ, Nandan S, Lansbury PT, Caughey B (1995) Non-genetic propagation of strain-specific properties of scrapie prion protein. Nature 375(6533):698–700

Bolton DC, McKinley MP, Prusiner SB (1982) Identification of a protein that purifies with the scrapie prion. Science 218(4579):1309–1311
Bruce ME (1993) Scrapie strain variation and mutation. Br Med Bull 49(4):822–838
Brundin P, Melki R, Kopito R (2010) Prion-like transmission of protein aggregates in neurodegenerative diseases. Nat Rev Mol Cell Biol 11(4):301–307. doi:nrm2873 [pii] 10.1038/nrm2873
Cancellotti E, Wiseman F, Tuzi NL, Baybutt H, Monaghan P, Aitchison L, Simpson J, Manson JC (2005) Altered glycosylated PrP proteins can have different neuronal trafficking in brain but do not acquire scrapie-like properties. J Biol Chem 280(52):42909–42918
Carlson GA (1996) Prion strains. Curr Top Microbiol Immunol 207:35–47
Castilla J, Saa P, Hetz C, Soto C (2005) In vitro generation of infectious scrapie prions. Cell 121(2):195–206
Castilla J, Morales R, Saa P, Barria M, Gambetti P, Soto C (2008) Cell-free propagation of prion strains. EMBO J 27(19):2557–2566
Chandler RL (1961) Encephalopathy in mice produced by inoculation with scrapie brain material. Lancet 1:1378–1379
Cohen FE (1999) Protein misfolding and prion diseases. J Mol Biol 293(2):313–320
Colby DW, Wain R, Baskakov IV, Legname G, Palmer CG, Nguyen HO, Lemus A, Cohen FE, DeArmond SJ, Prusiner SB (2010) Protease-sensitive synthetic prions. PLoS Pathog 6(1):e1000736. doi:10.1371/journal.ppat.1000736
Cordeiro Y, Silva JL (2005) The hypothesis of the catalytic action of nucleic acid on the conversion of prion protein. Protein Pept Lett 12(3):251–255
Cuillé J, Chelle PL (1939) Experimental transmission of trembling to the goat. C R Seances Acad Sci 208:1058–1060
Cushman M, Johnson BS, King OD, Gitler AD, Shorter J (2010) Prion-like disorders: blurring the divide between transmissibility and infectivity. J Cell Sci 123(Pt 8):1191–1201. doi:123/8/1191[pii]10.1242/jcs.051672
DeArmond SJ, Sanchez H, Yehiely F, Qiu Y, Ninchak-Casey A, Daggett V, Camerino AP, Cayetano J, Rogers M, Groth D, Torchia M, Tremblay P, Scott MR, Cohen FE, Prusiner SB (1997) Selective neuronal targeting in prion disease. Neuron 19(6):1337–1348
Deleault NR, Lucassen RW, Supattapone S (2003) RNA molecules stimulate prion protein conversion. Nature 425(6959):717–720
Deleault NR, Geoghegan JC, Nishina K, Kascsak R, Williamson RA, Supattapone S (2005) Protease-resistant prion protein amplification reconstituted with partially purified substrates and synthetic polyanions. J Biol Chem 280(29):26873–26879
Deleault NR, Harris BT, Rees JR, Supattapone S (2007) Formation of native prions from minimal componenets in vitro. Proc Natl Acad Sci USA 104(23):9741–9746
Deleault NR, Kascsak R, Geoghegan JC, Supattapone S (2010) Species-dependent differences in cofactor utilization for formation of the protease-resistant prion protein in vitro. Biochemistry 49(18):3928–3934
Gabizon R, McKinley MP, Prusiner SB (1987) Purified prion proteins and scrapie infectivity copartition into liposomes. Proc Natl Acad Sci USA 84(12):4017–4021
Gajdusek DC (1967) Slow-virus infections of the nervous system. N Engl J Med 276(7):392–400. doi:10.1056/NEJM196702162760708
Gajdusek DC, Gibbs CJ Jr, Alpers M (1966) Experimental transmission of a kuru-like syndrome to chimpanzees. Nature 209:794–796
Gale P (2006) The infectivity of transmissible spongiform encephalopathy agent at low doses: the importance of phospholipid. J Appl Microbiol 101(2):261–274. doi:JAM3110[pii]10.1111/j.1365-2672.2006.03110.x
Geoghegan JC, Valdes PA, Orem NR, Deleault NR, Williamson RA, Harris BT, Supattapone S (2007) Selective incorporation of polyanionic molecules into hamster prions. J Biol Chem 282(50):36341–36353
Geoghegan JC, Miller MB, Kwak AH, Harris BT, Supattapone S (2009) Trans-dominant inhibition of prion propagation in vitro is not mediated by an accessory cofactor. PLoS Pathog 5(7):e1000535

Green KM, Castilla J, Seward TS, Napier DL, Jewell JE, Soto C, Telling GC (2008) Accelerated high fidelity prion amplification within and across prion species barriers. PLoS Pathog 4(8):e1000139. doi:10.1371/journal.ppat.1000139
Griffith JS (1967) Self-replication and scrapie. Nature 215(105):1043–1044
Grossman A, Zeiler B, Sapirstein V (2003) Prion protein interactions with nucleic acid: possible models for prion disease and prion function. Neurochem Res 28(6):955–963
Hijazi N, Shaked Y, Rosenmann H, Ben-Hur T, Gabizon R (2003) Copper binding to PrPC may inhibit prion disease propagation. Brain Res 993(1–2):192–200
Hsiao KK, Groth D, Scott M, Yang SL, Serban H, Rapp D, Foster D, Torchia M, Dearmond SJ, Prusiner SB (1994) Serial transmission in rodents of neurodegeneration from transgenic mice expressing mutant prion protein. Proc Natl Acad Sci USA 91(19):9126–9130
Hunter GD (1969) The size and intracellular location of the scrapie agent. Biochem J 114(2):22P–23P
Hunter GD, Collis SC, Millson GC, Kimberlin RH (1976) Search for scrapie-specific RNA and attempts to detect an infectious DNA or RNA. J Gen Virol 32(2):157–162
Jones EM, Surewicz WK (2005) Fibril conformation as the basis of species- and strain-dependent seeding specificity of mammalian prion amyloids. Cell 121(1):63–72
Kaneko K, Zulianello L, Scott M, Cooper CM, Wallace AC, James TL, Cohen FE, Prusiner SB (1997) Evidence for protein X binding to a discontinuous epitope on the cellular prion protein during scrapie prion propagation. Proc Natl Acad Sci USA 94(19):10069–10074
Kazlauskaite J, Sanghera N, Sylvester I, Venien-Bryan C, Pinheiro TJ (2003) Structural changes of the prion protein in lipid membranes leading to aggregation and fibrillization. Biochemistry 42(11):3295–3304
Khalili-Shirazi A, Summers L, Linehan J, Mallinson G, Anstee D, Hawke S, Jackson GS, Collinge J (2005) PrP glycoforms are associated in a strain-specific ratio in native PrPSc. J Gen Virol 86(Pt 9):2635–2644
Kim J, Holtzman DM (2010) Medicine. Prion-like behavior of amyloid-beta. Science 330(6006):918–919. doi:330/6006/918[pii]10.1126/science.1198314
Kim JI, Cali I, Surewicz K, Kong Q, Raymond GJ, Atarashi R, Race B, Qing L, Gambetti P, Caughey B, Surewicz WK (2010) Mammalian prions generated from bacterially expressed prion protein in the absence of any mammalian cofactors. J Biol Chem 285(19):14083–14087. doi:C110.113464[pii]10.1074/jbc.C110.113464
King CY, Diaz-Avalos R (2004) Protein-only transmission of three yeast prion strains. Nature 428(6980):319–323
Koch R (1893) Über den augenblicklichen Stand der bakteriologischen Choleradiagnose. Zeitschrift für Hygiene und Infectionskrankheiten 14:319–333
Kocisko DA, Come JH, Priola SA, Chesebro B, Raymond GJ, Lansbury PT, Caughey B (1994) Cell-free formation of protease-resistant prion protein. Nature 370(6489):471–474
Kuczius T, Buschmann A, Zhang W, Karch H, Becker K, Peters G, Groschup MH (2004) Cellular prion protein acquires resistance to proteolytic degradation following copper ion binding. Biol Chem 385(8):739–747. doi:10.1515/BC.2004.090
Lee CI, Yang Q, Perrier V, Baskakov IV (2007) The dominant-negative effect of the Q218K variant of the prion protein does not require protein X. Protein Sci 16(10):2166–2173
Lee SJ, Desplats P, Sigurdson C, Tsigelny I, Masliah E (2010) Cell-to-cell transmission of non-prion protein aggregates. Nat Rev Neurol 6(12):702–706. doi:nrneurol.2010.145[pii]10.1038/nrneurol.2010.145
Legname G, Baskakov IV, Nguyen HO, Riesner D, Cohen FE, DeArmond SJ, Prusiner SB (2004) Synthetic mammalian prions. Science 305(5684):673–676
Leucht C, Simoneau S, Rey C, Vana K, Rieger R, Lasmezas CI, Weiss S (2003) The 37 kDa/67 kDa laminin receptor is required for PrP(Sc) propagation in scrapie-infected neuronal cells. EMBO Rep 4(3):290–295. doi:10.1038/sj.embor.embor768embor768[pii]
Li J, Browning S, Mahal SP, Oelschlegel AM, Weissmann C (2009) Darwinian evolution of prions in cell culture. Science 327(5967):869–872
Lucassen R, Nishina K, Supattapone S (2003) In vitro amplification of protease-resistant prion protein requires free sulfhydryl groups. Biochemistry 42(14):4127–4135

Mahal SP, Baker CA, Demczyk CA, Smith EW, Julius C, Weissmann C (2007) Prion strain discrimination in cell culture: the cell panel assay. Proc Natl Acad Sci USA 104(52):20908–20913

Makarava N, Baskakov IV (2008) The same primary structure of the prion protein yields two distinct self-propagating states. J Biol Chem 283(23):15988–15996

Makarava N, Kovacs GG, Bocharova O, Savtchenko R, Alexeeva I, Budka H, Rohwer RG, Baskakov IV (2010) Recombinant prion protein induces a new transmissible prion disease in wild-type animals. Acta Neuropathol 119(2):177–187

Manuelidis L (2011) Nuclease resistant circular DNAs copurify with infectivity in scrapie and CJD. J Neurovirol 17(2):131–145. doi:10.1007/s13365-010-0007-0

Manuelidis L, Yu ZX, Barquero N, Mullins B (2007) Cells infected with scrapie and Creutzfeldt–Jakob disease agents produce intracellular 25-nm virus-like particles. Proc Natl Acad Sci USA 104(6):1965–1970. doi:0610999104[pii]10.1073/pnas.0610999104

Mays CE, Ryou C (2010) Plasminogen stimulates propagation of protease-resistant prion protein in vitro. FASEB J 24(12):5102–5112. doi:fj.10-163600[pii]10.1096/fj.10-163600

Mead S, Whitfield J, Poulter M, Shah P, Uphill J, Campbell T, Al-Dujaily H, Hummerich H, Beck J, Mein CA, Verzilli C, Whittaker J, Alpers MP, Collinge J (2009) A novel protective prion protein variant that colocalizes with kuru exposure. N Engl J Med 361(21):2056–2065. doi:361/21/2056[pii]10.1056/NEJMoa0809716

Merz PA, Kascsak RJ, Rubenstein R, Carp RI, Wisniewski HM (1987) Antisera to scrapie-associated fibril protein and prion protein decorate scrapie-associated fibrils. J Virol 61(1):42–49

Meyer N, Rosenbaum V, Schmidt B, Gilles K, Mirenda C, Groth D, Prusiner SB, Riesner D (1991) Search for a putative scrapie genome in purified prion fractions reveals a paucity of nucleic acids. J Gen Virol 72(Pt 1):37–49

Meyer-Luehmann M, Coomaraswamy J, Bolmont T, Kaeser S, Schaefer C, Kilger E, Neuenschwander A, Abramowski D, Frey P, Jaton AL, Vigouret JM, Paganetti P, Walsh DM, Mathews PM, Ghiso J, Staufenbiel M, Walker LC, Jucker M (2006) Exogenous induction of cerebral beta-amyloidogenesis is governed by agent and host. Science 313(5794):1781–1784. doi:313/5794/1781[pii]10.1126/science.1131864

Miller MB, Geoghegan JC, Supattapone S (2011) Dissociation of infectivity from seeding ability in prions with alternate docking mechanism. PLoS Pathog 7(7):e1002128. doi:10.1371/journal.ppat.1002128PPATHOGENS-D-10-00102[pii]

Morillas M, Swietnicki W, Gambetti P, Surewicz WK (1999) Membrane environment alters the conformational structure of the recombinant human prion protein. J Biol Chem 274(52):36859–36865

Orem NR, Geoghegan JC, Deleault NR, Kascsak R, Supattapone S (2006) Copper (II) ions potently inhibit purified PrPres amplification. J Neurochem 96(5):1409–1415

Perrier V, Kaneko K, Safar J, Vergara J, Tremblay P, DeArmond SJ, Cohen FE, Prusiner SB, Wallace AC (2002) Dominant-negative inhibition of prion replication in transgenic mice. Proc Natl Acad Sci USA 99(20):13079–13084. doi:10.1073/pnas.182425299182425299[pii]

Piro JR, Harris BT, Nishina K, Soto C, Morales R, Rees JR, Supattapone S (2009) Prion protein glycosylation is not required for strain-specific neurotropism. J Virol 83(11):5321–5328

Piro JR, Harris BT, Supattapone S (2011) In situ photodegradation of incorporated polyanion does not alter prion infectivity. PLoS Pathog 7(2):e1002001. doi:10.1371/journal.ppat.1002001

Prusiner SB (1982) Novel proteinaceous infectious particles cause scrapie. Science 216(4542):136–144

Prusiner SB (1997) Prion diseases and the BSE crisis. Science 278(5336):245–251

Prusiner SB (1998) Prions. Proc Natl Acad Sci USA 95(23):13363–13383

Prusiner SB, Groth DF, Bolton DC, Kent SB, Hood LE (1984) Purification and structural studies of a major scrapie prion protein. Cell 38(1):127–134

Quaglio E, Chiesa R, Harris DA (2001) Copper converts the cellular prion protein into a protease-resistant species that is distinct from the scrapie isoform. J Biol Chem 276(14):11432–11438

Robertson J, Gomersall M, Gill P (1975) Mycoplasma hominis: growth, reproduction, and isolation of small viable cells. J Bacteriol 124(2):1007–1018

Saborio GP, Permanne B, Soto C (2001) Sensitive detection of pathological prion protein by cyclic amplification of protein misfolding. Nature 411(6839):810–813

Safar JG, Kellings K, Serban A, Groth D, Cleaver JE, Prusiner SB, Riesner D (2005) Search for a prion-specific nucleic acid. J Virol 79(16):10796–10806

Sandberg MK, Al-Doujaily H, Sharps B, Clarke AR, Collinge J (2011) Prion propagation and toxicity in vivo occur in two distinct mechanistic phases. Nature 470(7335):540–542. doi:nature09768|pii|10.1038/nature09768

Sigurdsson B (1954) Rida, a chronic encephalitis of sheep with general remarks on infections which develop slowly and some of their special characteristics. Br Vet J 110:341–354

Silveira JR, Raymond GJ, Hughson AG, Race RE, Sim VL, Hayes SF, Caughey B (2005) The most infectious prion protein particles. Nature 437(7056):257–261

Supattapone S (2010) Biochemistry. What makes a prion infectious? Science 327(5969):1091–1092. doi:327/5969/1091|pii|10.1126/science.1187790

Tanaka M, Chien P, Naber N, Cooke R, Weissman JS (2004) Conformational variations in an infectious protein determine prion strain differences |see comment|. Nature 428(6980):323–328

Tanaka M, Chien P, Yonekura K, Weissman JS (2005) Mechanism of cross-species prion transmission: an infectious conformation compatible with two highly divergent yeast prion proteins. Cell 121(1):49–62

Taylor DR, Whitehouse IJ, Hooper NM (2009) Glypican-1 mediates both prion protein lipid raft association and disease isoform formation. PLoS Pathog 5(11):e1000666. doi:10.1371/journal.ppat.1000666

Telling GC, Scott M, Mastrianni J, Gabizon R, Torchia M, Cohen FE, DeArmond SJ, Prusiner SB (1995) Prion propagation in mice expressing human and chimeric PrP transgenes implicates the interaction of cellular PrP with another protein. Cell 83(1):79–90

Tuzi NL, Cancellotti E, Baybutt H, Blackford L, Bradford B, Plinston C, Coghill A, Hart P, Piccardo P, Barron RM, Manson JC (2008) Host PrP glycosylation: a major factor determining the outcome of prion infection. PLoS Biol 6(4):e100

Vorberg I, Priola SA (2002) Molecular basis of scrapie strain glycoform variation. J Biol Chem 277(39):36775–36781

Wang F, Yang F, Hu Y, Wang X, Wang X, Jin C, Ma J (2007) Lipid interaction converts prion protein to a PrP(Sc)-like proteinase K-resistant conformation under physiological conditions. Biochemistry 46(23):7045–7053

Wang F, Wang X, Yuan CG, Ma J (2010) Generating a prion with bacterially expressed recombinant prion protein. Science 327(5969):1132–1135

Wickner RB, Masison DC, Edskes HK (1995) |PSI| and |URE3| as yeast prions. Yeast 11(16):1671–1685

Wong C, Xiong LW, Horiuchi M, Raymond L, Wehrly K, Chesebro B, Caughey B (2001) Sulfated glycans and elevated temperature stimulate PrP(Sc)-dependent cell-free formation of protease-resistant prion protein. EMBO J 20(3):377–386

Chapter 8
Prion Protein Conversion and Lipids

Jiyan Ma

Abstract The conversion of α-helical rich normal prion protein to a β-sheeted pathogenic isoform is central to the pathogenesis of prion disease. Recent studies have revealed the importance of cofactors in prion protein conformal change and in generating prion infectivity. Lipid appears to be a critical cofactor because of its unique biophysical properties and its ability to induce protein conformational changes. Biophysical and biochemical analyses of lipid–prion protein interactions and the resulting prion protein conformational changes revealed a huge impact of lipids on prion protein conformation. Studies of disease-associated mutations and the generation of highly infectious prions with bacterially expressed recombinant prion protein in the presence of lipid support the relevance of lipid interaction to prion disease. The hypothesized roles of lipid in prion protein conversion require rigorous future researches, which are essential for unveiling the molecular mechanism of prion infectivity.

Keywords Prion protein • Prion protein conversion • Lipids • TSEs • Prion infectivity

8.1 Introduction

Transmissible spongiform encephalopathies (TSEs), or prion diseases, are a large group of infectious neurodegenerative disorders characterized by an unusual infectious agent (Prusiner 1998; Caughey et al. 2009; Aguzzi et al. 2008; Collinge 2001; Collinge and Clarke 2007; Weissmann 2004). Prion hypothesis postulates that the infectious agent, PrP^{Sc}, is an altered conformational isoform of host-encoded prion

J. Ma, Ph.D. (✉)
Department of Molecular and Cellular Biochemistry, Ohio State University, 1645 Neil Avenue, Columbus, OH 43210, USA
e-mail: ma.131@osu.edu

W.-Q. Zou and P. Gambetti (eds.), *Prions and Diseases: Volume 1, Physiology and Pathophysiology*, DOI 10.1007/978-1-4614-5305-5_8,

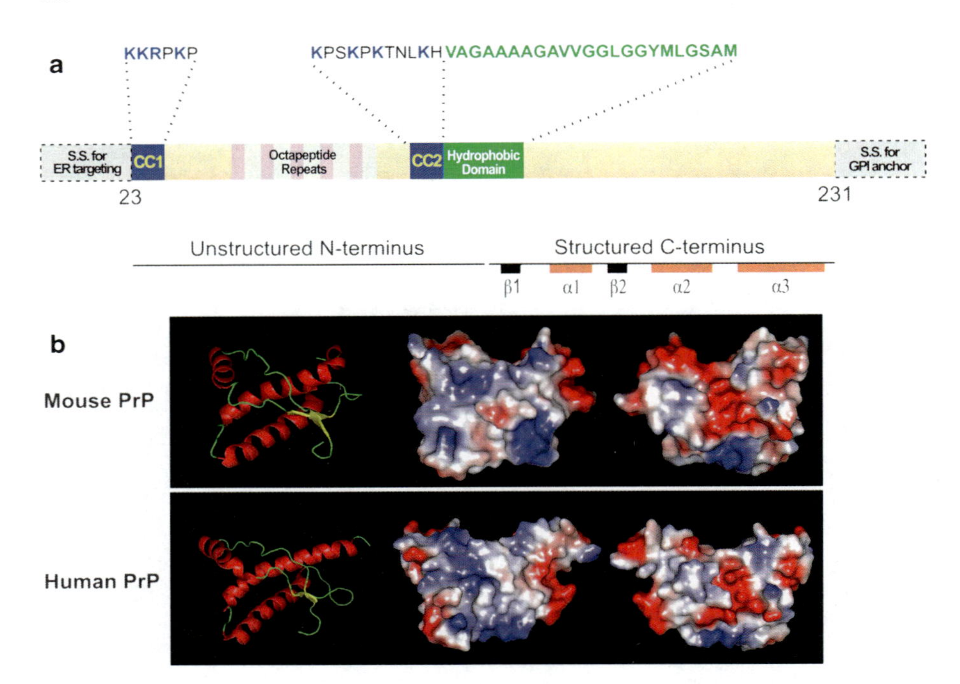

Fig. 8.1 (**a**) PrP contains two positively charged clusters CC1 and CC2 (*blue*) and a hydrophobic domain (*green*). The unstructured and structured regions of PrP are indicated. *S.S.*, signal sequence. (**b**) Surface charges of human and mouse PrPs' structured regions (amino acid 121–231) are colored according to electrostatic potential with *blue* for positive charges and *red* for negative charges. The images in the *middle* show the surface charges of PrP structures on the *left*. The images on the *right* are the surfaces after a 180° rotation around the vertical axis of the images in the *middle*. Images were generated by the PyMOL software

protein (PrP) (Prusiner 1982). PrP is a cell surface localized *N*-linked glycoprotein tethered to lipid membranes through a glycosylphosphatidylinositol (GPI) anchor. The normal form of PrP, PrP^C, is an α-helical rich protein containing an unstructured N terminus and a well-structured C terminus with three helices and a short β-sheet (Fig. 8.1) (Riek et al. 1996, 1997; Donne et al. 1997). During prion disease, a significant portion of PrP molecules converts to the β-sheeted PrP^{Sc} conformation (Smirnovas et al. 2011; Caughey et al. 1991; Gasset et al. 1993). The two conformational states can be differentiated by biochemical measures such as solubility and protease sensitivity. PrP^C is soluble in mild detergents and sensitive to proteinase K (PK) digestion, while PrP^{Sc} is aggregated and the C terminus of PrP^{Sc} is highly resistant to PK digestion (Prusiner 1998; Caughey et al. 2009; Aguzzi et al. 2008; Collinge 2001; Collinge and Clarke 2007; Weissmann 2004).

Prion hypothesis posits that, due to its self-perpetuating property, PrP^{Sc} serves as a template and converts PrP^C into the pathogenic PrP^{Sc} conformation (Prusiner 1998). This prediction was first confirmed by the cell-free conversion assay, in which highly purified PrP^{Sc} seeded purified PrP^C into PK-resistant PrP^{Sc} conformation

(Kocisko et al. 1994; Bessen et al. 1995), demonstrating the seeding capability of PrP^{Sc}. The more recently developed protein misfolding cyclic amplification (PMCA) technique, that processes whole brain homogenates with cycles of alternating sonication and incubation, is much more efficient in propagating the PrP^{Sc} conformation. This high efficiency has lead to the landmark study demonstrating simultaneous propagation of PK-resistant PrP^{Sc} and prion infectivity in a test tube (Castilla et al. 2005). Although it is still not completely understood why the efficiency differs so much in these two assays, it has been shown that in vivo factor(s) in the brain homogenate plays a role in facilitating PrP conversion and/or stabilizing the resulting PrP^{Sc} conformation (Deleault et al. 2003). Deleault et al. revealed that a variety of polyanions are able to enhance PrP^{Sc}-templated conversion and RNA appears to be the most potent stimulator (Deleault et al. 2003, 2005).

The requirement of factors other than PrP in PrP^{C}-to-PrP^{Sc} conversion is consistent with the notion that there is an energy barrier between the two conformational states (Baskakov et al. 2001). In vivo, PrP conversion mainly occurs on the cell surface or in the endocytic pathway (Caughey and Raymond 1991; Borchelt et al. 1992), indicating that the conversion starts with fully folded α-helical rich PrP^{C} conformation. A chaperone-like activity would help PrP^{C} to overcome the energy barrier and convert to the β-sheeted PrP^{Sc} conformation. Since there is little evidence supporting the involvement of another protein in PrP conversion, other biological molecules such as lipids, oligosaccharides, nucleic acids, or proteoglycans need to be considered for this activity.

8.2 Supporting Evidence for the Involvement of Lipid in PrP Conversion

Lipid appears to be a good candidate because of its proximity to the GPI-anchored PrP and the unique impact of lipid interaction on protein structure. The PrP^{C}-to-PrP^{Sc} conversion requires both the unfolding of α-helical rich PrP^{C} and the formation of β-sheeted PrP^{Sc}. It is well established that protein–lipid membrane interaction is able to unfold structured proteins (van der Goot et al. 1991; Muga et al. 1993; Pinheiro and Watts 1994; Banuelos and Muga 1995; Fisher and Ryan 1999); this effect would lower the energy barrier and remove the first thermodynamic obstacle in PrP conversion. Moreover, the interfacial region of lipid bilayer is known to have a potent capability of inducing secondary protein structures, either α-helices or β-sheets (White et al. 2001; Wimley et al. 1998). Thus, the PrP–lipid membrane interaction would facilitate both steps in converting α-helical PrP^{C} to β-sheeted PrP^{Sc}.

The involvement of lipid in PrP conversion is also consistent with previous experimental observations. First, GPI-anchored PrP^{C} can be released from cell surface by phospholipase C (PI-PLC) digestion, whereas the converted PrP^{Sc} resists PI-PLC digestion (Caughey and Raymond 1991; Borchelt et al. 1992). A GPI-anchor independent lipid membrane interaction by PrP^{Sc} is one of the plausible explanations for the development of PI-PLC resistance, which is also consistent with the result that a GPI-independent lipid interaction is essential for PrP conversion in cell-free conversion assay (Baron and Caughey 2003). Second, cell biological studies reveal that changing lipid contents in prion-infected cells markedly alters PrP^{Sc} production (Taraboulos et al. 1995; Naslavsky et al. 1999), which could be due to the alteration of PrP maturation or trafficking (Sarnataro et al. 2004). However, it is also possible that changing lipid membrane composition may alter its interaction with PrP, which could consequently influence the production of PrP^{Sc}. Third, various lipid molecules have been identified in "prion rod," one of the most pure preparations of the infectious particle (Klein et al. 1998). Removing lipids from "prion rod" by treatments of SDS, sonication, and SDS-PAGE results in the complete loss of prion infectivity (Leffers et al. 2005). This observation could be explained by altering PrP conformation during these treatments. Alternatively, SDS and sonication treatments may disrupt PrP–lipid interaction and destabilize the infectious PrP^{Sc} conformation, which would also leads to the loss of infectivity. Consistent with the latter explanation, reincorporation of purified "prion rod" into lipid vesicles results in higher infectivity (Gabizon et al. 1987) and PrP^{Sc}-containing microsomes infect cultured cells with a higher efficiency than detergent-purified PrP^{Sc} (Baron et al. 2006). Collectively, these observations support that the PrP–lipid interaction is involved in the PrP^{C}-to-PrP^{Sc} conversion.

8.3 Biophysical Studies of PrP–Lipid Interaction

Definitive evidence supporting a GPI-anchor independent PrP–lipid interaction comes from in vitro analyses with purified bacterially expressed recombinant PrP (rPrP) and model lipid membranes. Using spectroscopic approaches, Morillas et al. showed that human rPrP binds to anionic lipid-containing membranes and the rPrP–lipid-binding destabilizes the structured C-terminal domain of PrP (Morillas et al. 1999). The facts that rPrP–lipid interaction is highly pH-dependent and rPrP only binds to anionic lipids indicate a critical role of electrostatic interaction. Since electrostatic interactions are critical for PrP stability and the presence of salts destabilizes rPrP (Apetri and Surewicz 2003), the lipid-binding induced PrP destabilization could be, at least partly, due to the disruption of salt bridges in the folded C terminus by electrostatic rPrP–lipid interaction.

Using similar biophysical approaches, Pinheiro and colleagues confirmed the binding of PrP to anionic lipids using hamster rPrP(90–231) (Sanghera and Pinheiro 2002; Critchley et al. 2004). Interestingly, they also reported that hamster rPrP(90–231) could bind to either zwitterionic DPPC (1,2-dipalmitoyl-*sn*-glycero-3-phosphocholine) or a mixture of DPPC, cholesterol and sphingomyelin (molar ratio at 50:30:20) at pH 7, but not at pH 5 (Sanghera and Pinheiro 2002). The binding of hamster rPrP(90–231) to DPPC or DPPC/cholesterol/sphingomyelin is believed to be driven by hydrophobic lipid–protein interactions, which increases the α-helical content of hamster rPrP(90–231) (Sanghera and Pinheiro 2002). DPPC has a phase transition temperature of 41°C and it is in a gel phase at room temperature with fully extended and closely packed acyl chains. In contrast, all other lipids used to study rPrP–lipid interaction are in a liquid crystalline phase in which the acyl chains are randomly oriented and in a more fluid state. Notably, the GPI-anchored PrP is localized in the lipid rafts, which are specialized membrane microdomains of tightly packed lipids. The gel phase DPPC may resemble the rigidity of lipid rafts to certain extent, although DPPC is not a major component of PrP associated lipid raft (Brugger et al. 2004). Whether this particular physical property of DPPC contributes to the binding of hamster rPrP(90–231) remains unclear.

8.4 Analysis of PrP–Lipid Interaction Using Density Gradient and Protease Digestion

Besides spectroscopic approaches, density gradient analysis is a straightforward approach to directly measure protein–lipid interaction. Using this approach, we show that full-length α-helical rich mouse rPrP binds to anionic lipids, but not to zwitterionic or cationic lipids (Wang et al. 2007). The interaction between mouse rPrP and anionic lipids initiates with electrostatic contacts, a process that can be blocked by high concentrations of salt. Once electrostatic interaction brings rPrP to the vicinity of lipid bilayer, the hydrophobic domain of rPrP interacts with acyl chains of lipid membrane hydrophobically. The strength of hydrophobic interaction can be analyzed by extraction of the rPrP–lipid complex using a buffer with a high concentration of salt and/or high pH.

The binding of rPrP to anionic POPG (1-palmitoyl-2-oleoylphosphatidylglycerol) increases the β-sheet content of rPrP and results in two C-terminal proteinase K (PK)-resistant bands with apparent molecular weights at 15 and 14.5 kDa. Addition of salt at physiological concentration (150 mM NaCl) to the system induces further rPrP conformational change which is reflected by a further change in far-UV circular dichroism (CD) spectrum, a significantly enhanced PK resistance and the detection of a single C-terminal 15 kDa PK-resistant band by immunoblot analysis (Wang et al. 2007). Interestingly, the binding of rPrP to anionic lipid does not always leads to PK resistance. For example, little PK resistance was detected when rPrP binds to anionic POPS (1-palmitoyl-2-oleoylphosphatidylserine). However, when rPrP binds to vesicles consisting of 1:1 molar ratio of POPS and zwitterionic POPC (1-palmitoyl-2-oleoyl-*sn*-glycero-3-phosphocholine), a strong 15 kDa C-terminal PK-resistant band was detected (Wang et al. 2007). This observation clearly demonstrates that the PK resistance is not simply due to the binding of rPrP to anionic lipid-containing vesicles. Instead, it is due to lipid induced PrP conformational change, which is influenced by the polar headgroup of phospholipids and the distribution of these headgroups on the lipid membranes.

In addition to the C-terminal 15 kDa PK-resistant band, the rPrP binding to anionic lipid-containing membranes also results in a 13.5-kDa N-terminal PK-resistant band (Wang et al. 2007). The simultaneous appearance of both N- and C-terminal PK-resistant fragments and the fact that the sum of these two fragments is greater than the molecular weight of rPrP suggest that rPrP binds to anionic lipid-containing membranes in two different modes. This interpretation is consistent with the observation that, when lipid bilayer is disrupted by a detergent, only the C-terminal 15 kDa PK-resistant fragment can be maintained by protein aggregation (Wang et al. 2007).

The density gradient analyses provide the tools to dissect different aspects of rPrP–lipid interaction, including: the initial electrostatic interaction that can be inhibited by high concentrations of salt; the ensuing hydrophobic interaction that can be analyzed by extracting rPrP–lipid complex with an alkaline buffer containing high concentrations of salt; and the lipid induced rPrP conformational changes that can be analyzed by PK digestion. These tools allow us to characterize the influences of various PrP domains, mutations, and polymorphism on PrP–lipid interaction.

8.5 The Influence of PrP Mutations on rPrP–Lipid Interaction

After removing the N-terminal signal sequence for endoplasmic reticulum targeting and C-terminal signal sequence for GPI anchor addition, the primary amino acid sequence of mature fragment of PrP (Fig. 8.1a) contains two clusters of positively charged amino acid residues at the N terminus (amino acid 23–27, designated as CC1; for clarity, amino acids are numbered according to human PrP throughout the chapter) and in the middle region (amino acid 101–110, designated as CC2). A hydrophobic domain (amino acid 112–134, designated as HD) is located next to the CC2 region. Besides the clusters of positively charged amino acids, the structured C-terminal domain also contains positively or negatively charged surface patches (Fig. 8.1b), which may also contribute to PrP–lipid interaction.

Mutant rPrP without the hydrophobic domain (designated as: ΔH) still binds to anionic lipids, but unlike wild-type rPrP, the ΔH mutant can be extracted from rPrP–lipid complex by an alkaline salt buffer and is without either N- or C-terminal PK-resistant fragments (Wang et al. 2010a; Wang and Ma Unpublished results). These results show that the hydrophobic rPrP–lipid interaction is largely mediated by the HD domain and the development of both N- and C-terminal PK resistance depends on the hydrophobic rPrP–lipid interaction.

For electrostatic PrP–lipid interaction, CC1, CC2, and the positively charged surface patches in the structured C-terminal domain all play a role. The electrostatic PrP–lipid interaction mediated by different PrP regions may orient PrP in such a way that it leads to a difference in hydrophobic PrP–lipid interaction and PrP conformation. Since the N terminus of PrP is highly flexible, there is little conformational restraint to prevent CC1 and CC2 regions from orienting PrP in a variety of ways on the surface of lipid bilayer, which could potentially lead to a great variety of stable PrP conformations.

The complexity in the electrostatic PrP–lipid interaction is reflected in the analyses of different rPrP mutants (Wang et al. 2010a). Deletion of N-terminal CC1 region reduces electrostatic interaction between rPrP and anionic POPG, leading to a reduced C-terminal PK resistance. This effect can be attributed to the loss of positive charges of rPrP. In contrast, the rPrP mutant, in which four positively charged lysines in CC2 region are replaced by isoleucine (designated as: K/I mutant), does not appear to alter the strength of either electrostatic or hydrophobic PrP–POPG interaction. However, the C-terminal PK resistance of K/I mutant is significantly reduced. Therefore, although the positive charges in CC2 region minimally affect PrP's initial electrostatic contact with anionic lipids, the interaction between these lysines and the negatively charged phospholipid headgroups appear to play a role in orienting rPrP and assisting in the formation of PK-resistant PrP.

Two biochemically similar disease-associated mutants, P102L and P105L, are both located in the CC2 regions and flanked by lysines. Since proline is conformationally restrained, replacing proline with leucine would alter the spatial arrangement of positively charged lysines. Interestingly, the P102L mutation does not affect rPrP–POPG binding, but completely eliminates the anionic lipid-induced PK resistance. In contrast, the P105L mutant significantly reduces the electrostatic rPrP–POPG interaction and the anionic lipid-induced PK resistance. When both rPrP mutants are allowed to bind to anionic POPG, neither P102L nor P105L alters the strength of hydrophobic rPrP–POPG interaction.

Considering all three CC2 mutants analyzed, it can be concluded that, despite the cluster of positively charged lysines, the CC2 region minimally affects the electrostatic interaction between rPrP and anionic POPG. The reduction of electrostatic rPrP–POPG binding caused by P105L is likely due to its influence on the global PrP structure, which alters the positively charged surface patches in the C-terminal structured region or the presentation of N-terminal CC1 region. Since all three mutants reduce anionic lipid-induced PK resistance, it is likely that the CC2 region is important for orienting rPrP on lipid membranes, which leads to the PK resistance.

The hydrophobic region localized 129 methionine (129M) and valine (129V) polymorphisms significantly affect the susceptibility and pathogenesis of prion disease (Ironside et al. 2005), yet very few biochemical differences between the two PrP variants can be detected. Analysis of the two variants reveals a stronger hydrophobic interaction between the 129M variant and total mouse brain lipids. This result seems to be counterproductive since valine is more hydrophobic than methionine. However, all amino acids in the hydrophobic domain are capable of interacting with the hydrophobic acyl chains of lipids. Substituting methionine with valine increases the hydrophobicity, which likely results in a tighter binding of the hydrophobic acyl chains to residue 129 and alters the interaction between acyl chains and surrounding hydrophobic amino acids. Thus, the total strength of the hydrophobic lipid interaction is lower in 129V.

It is important to note that rPrP differs from native PrP^{C} in that it lacks *N*-linked oligosaccharides and a GPI anchor. Model lipid vesicles used in the in vitro studies also differ from in vivo lipid membranes in composition, curvature, and local environments. Therefore, one should not simply assume that these in vitro results could be directly extrapolated to the in vivo condition. However, two important observations from these in vitro studies support the relevance of PrP–lipid interaction to the pathogenesis of prion disease. First, disease-associated PrP mutants and the 129 polymorphisms clearly affect PrP–lipid interaction, indicating a role of altered PrP–lipid interaction in the pathogenesis of prion disease. Second, the lipid interaction is sufficient to convert fully folded α-helical rich rPrP into a conformation that is similar to the pathogenic PrP^{Sc} form, with increased β-sheet contents and a highly PK-resistant C terminus. Previous in vitro conversions of α-helical rich rPrP into various aggregated forms all required treatments of denaturant or reducing agent (Legname et al. 2004; Bocharova et al. 2005; Colby et al. 2010; Apetri et al. 2005; Jackson et al. 1999), yet, the lipid-mediated rPrP conformational change does not. This difference indicates that the lipid interaction is capable of overcoming the energy barrier and converting rPrP to a conformation similar to PrP^{Sc}.

8.6 Forming Recombinant Prions with Lipid as a Cofactor

The similarities between lipid-induced rPrP conformation and PrP^{Sc} suggest that lipid might be a necessary cofactor for the conversion of rPrP into an infectious conformation. Studies from Supattapone's laboratory revealed that polyanions, particularly RNA, facilitate PrP conversion in PMCA (Deleault et al. 2003, 2005). More importantly, they also showed that native prions can be formed de novo with native PrP^C purified from golden Syrian hamster brains, copurified lipid molecules, and polyriboadenylic acid (poly(rA)) (Deleault et al. 2007). This simplified PMCA system is ideal for testing the role of lipid in converting rPrP into an infectious conformation. Using this system, we tested whether rPrP is able to convert to the infectious conformation in the presence of synthetic phospholipids and total RNA isolated from normal mouse liver (Wang et al. 2010b). Indeed, the PK-resistant form of rPrP was generated and could be propagated indefinitely by serial PMCA. Because a portion of rPrP gained PK resistance after PMCA, we use "rPrP-res" to represent the rPrP conformational state(s) in the PMCA product, which does not necessarily imply that the infectious conformation has to be PK resistant.

The rPrP-res has all the signature characteristics of PrP^{Sc}: aggregated, C-terminal PK-resistance, the capability of converting endogenous PrP^C in the brain homogenate to PrP^{Sc} by serial PMCA, and the capability of infecting cultured cells (Wang et al. 2010b). Most importantly, it causes prion disease in wild-type mice with an incubation time similar to those have been observed with naturally occurring prions (Wang et al. 2010b). Therefore, rPrP-res is not only infectious but also contains relatively high prion infectivity.

Generating prion infectivity with bacterially expressed rPrP has also been reported by several groups (Legname et al. 2004; Makarava et al. 2010; Kim et al. 2010; Colby et al. 2010). The rPrP amyloid fibers have been shown to induce prion disease in transgenic mice overexpressing PrP, but not in wild-type mice (Legname et al. 2004; Colby et al. 2010). Full-length hamster rPrP fibers subjected to an "annealing" procedure (5 cycles of incubations at 80°C and 37°C in the presence of normal hamster brain homogenate or bovine serum albumin) are able to induce the formation of infectious prions in a subset of asymptomatic wild-type hamsters

(Makarava et al. 2010). Using PMCA seeded by PrP^{Sc} partially purified from 263 K scrapie-infected hamster brain, Kim et al. showed that the converted bacterially expressed hamster rPrP (designated as $rPrP^{PMCA}$) is able to cause prion disease in wild-type hamsters despite a relative large variability in incubation times and attack rates (Kim et al. 2010).

Compared to PMCA generated rPrP-res (Wang et al. 2010b) or PrP^{Sc}-seeded $rPrP^{PMCA}$ (Kim et al. 2010), rPrP amyloid fibers appear to have a much lower infectivity, which fails to induce prion disease in wild-type animal (Legname et al. 2004; Colby et al. 2010) or only induce infectious prion formation in asymptomatic wild-type hamsters (Makarava et al. 2010). The low infectivity of rPrP amyloid fibers suggests a possibility that, instead of mature fibers, the infectivity could be associated with some type of oligomeric rPrP structures, which can be on or off the amyloidogenic pathway. In vivo, the rPrP oligomer may be stabilized by binding to a cofactor. The "annealing" step may rearrange the quaternary rPrP structure to increase the formation and/or stabilization of infectious rPrP oligomers. This hypothesis accounts for the discrepancy of a large amount of fibers in the inoculum and a low infectivity in bioassay, and explains the differences between fibers with or without "annealing." It is also consistent with the observation that the most infectious prion particles are oligomers (Silveira et al. 2005) and the apparent lack of fibers in PMCA-generated rPrP-res (Piro et al. 2011).

Growing rPrP amyloid fiber requires chaotropic agents such as guanidinium hydrochloride or urea (Legname et al. 2004; Bocharova et al. 2005; Colby et al. 2010; Apetri et al. 2005). The chaotropic agents may play a role similar to the binding to lipid membranes, that is, unfolding α-helical rich rPrP to allow the formation of β-sheeted amyloid fibers. However, the condition used for PrP^{Sc}-seeded formation of $rPrP^{PMCA}$ is quite different, which does not require chaotropic agent or cofactors such as lipids or polyanions (Kim et al. 2010). The following two reasons may contribute to the success in generating infectious $rPrP^{PMCA}$. First, the sonication step in PMCA is drastically different from the amyloid fiber growing condition, which may provide activation energy needed for PrP conversion or directly affect rPrP conformation. In addition, the conversion buffer for PrP^{Sc}-seeded $rPrP^{PMCA}$ formation contains anionic detergent sodium dodecyl sulfate (SDS) and nonionic detergent Triton X-100. Both detergents contain a hydrophilic group and a hydrophobic moiety, which resembles the structural characteristics of lipid molecules. Moreover, the anionic SDS has been shown to promote the conversion from α-helical rich rPrP to β-sheeted conformations (Leffers et al. 2005). Thus, in PrP^{Sc}-seeded $rPrP^{PMCA}$ formation, SDS and Triton X-100 may partially replace the function of lipid molecules and/or polyanions in promoting rPrP conversion.

Among all in vitro-generated recombinant prions, rPrP-res produced by PMCA in the presence of phospholipid POPG and total RNA appears to contain the highest infectivity. Not only does it cause prion disease in wild-type mice with a relatively short incubation time and 100% attack rate, it also infects cultured cells and propagates the PK-resistant conformation to native PrP^{C} via PMCA (Wang et al. 2010b). The high infectivity associated with rPrP-res could be attributed to a variety of reasons, but the presence of lipid molecules, a distinct characteristic of this system, likely plays a role in generating the highly infectious rPrP-res.

8.7 Possible Roles of Lipid in Forming an Infectious Prion

Although the involvement of lipid molecules in generating infectious prions is supported by experimental results, many questions remain to be answered, such as: what type of lipid molecules or which combinations of lipids are the best cofactors for the formation of an infectious prion, or whether different lipid molecules can lead to distinct prion strains. The most fundamental question that needs to be addressed is whether or not lipid is an essential part of the infectious agent. Depending on whether lipid is or is not an essential part of the infectious agent, the following roles of lipids can be envisaged.

If the "protein-only" hypothesis is correct, then the converted PrP^{Sc} conformer itself should be sufficient to cause prion disease. In this scenario, lipid may act like a molecular chaperone that facilitates PrP conversion by unfolding α-helical rich PrP and/or promoting the formation of the β-sheeted PrP^{Sc} conformer. Alternatively, lipid molecules may simply enhance the in vivo retention time of PrP^{Sc}. In this case, the infectious agent is the PrP^{Sc} conformer, but its association with lipid molecules may prevent its clearance and thereby enhance the infectivity. The third possibility could be that the lipid molecules facilitate the binding of infectious particle to cellular membranes, where the pathogenic PrP^{Sc} conformer will encounter and convert membrane attached PrP^C. The latter two possibilities would account for the increased infectivity when PrP^{Sc} is associated with lipid membranes (Gabizon et al. 1987; Baron et al. 2006).

In case that lipid is an essential part of the infectious agent, lipid molecules may still play the roles proposed above, and in addition, they will contribute to the stabilization of infectious PrP^{Sc} conformation. Early studies by Alper et al. showed that oxygen greatly sensitizes the infectious agent to ultraviolet irradiation and such a large oxygen effect on ultraviolet irradiation is characteristic for the involvement of lipid molecules (Alper et al. 1978). This observation is in agreement with the notion that lipid is an integral part of the infectious agent and plays an essential role in stabilizing the infectious PrP^{Sc} conformation. Moreover, it is well known that PrP^{Sc} and PrP^C share the same amino acid sequence, but they have to reach different minimum-energy conformations. Similar to the conformational states of a receptor with or without ligand binding, the thermodynamically stable point of PrP^{Sc}–lipid complex would be different from that of free PrP^C molecule. Thus, the self-propagating PrP^{Sc} conformation could be stably maintained by forming a PrP^{Sc}–lipid complex, which would allow the unorthodox prion phenomenon to be simply explained within Christian Anfinsen's protein folding paradigm.

These hypotheses should be rigorously tested and results from these studies will help us to understand the molecular mechanism of prion infectivity. In addition to addressing a long-lasting question with intellectual significance, elucidating the role of lipid or other cofactors in PrP conformational change and in the formation of an infectious prion may lead to novel prophylactic, diagnostic, and therapeutic strategies against these fatal neurodegenerative disorders.

Acknowledgments Thanks to Xinhe Wang, Fei Wang, and Jessica Chadwick at Ohio State University for comments and thanks to Fei Wang for generating the images with PyMOL.

References

Aguzzi A, Baumann F, Bremer J (2008) The prion's elusive reason for being. Annu Rev Neurosci 31:439–477. doi:10.1146/annurev.neuro.31.060407.125620

Alper T, Haig DA, Clarke MC (1978) The scrapie agent: evidence against its dependence for replication on intrinsic nucleic acid. J Gen Virol 41(3):503–516

Apetri AC, Surewicz WK (2003) Atypical effect of salts on the thermodynamic stability of human prion protein. J Biol Chem 278(25):22187–22192. doi:10.1074/jbc.M302130200M302130200[pii]

Apetri AC, Vanik DL, Surewicz WK (2005) Polymorphism at residue 129 modulates the conformational conversion of the D178N variant of human prion protein 90–231. Biochemistry 44(48):15880–15888. doi:10.1021/bi051455+

Banuelos S, Muga A (1995) Binding of molten globule-like conformations to lipid bilayers Structure of native and partially folded alpha-lactalbumin bound to model membranes. J Biol Chem 270(50):29910–29915

Baron GS, Caughey B (2003) Effect of glycosylphosphatidylinositol anchor-dependent and -independent prion protein association with model raft membranes on conversion to the protease-resistant isoform. J Biol Chem 278(17):14883–14892

Baron GS, Magalhaes AC, Prado MA, Caughey B (2006) Mouse-adapted scrapie infection of SN56 cells: greater efficiency with microsome-associated versus purified PrP-res. J Virol 80(5):2106–2117

Baskakov IV, Legname G, Prusiner SB, Cohen FE (2001) Folding of prion protein to its native alpha-helical conformation is under kinetic control. J Biol Chem 276(23):19687–19690

Bessen RA, Kocisko DA, Raymond GJ, Nandan S, Lansbury PT, Caughey B (1995) Non-genetic propagation of strain-specific properties of scrapie prion protein. Nature 375(6533):698–700. doi:10.1038/375698a0

Bocharova OV, Breydo L, Parfenov AS, Salnikov VV, Baskakov IV (2005) In vitro conversion of full-length mammalian prion protein produces amyloid form with physical properties of PrP(Sc). J Mol Biol 346(2):645–659

Borchelt DR, Taraboulos A, Prusiner SB (1992) Evidence for synthesis of scrapie prion proteins in the endocytic pathway. J Biol Chem 267(23):16188–16199

Brugger B, Graham C, Leibrecht I, Mombelli E, Jen A, Wieland F, Morris R (2004) The membrane domains occupied by glycosylphosphatidylinositol-anchored prion protein and Thy-1 differ in lipid composition. J Biol Chem 279(9):7530–7536. doi:10.1074/jbc.M310207200M310207200[pii]

Castilla J, Saa P, Hetz C, Soto C (2005) In vitro generation of infectious scrapie prions. Cell 121(2):195–206

Caughey B, Raymond GJ (1991) The scrapie-associated form of PrP is made from a cell surface precursor that is both protease- and phospholipase-sensitive. J Biol Chem 266(27):18217–18223

Caughey BW, Dong A, Bhat KS, Ernst D, Hayes SF, Caughey WS (1991) Secondary structure analysis of the scrapie-associated protein PrP 27–30 in water by infrared spectroscopy. Biochemistry 30(31):7672–7680

Caughey B, Baron GS, Chesebro B, Jeffrey M (2009) Getting a grip on prions: oligomers, amyloids, and pathological membrane interactions. Annu Rev Biochem 78:177–204. doi:10.1146/annurev.biochem.78.082907.145410

Colby DW, Wain R, Baskakov IV, Legname G, Palmer CG, Nguyen HO, Lemus A, Cohen FE, DeArmond SJ, Prusiner SB (2010) Protease-sensitive synthetic prions. PLoS Pathog 6(1):e1000736. doi:10.1371/journal.ppat.1000736

Collinge J (2001) Prion diseases of humans and animals: their causes and molecular basis. Annu Rev Neurosci 24:519–550

Collinge J, Clarke AR (2007) A general model of prion strains and their pathogenicity. Science 318(5852):930–936

Critchley P, Kazlauskaite J, Eason R, Pinheiro TJ (2004) Binding of prion proteins to lipid membranes. Biochem Biophys Res Commun 313(3):559–567

Deleault NR, Lucassen RW, Supattapone S (2003) RNA molecules stimulate prion protein conversion. Nature 425(6959):717–720

Deleault NR, Geoghegan JC, Nishina K, Kascsak R, Williamson RA, Supattapone S (2005) Protease-resistant prion protein amplification reconstituted with partially purified substrates and synthetic polyanions. J Biol Chem 280(29):26873–26879

Deleault NR, Harris BT, Rees JR, Supattapone S (2007) Formation of native prions from minimal components in vitro. Proc Natl Acad Sci USA 104(23):9741–9746

Donne DG, Viles JH, Groth D, Mehlhorn I, James TL, Cohen FE, Prusiner SB, Wright PE, Dyson HJ (1997) Structure of the recombinant full-length hamster prion protein PrP(29–231): the N terminus is highly flexible. Proc Natl Acad Sci USA 94(25):13452–13457

Fisher CA, Ryan RO (1999) Lipid binding-induced conformational changes in the N-terminal domain of human apolipoprotein E. J Lipid Res 40(1):93–99

Gabizon R, McKinley MP, Prusiner SB (1987) Purified prion proteins and scrapie infectivity copartition into liposomes. Proc Natl Acad Sci USA 84(12):4017–4021

Gasset M, Baldwin MA, Fletterick RJ, Prusiner SB (1993) Perturbation of the secondary structure of the scrapie prion protein under conditions that alter infectivity. Proc Natl Acad Sci USA 90(1):1–5

Ironside JW, Ritchie DL, Head MW (2005) Phenotypic variability in human prion diseases. Neuropathol Appl Neurobiol 31(6):565–579. doi:NAN697|pii|10.1111/j.1365-2990.2005.00697.x

Jackson GS, Hosszu LL, Power A, Hill AF, Kenney J, Saibil H, Craven CJ, Waltho JP, Clarke AR, Collinge J (1999) Reversible conversion of monomeric human prion protein between native and fibrilogenic conformations. Science 283(5409):1935–1937

Kim JI, Cali I, Surewicz K, Kong Q, Raymond GJ, Atarashi R, Race B, Qing L, Gambetti P, Caughey B, Surewicz WK (2010) Mammalian prions generated from bacterially expressed prion protein in the absence of any mammalian cofactors. J Biol Chem 285(19):14083–14087. doi:C110.113464|pii|10.1074/jbc.C110.113464

Klein TR, Kirsch D, Kaufmann R, Riesner D (1998) Prion rods contain small amounts of two host sphingolipids as revealed by thin-layer chromatography and mass spectrometry. Biol Chem 379(6):655–666

Kocisko DA, Come JH, Priola SA, Chesebro B, Raymond GJ, Lansbury PT, Caughey B (1994) Cell-free formation of protease-resistant prion protein. Nature 370(6489):471–474

Leffers KW, Wille H, Stohr J, Junger E, Prusiner SB, Riesner D (2005) Assembly of natural and recombinant prion protein into fibrils. Biol Chem 386(6):569–580

Legname G, Baskakov IV, Nguyen HO, Riesner D, Cohen FE, DeArmond SJ, Prusiner SB (2004) Synthetic mammalian prions. Science 305(5684):673–676

Makarava N, Kovacs GG, Bocharova O, Savtchenko R, Alexeeva I, Budka H, Rohwer RG, Baskakov IV (2010) Recombinant prion protein induces a new transmissible prion disease in wild-type animals. Acta Neuropathol 119(2):177–187. doi:10.1007/s00401-009-0633-x

Morillas M, Swietnicki W, Gambetti P, Surewicz WK (1999) Membrane environment alters the conformational structure of the recombinant human prion protein. J Biol Chem 274(52):36859–36865

Muga A, Gonzalez-Manas JM, Lakey JH, Pattus F, Surewicz WK (1993) pH-dependent stability and membrane interaction of the pore-forming domain of colicin A. J Biol Chem 268(3):1553–1557

Naslavsky N, Shmeeda H, Friedlander G, Yanai A, Futerman AH, Barenholz Y, Taraboulos A (1999) Sphingolipid depletion increases formation of the scrapie prion protein in neuroblastoma cells infected with prions. J Biol Chem 274(30):20763–20771

Pinheiro TJ, Watts A (1994) Lipid specificity in the interaction of cytochrome c with anionic phospholipid bilayers revealed by solid-state 31P NMR. Biochemistry 33(9):2451–2458

Piro JR, Wang F, Walsh DJ, Rees JR, Ma J, Supattapone S (2011) Seeding specificity and ultrastructural characteristics of infectious recombinant prions. Biochemistry 50(33):7111–7116

Prusiner SB (1982) Novel proteinaceous infectious particles cause scrapie. Science 216(4542): 136–144

Prusiner SB (1998) Prions. Proc Natl Acad Sci USA 95(23):13363–13383

Riek R, Hornemann S, Wider G, Billeter M, Glockshuber R, Wuthrich K (1996) NMR structure of the mouse prion protein domain PrP(121–321). Nature 382(6587):180–182

Riek R, Hornemann S, Wider G, Glockshuber R, Wuthrich K (1997) NMR characterization of the full-length recombinant murine prion protein, mPrP(23–231). FEBS Lett 413(2):282–288

Sanghera N, Pinheiro TJ (2002) Binding of prion protein to lipid membranes and implications for prion conversion. J Mol Biol 315(5):1241–1256

Sarnataro D, Campana V, Paladino S, Stornaiuolo M, Nitsch L, Zurzolo C (2004) PrP(C) association with lipid rafts in the early secretory pathway stabilizes its cellular conformation. Mol Biol Cell 15(9):4031–4042

Silveira JR, Raymond GJ, Hughson AG, Race RE, Sim VL, Hayes SF, Caughey B (2005) The most infectious prion protein particles. Nature 437(7056):257–261

Smirnovas V, Baron GS, Offerdahl DK, Raymond GJ, Caughey B, Surewicz WK (2011) Structural organization of brain-derived mammalian prions examined by hydrogen-deuterium exchange. Nat Struct Mol Biol 18(4):504–506. doi:nsmb.2035|pii|10.1038/nsmb.2035

Taraboulos A, Scott M, Semenov A, Avrahami D, Laszlo L, Prusiner SB (1995) Cholesterol depletion and modification of COOH-terminal targeting sequence of the prion protein inhibit formation of the scrapie isoform. J Cell Biol 129(1):121–132

van der Goot FG, Gonzalez-Manas JM, Lakey JH, Pattus F (1991) A "molten-globule" membrane-insertion intermediate of the pore-forming domain of colicin A. Nature 354(6352):408–410. doi:10.1038/354408a0

Wang F, Yang F, Hu Y, Wang X, Jin C, Ma J (2007) Lipid interaction converts prion protein to a PrPSc-like proteinase K-resistant conformation under physiological conditions. Biochemistry 46(23):7045–7053

Wang F, Yin S, Wang X, Zha L, Sy MS, Ma J (2010a) Role of the highly conserved middle region of prion protein (PrP) in PrP-lipid interaction. Biochemistry 49(37):8169–8176. doi:10.1021/bi101146v

Wang F, Wang X, Yuan CG, Ma J (2010b) Generating a prion with bacterially expressed recombinant prion protein. Science 327(5969):1132–1135. doi:science.1183748|pii|10.1126/science.1183748

Weissmann C (2004) The state of the prion. Nat Rev Microbiol 2(11):861–871

White SH, Ladokhin AS, Jayasinghe S, Hristova K (2001) How membranes shape protein structure. J Biol Chem 276(35):32395–32398

Wimley WC, Hristova K, Ladokhin AS, Silvestro L, Axelsen PH, White SH (1998) Folding of beta-sheet membrane proteins: a hydrophobic hexapeptide model. J Mol Biol 277(5):1091–1110

Chapter 9
New Perspectives on Prion Conversion: Introducing a Mechanism of Deformed Templating

Ilia V. Baskakov

Abstract The transmissible agent of prion disease consists of a prion protein in its abnormal, β-sheet-rich state (PrP^{Sc}), which replicates itself according to the template-assisted mechanism. This mechanism postulates that the folding pattern of a newly recruited polypeptide chain accurately reproduces that of a PrP^{Sc} template. This chapter introduces a new mechanism of PrP^{Sc} formation and replication designated as "deformed templating." In contrast to classical templating, "deformed templating" postulates that PrP fibrils or particles with one cross-β-sheet structure can catalyze formation of PrP particles with fundamentally different structure of cross-β sheet. As a result, significant change in the PrP folding pattern can occur within cross-β spine. The mechanism of deformed templating predicts that PrP^{Sc} and transmissible prion diseases can be induced by cross-β prion protein structures substantially different from that of authentic PrP^{Sc}. The data on synthetic prions, i.e., inducing transmissible prion diseases with recombinant PrP amyloid fibrils strongly support the new mechanism. The possibility that a mechanism similar to deformed templating accounts for prion adaptation to new hosts is discussed. The new concept of deformed templating provides important new insight into genesis and evolution of the transmissible states of the prion protein and has numerous implications for understanding the etiology of prion and other neurodegenerative diseases.

Keywords Prion protein • Prion conversion • Template mechanism • Deformed templating • Cross-β folding pattern

I.V. Baskakov, Ph.D. (✉)
Department of Anatomy and Neurobiology, Center for Biomedical Engineering and Technology, University of Maryland School of Medicine,
725 W. Lombard St., Baltimore, MD 21201, USA
e-mail: baskakov@umaryland.edu

W.-Q. Zou and P. Gambetti (eds.), *Prions and Diseases: Volume 1, Physiology and Pathophysiology*, DOI 10.1007/978-1-4614-5305-5_9,

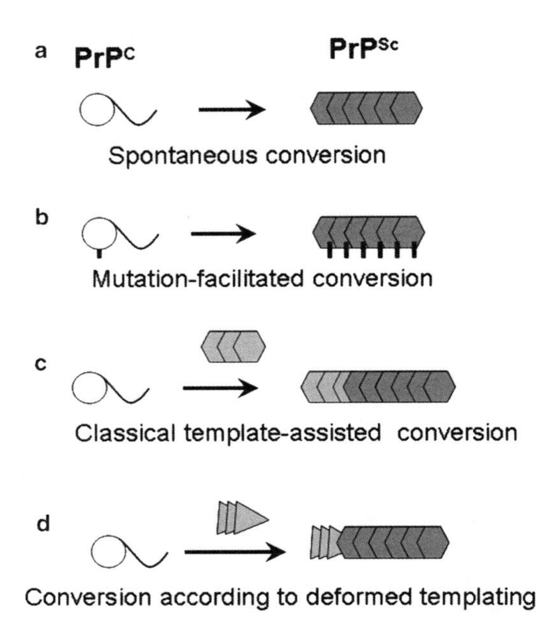

Fig. 9.1 Four mechanisms for PrP^{Sc} formation. (**a**) Spontaneous conversion of PrP^{C} into PrP^{Sc} is believed to underlie the sporadic forms of the prion diseases. (**b**) Disease-related mutations in prion protein can facilitate the conversion of PrP^{C} into PrP^{Sc}. (**c**) The template-assisted model postulates that PrP^{Sc} replicate its pathogenic structure by recruiting and converting PrP^{C}. According to this model, the folding pattern of a newly recruited polypeptide chain accurately reproduces that of a PrP^{Sc} template. (**d**) A new mechanism referred to as deformed templating postulates that the formation of PrP^{Sc} de novo can be seeded by abnormal PrP structures substantially different from that of authentic PrP^{Sc}. A transformation from one cross-β folding pattern present in a template to a significantly different folding pattern, the one specific for PrP^{Sc}, occurs during deformed templating

9.1 Introduction

Prion diseases, or transmissible spongiform encephalopathies, are fatal neurodegenerative disorders that can be sporadic, inherited, or infectious in origin. Misfolding and aggregation of the normal, cellular form of the prion protein (PrP^{C}) into an abnormal β-sheet-rich, disease-related conformation (PrP^{Sc}) underlie the pathogenic mechanisms of the prion diseases for all three origins (Prusiner 1996). Spontaneous conversion of PrP^{C} into PrP^{Sc} is believed to underlie the sporadic forms of prion diseases (Fig. 9.1a). The low occurrence rate of sporadic prion disease is likely to reflect the extremely low probability of spontaneous conversion of PrP^{C} into PrP^{Sc}. Inherited forms of the disease have been linked to a number of single-point mutations, truncation, or octarepeat expansion mutations in the *PRNP* gene (a gene that encodes prion protein), with more than 20 disease-inducing mutations identified so far (Prusiner and Scott 1997) (Fig. 9.1b). In addition to sporadic and inherited origins, prion diseases can be also acquired via transmission, just like other infectious diseases. The "protein only" hypothesis postulates that the transmissible agent of prion disease consists of a prion protein in its abnormal, β-sheet-rich, disease-related state (PrP^{Sc}), which is capable of propagating its abnormal conformation in an autocatalytic manner by recruiting and converting PrP^{C} (Prusiner 1982; Griffith 1967). According to the classical templating mechanism, during PrP^{Sc}-seeded conversion, the folding pattern of a newly recruited polypeptide chain accurately replicates that of a PrP^{Sc} template (Fig. 9.1c) (Cohen and Prusiner 1998). As a result, the PrP^{Sc}-specific folding pattern can be amplified endlessly and with a high precision, when homologous PrP^{C} molecules are provided.

This chapter introduces a new mechanism of PrP^{Sc} formation designated as "deformed templating." Deformed templating involves a switching from one cross-β folding pattern present in a template to a significantly different folding pattern, the one specific for PrP^{Sc} (Fig. 9.1d). Experimental data on synthetic prions accumulated over the past decade strongly support this hypothesis. The new concept on deformed templating offers a new perspective on genesis, evolution, and adaptation of transmissible prion structures.

9.2 Generating Transmissible Prion Diseases De Novo

The recent years witnessed a number of studies, where transmissible prion diseases were generated in animals de novo by inoculating prion infectious material produced in vitro (Legname et al. 2004; Colby et al. 2009, 2010; Makarava et al. 2010, 2011; Barria et al. 2009; Deleault et al. 2007; Wang et al. 2010). All studies on generating prion infectivity could be divided into two large groups, where the material for inoculating animals was produced either using (1) serial protein misfolding cyclic amplification (sPMCA) (Barria et al. 2009; Deleault et al. 2007; Wang et al. 2010) or (2) in vitro fibrillation protocols that utilized recombinant PrP (rPrP) (Legname et al. 2004; Colby et al. 2009, 2010; Makarava et al. 2010, 2011).

In the studies that employed the first approach, the application of sPMCA accomplished two purposes (1) generating PrP^{Sc} particles de novo and (2) amplification of newly formed PrP^{Sc} to the amounts that can effectively produce clinical disease in wild-type animals with 100% success rate (Barria et al. 2009; Deleault et al. 2007; Wang et al. 2010). De novo formation of PrP^{Sc} in PMCA showed stochastic behavior, i.e., less than 100% of the sPMCA reactions were positive, while the number of sPMCA rounds required for amplification of the newly formed PrP^{Sc} to the amount detectible by Western blot was variable even within the same experimental conditions. Therefore, it remains unclear whether the substantial infectivity produced after multiple rounds of sPMCA was a result of amplification of a few or even a single PrP^{Sc} particle.

The second approach involved conversion of rPrP into amyloid fibrils in vitro without application of sPMCA (Legname et al. 2004; Colby et al. 2009, 2010; Makarava et al. 2010, 2011). In these studies, transmissible diseases were generated either in transgenic animals with high levels of PrP^{C} expression or in wild-type animals. In transgenic animals, the disease was produced with 100% success rate in the first passage, although after relatively long incubation time (Legname et al. 2004; Colby et al. 2009, 2010). In wild-type animals, the disease was induced with less than 100% success rate, while two or even three serial passages required for appearance of clinical symptoms (Makarava et al. 2010, 2011). Critical concerns that rPrP amyloid fibrils did not induce the disease de novo but only accelerated an ongoing pathogenic process have been raised regarding the studies performed on transgenic mice (Caughey et al. 2009; Caughey and Baron 2006; Soto 2011). Indeed, the mice that overexpress PrP^{C} were found to develop a neurological disorder that was accompanied by PrP aggregation, although these disorders were not transmissible in serial passages (Colby et al. 2010). In contrast to sporadic formation of non-transmissible PrP aggregates, inoculation of rPrP fibrils triggered formation of authentic PrP^{Sc} that can transmit disease, a process that appears to compete with aggregation of non-transmissible PrP.

The experiments conducted using Syrian hamsters provided strong evidence that rPrP fibrils can induce transmissible prion disease de novo in wild-type animals (Makarava et al. 2010, 2011). However, when triggered by rPrP fibrils, only a small fraction of animals showed signs of infection. Furthermore, the clinical disease was observed only at the second or third serial passages. Less than a 100% success rate and long clinically silent stage raised a number of questions regarding the molecular mechanism underlying the genesis of transmissible prions. In the prevailing opinion, the preparations of rPrP amyloid fibrils must contain minuscule amounts of PrP^{Sc} or particles with the structure of PrP^{Sc}, and that this tiny subfraction was responsible for the disease.

9.3 The Mechanism of Triggering Prion Disease by rPrP Amyloid Fibrils

Before discussing models on triggering transmissible prion diseases by rPrP fibrils, it is useful to briefly review the data on the structure of rPrP fibrils and PrP^{Sc}. Several recent studies presented strong evidence that the structures of rPrP amyloid fibrils are fundamentally different from those of authentic PrP^{Sc}, which was either isolated from scrapie-infected animals or produced via sPMCA in vitro (Wille et al. 2009; Ostapchenko et al. 2010; Piro et al. 2011). X-ray diffraction experiments revealed substantial differences in equatorial diffraction patterns collected from rPrP fibrils and PrP^{Sc} purified from scrapie brains, illustrating that their folding patterns were significantly different (Wille et al. 2009; Ostapchenko et al. 2010). The results of X-ray diffraction analysis were consistent with the FTIR data, which also pointed to significant differences between conformations of PrP^{Sc} and rPrP fibrils (Spassov et al. 2006; Makarava and Baskakov 2008). For different prion strains, the maxima of the β-sheet absorption collected for PrP^{Sc} isolates varied between 1625 and 1637 cm^{-1} (Spassov et al. 2006), whereas the maxima of β-sheet absorption for rPrP fibrils was found to be at 1614 and 1626/28 cm^{-1} under the same solvent conditions (Makarava and Baskakov 2008; Ostapchenko et al. 2010). Substantial

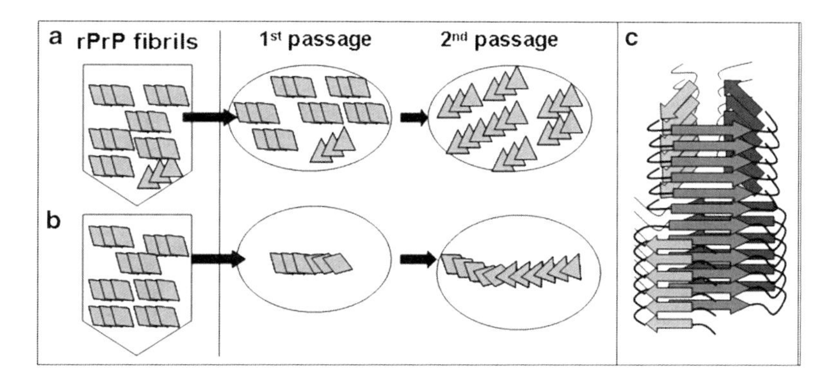

Fig. 9.2 Schematic representation of two mechanisms responsible for generating transmissible prion diseases de novo. According to the first mechanism, (**a**) the preparations of rPrP amyloid fibrils (schematically shown as *parallelograms*) contain very small amounts of PrP^{Sc} (shown as *triangles*). The silent stage of the disease is attributed to the long time required for amplification of this extremely small amount of PrP^{Sc}. A second mechanism referred to as deformed templating postulates that there are no PrP^{Sc} particles in the preparations of amyloid fibrils. (**b**) Instead, when inoculated into animals, amyloid fibrils can seed conversion of PrP^{C} into PrP^{Sc}-like structures, although with a low efficiency. The process of transformation of rPrP fibrils into PrP^{Sc} might involve several stages, which take place during long clinically silent stage. (**c**) A schematic diagram illustrating conformational switch within individual fibril (Makarava et al. 2009). The hybrid fibril consists of two segments with different global folding patterns. In both segments, common β-strand structure is formed within the same PrP region that links two segments together. This diagram does not intend to model PrP folding pattern within amyloid fibrils or PrP^{Sc}

differences in positions of the absorption maxima support the notion that the cross-β-sheet folding patterns in PrP^{Sc} and rPrP fibrils are fundamentally different. Furthermore, as judged from AFM and EM imaging, PrP^{Sc} produced in vitro displayed a spherical but not the elongated shape typical for amyloid fibrils (Piro et al. 2011). rPrP fibrils were not capable of seeding PrP^{Sc} formation in sPMCA further supporting the view that the structures of PrP^{Sc} and rPrP fibrils are different (Piro et al. 2011). If the structures of rPrP fibrils and PrP^{Sc} are fundamentally different, how can the first seed the last one?

Bearing in mind the results of structural studies, two alternative models can be put forward. According to the first model, the preparations of rPrP amyloid fibrils contain very small amounts of PrP^{Sc} or particles with a structure similar to authentic PrP^{Sc} (Fig. 9.2a). If this is correct, then the low success rate in infecting the animals and the long clinically silent stage can be attributed to the miniscule amounts of PrP^{Sc} in preparation of the fibrils. The second model proposes that formation of PrP^{Sc} and transmissible prion diseases in wild-type animals are triggered by rPrP seeding material that lacks PrP^{Sc} (Fig. 9.2b). Despite substantial differences in folding patterns, rPrP fibrils are capable of triggering formation of PrP^{Sc}. Because of the structural differences, rPrP fibril-induced seeding of PrP^{Sc} is not efficient, which explains the low rate of infection in the first passage. For the same reasons, transformation of rPrP amyloid structure into a structure of PrP^{Sc} might involve several steps before authentic PrP^{Sc} emerges, a process which is accompanied by a long clinically silent stage.

9.4 Experimental Evidence in Support of the Second Model

Experimental data accumulated to date strongly support the second model. First, no PrP^{Sc} could be detected in the preparations of rPrP amyloid fibrils in the sPMCAb format that detects single PrP^{Sc} particles (Makarava et al. 2011). If one assumes that the first model is correct, the amount of infectivity should be equivalent to approximately 0.5–1 infectious dose to account for the less than 100% infection rate in the first passage. This amount of infectivity is equivalent to ~10,000–100,000 PrP molecules or to ~100–1,000 PrP^{Sc} particles, assuming that an average PrP^{Sc} particle consists of ~100 PrP molecules (Saa et al. 2006). This amount of PrP^{Sc} was well above the detection limits of sPMCAb and should have been easily detected if present in preparations of rPrP fibrils.

Second, the experimental protocol used for producing rPrP amyloid fibrils employs denaturants (a mixture of 1 M GdnHCl and 3 M urea)—the solvent conditions under which PrP^{Sc} is largely unfolded. Because rPrP fibrils are much more stable than PrP^{Sc} (Makarava et al. 2010; Peretz et al. 2001; Sun et al. 2007), rPrP fibrils can be formed under solvent conditions where PrP^{Sc} is largely denatured. Furthermore, formation of authentic PrP^{Sc} in vitro requires RNA and lipids (Deleault et al. 2007; Wang et al. 2010), whereas rPrP amyloid fibrils were formed in the absence of any cellular cofactors. Therefore, it is highly unlikely that PrP^{Sc} particles with authentic structures could be formed during preparation of rPrP fibrils conducted in the absence of cofactors essential for authentic PrP^{Sc} structures and under solvent conditions that promote PrP^{Sc} denaturation.

Third, in recent studies on synthetic prions, a strong correlation between conformational stability of rPrP amyloid fibrils, the stability of PrP^{Sc} produced in animals upon inoculating rPrP fibrils, and the incubation time to disease were described (Colby et al. 2009). If a miniscule fraction in the preparation of rPrP fibrils is responsible for the disease, the correlation between stability of rPrP amyloid, which is a bulk property of fibril preparation, and PrP^{Sc} would be challenging to explain. Again, these results provide a strong support toward the second model.

Fourth, when transmissible prion disease is triggered by rPrP amyloid fibrils, a decrease in PrP^{Sc} conformational stability was observed during serial passages of synthetic prions (Legname et al. 2005; Makarava et al. 2010; Colby et al. 2009). Similar dynamics in PrP^{Sc} conformational stability was found regardless of whether transgenic mice or Syrian hamsters were inoculated with rPrP fibrils, suggesting that a common pathway in genesis and evolution of infectious structures might exist (Legname et al. 2005; Makarava et al. 2010; Colby et al. 2009). Observed changes in physical properties illustrate that the PrP^{Sc} structure undergoes transformation during serial transmission providing a direct support for the second model.

Fifth, as judged from the clinical and neuropathological features, the synthetic strains generated in Syrian hamsters by rPrP fibrils were remarkably different from all previously known hamster-adapted strains or strains generated by sPMCA (Deleault et al. 2007; Barria et al. 2009; Wang et al. 2010; Makarava et al. 2010, 2011). Among the most distinguishing features were slow progression of clinical disease, accumulation of large plaques in subpial and subependymal areas, distinctive lesion and PrP immunoreactivity profiles, and unusual clinical phenotype (obesity, hair loss) (Makarava et al. 2010). The fact that rPrP fibrils produced a disease phenotype remarkably different from the phenotype expressed by strains generated in sPMCA or isolated from animals is consistent with the hypothesis that rPrP fibrils gave rise to PrP^{Sc} with unique physical and biological features.

9.5 The New Mechanism on Deformed Templating

All the data on synthetic prions accumulated to date support the hypothesis that transmissible prion diseases can be triggered by PrP structures substantially different from that of authentic PrP^{Sc} (Fig. 9.2b). This hypothesis assumes that only partial overlap or distant similarities in structures of fibrillar rPrP and PrP^{Sc} are sufficient for triggering transmissible prion diseases. The precise mechanistic details for PrP^{Sc} formation by conformationally different structures via deformed templating mechanism remain to be elucidated. Nevertheless, recently discovered phenomenon of conformational switching within individual amyloid fibrils or particles provides one possible explanation of how such transformation might occur (Makarava et al. 2009; Baskakov 2009) (Fig. 9.2c). As a result of conformational switching, hybrid fibrils can be produced, where polypeptide folding pattern changes considerably along cross-β spine of individual fibril. According to the mechanism on deformed templating, the global folding patterns of PrP molecules within amyloid fibrils and PrP^{Sc} are different, yet nevertheless, they share common structural motifs. For instance, a common β-strand that can link two structures provides opportunity for limited templating (Fig. 9.2c). This model on conformational switching is consistent with experimental observations that the global structures of PrP^{Sc} and rPrP fibrils are different (Wille et al. 1996, 2009; Ostapchenko et al. 2010; Piro et al. 2011), while it explains a correlation between conformational stability of two structures (Colby et al. 2009). Because there is only partial overlap between two structures, the seeding of PrP^{Sc} by rPrP fibrils is not efficient, which explains the low infection rate in the first passage. Previous studies on molecular imaging of single amyloid fibrils provided a proof of principle that the conformational switching between two alternative PrP folding patterns can occur within an individual PrP fibril or particle (Makarava et al. 2009).

In classical templating, the folding pattern of a newly recruited polypeptide chain accurately reproduces that of a template. In deformed templating, while the template provides limited seeding, a newly recruited polypeptide chain acquires a new folding pattern which only partially overlaps with the folding pattern of a template.

Two glycosyl groups and GPI anchor present in PrP^{C} might impose spatial constraints on the spectrum of folding patterns available to PrP^{C} (Breydo et al. 2007), thus providing a driving force behind switching the recruited PrP folding pattern from rPrP fibril-specific to PrP^{S}-specific.

If the hypothesis that the fibril-specific PrP folding pattern can template a folding pattern typical for authentic PrP^{Sc} is correct, one can assume that the opposite reaction, i.e., the seeding of rPrP fibrils by PrP^{Sc}, is also possible. Indeed, several assays exploited the phenomenon of PrP^{Sc}-seeded conversion of α-rPrP into amyloid fibrils for detecting miniscule amounts of PrP^{Sc} (Colby et al. 2007; Atarashi et al. 2007). Interestingly, the structure of rPrP fibrils produced as a result of seeding by PrP^{Sc} only distantly resembled those of authentic PrP^{Sc} structure and had limited infectivity (Kim et al. 2010).

9.6 Conformational Switching Within Individual Amyloid Fibrils and Strain Adaptation Phenomenon

According to the prevailing view, multiple amyloid structures could be produced within the same amino acid sequence (Petkova et al. 2005; Makarava and Baskakov 2008). However, the folding pattern of a polypeptide chain within individual amyloid fibrils or PrP^{Sc} particles is believed to be uniform. In amyloid fibrils or PrP^{Sc} particles, β-strands are arranged perpendicularly to the axis of the cross-β spine (Wille et al. 2009; Ostapchenko et al. 2010), and their individual, strain-specific folding pattern provides a template for recruiting and converting a monomeric precursor at the growing edge. Faithful templating of cross-β structures is based on self-complementation of a polypeptide chain involved in cross-β assembly (Eisenberg et al. 2006). Self-complementation can be achieved through several mechanisms including tight complementarity of amino acid side chains in the steric zippers of the cross-β spine; the stacking of side chains in so-called polar zippers, where the side chain hydrogen bonds are formed between β-strands along the fibrillar axis; or domain swapping (Eisenberg et al. 2006).

The recent studies that employed single-fibril microscopy imaging revealed that the elongation of fibrils does not always support uniformity in cross-β structures within individual fibrils (Makarava et al. 2007, 2009). The cross-seeding reaction, where hamster rPrP fibrils were used as seeds while mouse rPrP was used as a substrate, was shown to produce hybrid fibrils consisted of two segments: one composed of hamster and another mouse rPrP (Makarava et al. 2009). Most importantly, within individual fibrils, the folding pattern within the mouse segment was considerably different from that within the hamster segment. The switch from hamster- to mouse-specific folding patterns within hybrid fibrils occurred presumably because the amino acid sequence of mouse rPrP was not compatible with the hamster-specific folding pattern. In addition to species-specific variations in amino acid sequence within prion-folding C-terminal domain, a deletion of a few N-terminal amino acid residues outside of the prion-folding domain can dramatically alter the folding pattern of the prion-folding domain (Ostapchenko et al. 2008).

To form a hybrid structure, two fibrillar segments with different global folds have to share a common local motif which will be responsible for the integrity of the hybrid structure (Fig. 9.2c) (Baskakov 2009). To satisfy this requirement, the same polypeptide region must adopt identical parallel β-strand conformation within two fundamentally different folding structures. Because the region that acquires the common β-strand conformation is connected by hydrogen bonds to the same region in the polypeptide molecules along the cross-β spine, the parallel β-sheet propagates along the whole length of the fibril despite being part of two different global folds. Hydrogen bonds running up and down the common β-sheet provide conformational stability for the whole hybrid structure. This model proposes that the catalytic activity in recruiting and converting monomeric precursors into an alternative cross-β folding pattern is due to partial overlap in folding patterns.

The observation of a conformational switch within individual fibrils highlights high adaptation potential for amyloid structures. Adaptive conformational switching permits recruitment of nonidentical but highly homologous polypeptide chains which otherwise are not compatible with the existing structure. Adaptive conformational switching within individual fibrils may provide a mechanistic explanation for strain mutation or modification, phenomena that have been frequently observed upon transmission of prions across species (Peretz et al. 2002; Castilla et al. 2008; Green et al. 2008). Notably, the process of triggering transmissible disease by rPrP fibrils displays features similar to those observed in the course of prion adaptation to a new host. Both processes are characterized by a long clinically silent stage (Hill and Collinge 2003; Hill et al. 2000; Race et al. 2001; Makarava et al. 2010, 2011), a transformation in the physical properties of PrP^{Sc} (Peretz et al. 2002; Makarava et al. 2010; Legname et al. 2005; Colby et al. 2009), and sometimes, a change in the accumulation pattern of PrP^{Sc} during serial transmission (Kimura et al. 2000; Makarava et al. 2011). Therefore, both phenomena, prion adaptation to a new host and triggering transmissible disease by rPrP fibrils, might share a common mechanism.

The studies on interspecies transmission using PMCA reactions or bioassays revealed that prion strains have variable adaptation potential. Some strains are capable of maintaining their individual strain-specific properties within a range of PrP amino acid sequences, whereas other strains can faithfully replicate only within a single or closely homologous PrP sequence (Peretz et al. 2002, ; Castilla et al. 2008; Green et al. 2008; Capobianco et al. 2007). The work on cross-seeding of rPrP amyloids provided direct illustration that self-replicating amyloid structures are not equally selective with respect to the amino acid sequence of the substrate molecules that can be recruited for their replication (Makarava et al. 2007). This difference in selectivity between amyloid strains specifies a direction in which adaptation or evolution of amyloid structures or prion strains occurs upon interspecies transmission. Conformational adaptation is expected to proceed from highly selective or species-specific structures toward promiscuous ones. A switch from a species-specific to promiscuous strain presumably occurs when a species-specific strain faces a heterologous substrate that is not compatible with the conformation of the original strain. Therefore, one can predict that interspecies transmissions could lead to formation of new promiscuous strains.

9.7 Cross talk Between Amyloidogenic Proteins

The hypothesis that transmissible prion diseases can be triggered by cross-β PrP structures substantially different from that of authentic PrP^{Sc} has large implications for understanding the etiology of prion and other neurodegenerative diseases. A growing number of studies have documented that amyloid forms of several proteins linked to neurodegenerative diseases were capable of seeding their own aggregation in a prion-like manner in a cell and spreading from cell to cell through the nervous system reviewed in Miller (2009), Frost and Diamond (2010), and Aguzzi and Rajendran (2009). One recent study provided strong evidence that pathological changes associated with non-prion neurodegenerative diseases could be induced or transmitted through inoculation of the aggregated forms of non-prion proteins such as Aβ (Eisele et al. 2010). Furthermore, recent study illustrated that amyloids can template structures different from their own (Makarava et al. 2009). It is generally assumed that self-perpetuating aggregation requires identity in amino acid sequence between seeds and substrate. Recent work, however, suggested the possibility of cross talk between non-related amyloidogenic proteins (Jean et al. 2007; Yan et al. 2007; Morales et al. 2010). In vivo, amyloidosis of one protein was found to be triggered by fibrils of an unrelated protein in a manner similar to cross-seeded polymerization (Jean et al. 2007; Yan et al. 2007; Morales et al. 2010). Cross talk between several yeast prion proteins provides another example of how direct interactions between newly forming and preexisting heterologous fibrils might take place in a cell (Derkatch et al. 1997, 2001, 2004). Moreover, pathological studies revealed that protein aggregates produced from two different proteins or peptides, including PrP, Aβ, α-synuclein, immunoglobulin light chain λ, and β_2 microglobulin, often colocalize within the same amyloid plaque in a variety of organs or tissues (Haik et al. 2002; Adjou et al. 2007; Takahashi et al. 1996; Miyazono et al. 1992; Galuske et al. 2004). The promiscuous nature of the propagating activity of amyloid structures can lead to devastating consequences for cellular health. For instance, the cross talk between non-related amyloidogenic proteins may offer a possible explanation for the development of age-related conformational disorders that are considered to be sporadic. In future studies, it would be interesting to define the spectrum of structures and sequences capable of triggering the PrP^C to PrP^{Sc} conversion and inducing transmissible prion diseases.

Acknowledgments This work was supported by NIH grant NS045585.

References

Adjou KT, Allix S, Ouidja MO, Backer S, Couquet C, Cornuejols MJ, Deslys JP, Bruqere H, Bruqere-Picoux J, El-Hachimi KH (2007) Alpha-synuclein accumulates in the brain of scrapie-affected sheep and goats. J Comp Pathol 137:78–81

Aguzzi A, Rajendran L (2009) The transcellular spread of cytosolic amyloids, prions, and prionoids. Neuron 64:783–790

Atarashi R, Moore RA, Sim VL, Hughson AG, Dorward DW, Onwubiko HA, Priola SA, Caughey B (2007) Ultrasensitive detection of scrapie prion protein using seeded conversion of recombinant prion protein. Nat Methods 4:645–650

Barria MA, Mukherjee A, Gonzalez-Romero D, Morales R, Soto C (2009) De novo generation of infectious prions in vitro produces a new disease phenotype. PLoS Pathog 5:e1000421

Baskakov IV (2009) Switching in amyloid structure within individual fibrils: implication for strain adaptation, species barrier and strain classification. FEBS Lett 583:2618–2622

Breydo L, Sun Y, Makarava N, Lee C-I, Novitskaia V, Bocharova OV, Kao JPY, Baskakov IV (2007) Nonpolar substitution at the C-terminus of the prion protein, a mimic of the glycosylphosphatidylinositol anchor, partially impairs amyloid fibril formation. Biochemistry 46:852–861

Capobianco R, Casalone C, Suardi S, Mangieri M, Miccolo C, Limido L, Catania M, Rossi G, Di Fede G, Giaccone G, Bruzzone MG, Minati L, Corona C, Acutis P, Gelmetti D, Lombardi G, Groschup MH, Buschmann A, Zanusso G, Monaco S, Caramelli M, Tagliavini F (2007) Conversion of the BASE prion strain into BSE strain: the origin of BSE? PLoS Pathog 3:e31

Castilla J, Gonzalez-Romero D, Saa P, Morales R, De Castro J, Soto C (2008) Crossing the species barrier by PrPSc replication in vitro generates unique infectious prions. Cell 134:757–768

Caughey B, Baron GS (2006) Prions and their partners in crime. Nature 443:803–810

Caughey B, Baron GS, Chesebro B, Jeffrey M (2009) Getting a grip on prions: oligomers, amyloids, and pathological membrane interactions. Annu Rev Biochem 78:177–204

Cohen FE, Prusiner SB (1998) Pathologic conformations of prion proteins. Annu Rev Biochem 67:793–819

Colby DW, Zhang Q, Wang S, Groth D, Legname G, Riesner D, Prusiner SB (2007) Prion detection by an amyloid seeding assay. Proc Acad Natl Sci USA 104:20914–20919

Colby DW, Giles K, Legname G, Wille H, Baskakov IV, DeArmond SJ, Prusiner SB (2009) Design and construction of diverse mammalian prion strains. Proc Acad Natl Sci USA 106: 20417–20422

Colby DW, Wain R, Baskakov IV, Legname G, PAlmer CG, Nguyen H-O, Lemus A, Cohen FE, DeArmond SJ, Prusiner SB (2010) Protease-sensitive synthetic prions. PLoS Pathog 6:e1000736

Deleault NR, Harris BT, Rees JR, Supattapone S (2007) Formation of native prions from minimal components in vitro. Proc Acad Natl Sci USA 104:9741–9746

Derkatch IL, Bradley ME, Zhou P, Chernoff YO, Liebman SW (1997) Genetic and environmental factors affecting the de novo appearance of the [PSI+] prion in *Saccharomyces cerevisiae*. Genetics 147:507–519

Derkatch IL, Bradley ME, Hong JY, Liebman SW (2001) Prions affect the appearance of other prions: the story of [PIN(+)]. Cell 106:171–182

Derkatch IL, Uptain SM, Quteiro TF, Krishnan R, Lindquist SL, Liebman SW (2004) Effects of Q/N-rich, polyQ, and non-polyQ, amyloids on the de novo formation of the [PSI+] prion in yeast and aggregation of Sup35 in vitro. Proc Acad Natl Sci USA 101:12934–12939

Eisele YS, Obermuller U, Heilbronner G, Baumann F, Kaeser SA, Wolburg H, Walker LC, Staufenbiel M, Heikenwalder M, Jucker M (2010) Peripherally applied Abeta-containing inoculates induce cerebral beta-amyloidosis. Science 330:980–982

Eisenberg D, Nelson R, Sawaya MR, Balbirnie M, Sambashivan S, Ivanova MI, Madsen AO, Riekel C (2006) The structural biology of protein aggregation diseases: fundamental questions and some answers. Acc Chem Res 39:568–575

Frost B, Diamond MI (2010) Prion-like mechanisms on neurodegenerative diseases. Nat Rev Neuroscience 11:155–159

Galuske RA, Drach LM, Nichtweiss M, Marquardt G, Franz K, Bohl J, Scholote W (2004) Colocalization of different types of amyloid in the walls of cerebral blood vessels of patients from cerebral amyloid angiopathy and spontaneous intracranial hemorrhage: a report of 5 cases. Clin Neuropathol 23:113–119

Green KM, Castilla J, Seward TS, Napier DL, Jewell JE, Soto C, Telling GC (2008) Accelerated high fidelity prion amplification within and across prion species barriers. PLoS Pathog 4:e1000139

Griffith JS (1967) Self-replication and scrapie. Nature 215:1043–1044

Haik S, Privat N, Adjou KT, Sazdovitch V, Dormont D, Duyckaerts C, Hauw JJ (2002) Alpha-synuclein-immunoreactive deposits in human and animal prion diseases. Acta Neuropathol 103:516–520
Hill AF, Collinge J (2003) Subclinical prion infection. Trends Microbiol 11:578–584
Hill AF, Joiner S, Linehan J, Desbruslais M, Lantos PL, Collinge J (2000) Species-barrier-independent prion replication in apparently resistant species. Proc Natl Acad Sci USA 97:10248–10253
Jean L, Thomas B, Tahiri-Alaoui A, Shaw M, Vaux DJ (2007) Heterologous amyloid seeding: revisiting the role of acetylcholinesterase in Alzheimer's disease. PLoS One 2:e652
Kim JI, Cali I, Surewicz K, Kong Q, Raymond GJ, Atarashi R, Race B, Qing L, Gambetti P, Surewicz WK (2010) Mammalian prions generated from bacterially expressed prion protein in the absence of any mammalian cofactors. J Biol Chem 285:14083–14087
Kimura K, Kubo M, Yokoyama T (2000) Characteristics of prion protein (PrP(Sc)) in the brains of hamsters inoculated serially with a mouse-passaged scrapie strain. J Comp Pathol 122:123–130
Legname G, Baskakov IV, Nguyen H-OB, Riesner D, Cohen FE, DeArmond SJ, Prusiner SB (2004) Synthetic mammalian prions. Science 305:673–676
Legname G, Nguyen H-OB, Baskakov IV, Cohen FE, DeArmond SJ, Prusiner SB (2005) Strain-specified characteristics of mouse synthetic prions. Proc Natl Acad Sci USA 102:2168–2173
Makarava N, Baskakov IV (2008) The same primary structure of the prion protein yields two distinct self-propagating states. J Biol Chem 283:15988–15996
Makarava N, Lee CI, Ostapchenko VG, Baskakov IV (2007) Highly promiscuous nature of prion polymerization. J Biol Chem 282:36704–36713
Makarava N, Ostapchenko VG, Savtchenko R, Baskakov IV (2009) Conformational switching within individual amyloid fibrils. J Biol Chem 284:14386–14395
Makarava N, Kovacs GG, Bocharova OV, Savtchenko R, Alexeeva I, Budka H, Rohwer RG, Baskakov IV (2010) Recombinant prion protein induces a new transmissible prion disease in wild type animals. Acta Neuropathol 119:177–187
Makarava N, Kovacs GG, Savtchenko R, Alexeeva I, Budka H, Rohwer RG, Baskakov IV (2011) Genesis of mammalian prions: from non-infectious amyloid fibrils to a transmissible prion disease. PLoS Pathog 7(12):e1002419
Miller G (2009) Could they all be prion diseases? Science 326:1337–1339
Miyazono M, Kitamoto T, Doh-ura K, Iwaki T, Tateishi J (1992) Creutzfeldt–Jakob disease with codon 129 polymorphism (Valine): a comparative study of patients with codon 102 point mutation or without mutations. Acta Neuropathol 84:349–354
Morales R, Estrada LD, Diaz-Espinoza R, Morales-Scheihing D, Jara MC, Castilla J, Soto C (2010) Molecular cross talk between misfolded proteins in animal models of Alzheimer's and prion diseases. J Neurosci 30:4528–4535
Ostapchenko VG, Makarava N, Savtchenko R, Baskakov IV (2008) The polybasic N-terminal region of the prion protein controls the physical properties of both the cellular and fibrillar forms of PrP. J Mol Biol 383:1210–1224
Ostapchenko VG, Sawaya MR, Makarava N, Savtchenko R, Nilsson KP, Eisenberg D, Baskakov IV (2010) Two amyloid states of the prion protein display significantly different folding patterns. J Mol Biol 400:908–921
Peretz D, Scott M, Groth D, Williamson A, Burton D, Cohen FE, Prusiner SB (2001) Strain-specified relative conformational stability of the scrapie prion protein. Protein Sci 10:854–863
Peretz D, Williamson RA, Legname G, Matsunaga Y, Vergara J, Burton D, DeArmond S, Prusiner S, Scott MR (2002) A change in the conformation of prions accompanies the emergence of a new prion strain. Neuron 34:921–932
Petkova AT, Leapman RD, Gua Z, Yau W-M, Mattson MP, Tycko R (2005) Self-propagating, molecular-level polymorphism in Alzheimer's b-amyloid fibrils. Science 307:262–265
Piro JR, Wang F, Walsh DJ, Rees JR, Ma J, Supattapone S (2011) Seeding specificity and ultrastructural characteristics of infectious recombinant prions. Biochemistry 50:7111–7116
Prusiner SB (1982) Novel proteinaceous infectious particles cause scrapie. Science 216:136–144

Prusiner SB (1996) Prion diseases. In: Nathanson N, Ahmed R, Gonzalez-Scarano F, Griffin D, Holmes K, Murphy FA, Robinson HL (eds) Viral pathogenesis. Raven, New York, pp 855–911
Prusiner SB, Scott MR (1997) Genetics of prions. Annu Rev Genet 31:139–175
Race R, Rainse A, Raymond G, Caughey B, Chesebro B (2001) Long-term subclinical carrier state precedes scrapie replication and adaptation in a resistant species: analogies to bovine spongiform encephalopathy and variant Creutzfeldt–Jakob disease in humans. J Virol 2001: 10106–10112
Saa P, Castilla J, Soto C (2006) Ultra-efficient replication of infectious prions by automated protein misfolding cyclic amplification. J Biol Chem 281:35245–35252
Soto C (2011) Prion hypothesis: the end of the controversy? Trends Biochem Sci 36:151–158
Spassov S, Beekes M, Naumann D (2006) Structural differences between TSEs strains investigated by FT-IR spectroscopy. Biochim Biophys Acta 1760:1138–1149
Sun Y, Breydo L, Makarava N, Yang Q, Bocharova OV, Baskakov IV (2007) Site-specific conformational studies of PrP amyloid fibrils revealed two cooperative folding domain within amyloid structure. J Biol Chem 282:9090–9097
Takahashi M, Hoshii Y, Kawano H, Gondo T, Ishihara T, Isobe T (1996) Ultrastructural evidence for colocalization of kappa light chain- and beta 2-microglobulin -derived amyloid using double labelling immunogold electron microscopy. Virchows Arch 429:383–388
Wang F, Wang X, Yuan C-G, Ma J (2010) Generating a prion bacterially expressed recombinant prion protein. Science 327:1132–1135
Wille H, Zhang G-F, Baldwin MA, Cohen FE, Prusiner SB (1996) Separation of scrapie prion infectivity from PrP amyloid polymers. J Mol Biol 259:608–621
Wille H, Bian W, McDonald M, Kendall A, Colby DW, Bloch L, Ollesh J, Borovinskiy AL, Cohen FE, Prusiner SB, Stubbs G (2009) Natural and synthetic prion structure from X-ray fiber diffraction. Proc Acad Natl Sci USA 106:16990–16995
Yan J, Fu X, Ge F, Zhang B, Yao J, Zhang H, Qian J, Tomozawa H, Naiki H, Sawashita J, Mori M, Higuchi K (2007) Cross-seeding and cross-competition in mouse apolipoprotein A-II amyloid fibrils and protein A amyloid fibrils. Am J Pathol 171:172–180

Chapter 10
Infectious and Pathogenic Forms of PrP

Emiliano Biasini and David A. Harris

Abstract Prion diseases are transmitted by unconventional infectious agents (prions) generated by the conformational conversion of PrP^C, a normal, cell-surface glycoprotein, into PrP^{Sc}, a misfolded isoform that propagates itself by a self-templating mechanism. Although PrP^{Sc} has commonly been considered the primary neurotoxic species in prion diseases, strong experimental evidence now challenges this dogma and suggests that alternative pathogenic forms of PrP may operate by altering the normal physiological function of PrP^C. In the past 15 years, we and others have generated cellular and animal models for studying prion diseases that shed light on important aspects of PrP infectivity, aggregation, and toxicity. In this chapter, we review some of these results and discuss our current understanding of the molecular processes responsible for the formation of aberrant forms of PrP and their acquisition of infectious and toxic properties.

Keywords Prion • Mutant PrP • Inherited prion diseases • Transfected cells • Transgenic mice

10.1 Introduction

Prion diseases, including Creutzfeldt–Jakob disease (CJD), Gerstmann–Sträussler–Scheinker syndrome (GSS), and fatal familial insomnia (FFI), are fatal disorders characterized by dementia, motor dysfunction, and cerebral amyloidosis (Collinge 2001; Prusiner 1998). Prion diseases can be acquired from horizontal transmission and occur sporadically or originate genetically in an autosomal-dominant fashion

E. Biasini, Ph.D. • D.A. Harris, M.D.,Ph.D. (✉)
Department of Biochemistry, Boston University School of Medicine,
72 East Concord Street, Boston, MA 02118, USA
e-mail: daharris@bu.edu

W.-Q. Zou and P. Gambetti (eds.), *Prions and Diseases: Volume 1, Physiology and Pathophysiology*, DOI 10.1007/978-1-4614-5305-5_10,

(Collinge and Palmer 1994). Current evidence indicates that the key event in the pathogenesis of all forms of prion diseases is the conformational conversion of the normal prion protein (PrP^{C}) into a pathogenic isoform (PrP^{Sc}) that has a high content of β-sheets. PrP^{Sc} accumulates in the central nervous system in an aggregated, protease-resistant form that is believed to propagate itself by impressing its abnormal conformation onto PrP^{C} molecules (Weissmann 2004). This phenomenon of protein-based inheritance has been extended to several non-Mendelian traits in yeast (Wickner et al. 2011) and, more recently, has provided an explanation for the spreading of several other disease-associated proteins within the nervous system (Frost and Diamond 2010).

Familial prion diseases display autosomal-dominant inheritance and are linked to insertional and point mutations in the PrP gene on chromosome 20 (Collinge 1993). Although these mutations are presumed to favor the spontaneous conversion of PrP^{C} into PrP^{Sc}, transmissibility has been formally demonstrated only for few of them, raising the possibility that at least some mutant PrP molecules may be pathogenic but not infectious (Collins and Masters 1995). This conclusion is consistent with other data indicating that several PrP isoforms, distinct from infectious PrP^{Sc}, possess neurotoxic properties (Chiesa and Harris 2001).

The presence of functional PrP^{C} molecules on the neuronal surface has also been recognized as a key factor for the pathogenesis of prion diseases. Strong experimental evidence for this conclusion is provided by the observation that grafted, prion-infected brain tissue is not toxic to surrounding nerve cells lacking endogenous PrP^{C} (Brandner et al. 1996). Moreover, targeted depletion of neuronal PrP in prion-infected mice rescues both neuronal loss and clinical signs, despite the continuous production of PrP^{Sc} by surrounding glial cells (Mallucci et al. 2003, 2007). Finally, recent data suggest that PrP^{C} may play a role in the toxicity of misfolded proteins associated with other neurodegenerative diseases, such as the Aβ oligomers involved in Alzheimer's disease (Laurén et al. 2009). Collectively, these data suggest that PrP^{C} may act as a transducer of toxic stimuli deriving from infectious PrP^{Sc}, noninfectious misfolded conformers of PrP, or other β-rich protein aggregates (Resenberger et al. 2011).

In this chapter, we will critically review the available experimental data regarding the existence of noninfectious, pathogenic forms of PrP and how these molecules could exert their toxicity by activating multiple neurotoxic pathways, some of which may involve the physiological function of PrP^{C}.

10.2 Modeling Familial Prion Diseases in Cultured Cells

Infectious and inherited forms of prion disease are both characterized by the accumulation of protease-resistant PrP aggregates in the brain. Cell culture models permissive for propagation of different prion strains have illuminated some of the basic principles underlying the formation and replication of PrP^{Sc}. In order to develop cellular models for studying the mechanisms that lie at the root of mutant

PrP misfolding and aggregation, we and others have employed various cell lines, including Chinese hamster ovary (CHO), neuroblastoma 2A (N2A), baby hamster kidney (BHK), human embryonic kidney 293 (HEK293) cells, and several others (Harris 1999). Each of these cell lines was used to generate stable clones expressing PrPs carrying mouse homologues of several human, disease-linked mutations, including P102L, D178N, V180I, D198N, E200K, V210I, as well as octarepeat insertions in the N-terminus of PrP. In many cases, mutant PrP molecules expressed in transfected cells undergo spontaneous misfolding and conversion into an aggregated state, which displays several biochemical features reminiscent of PrP^{Sc} (Lehmann and Harris 1996). These include insolubility in non-denaturing detergents, resistance to low concentrations of proteinase K (PK), and resistance to cleavage of their glycosyl–phosphatidyl–inositol (GPI) anchor by the enzyme phospholipase C (PIPLC). Mutant molecules also undergo structural rearrangements that involve the central hydrophobic domain region (Biasini et al. 2010), resulting in lack of reactivity with several PrP^{C}-directed antibodies, and display of conformational epitopes that are recognized by PrP^{Sc}-directed antibodies (Biasini et al. 2008a). Taken together, these studies indicate that disease-associated mutations can promote the spontaneous misfolding of the PrP molecule and generate a form that structurally resembles infectious PrP^{Sc}.

In order to gain information on the biology of PrP mutants, several laboratories, including our own, have studied the intracellular trafficking of these molecules by using immunolabeling techniques, as well as mutant PrP constructs tagged with specific antibody epitopes, probe-acceptor sequences, or green fluorescent proteins (GFP) (Campana et al. 2005). Collectively, these experiments revealed that mutant PrPs are often impaired in their trafficking to the cell surface, as a result of their accumulation in intracellular compartments such as the endoplasmic reticulum (ER) and the Golgi apparatus (Ivanova et al. 2001) or their selective rerouting to acidic lysosomal compartments (Ashok and Hegde 2009). Some of these mutant PrP molecules also show delayed maturation of their polysaccharide chains, which can be detected by treatment with endoglycosidase-H, indicating impaired transit through the mid-Golgi (Daude et al. 1997). These results suggest that misfolding of mutant PrP molecules occurs in the early secretory pathway, which may induce activation of the unfolded protein response (UPR) (Nunziante et al. 2011). However, the latter conclusion has not been consistently supported by experimental evidence (Quaglio et al. 2011). Other results have suggested that a subpopulation of PrP molecules can be substrates for the ER-associated degradation (ERAD) pathway, resulting in retrotranslocation from the ER lumen to the cytosol, deglycosylation by *N*-glycanases, and degradation by the proteasome machinery (Ma and Lindquist 2001). Cytoplasmic accumulation of these aberrant PrP isoforms (called cyPrP), which lack the N-terminal signal peptide, could be caused by disease-associated mutations or other pathological conditions, resulting in an impairment of proteasome function.

Several kinds of data indicate that cyPrP is cytotoxic. For example, ectopic expression of cyPrP in transgenic (Tg) mice induces cerebellar granule neuron (CGN) degeneration as well as behavioral and neuropathological abnormalities in the forebrain (Ma et al. 2002). cyPrP has also been shown to exert a cytotoxic activity by inactivating the E3 ubiquitin ligase mahogunin, providing a possible mechanism by which mislocalized PrP species could be pathogenic (Chakrabarti and Hegde 2009). However, some of these conclusions have been challenged by other data showing that, under physiological conditions, neither WT nor mutant PrPs are subjected to proteasomal degradation (Drisaldi et al. 2003). Under conditions of supraphysiological expression of PrP or pharmacological inhibition of the proteasome, a small percentage of PrP molecules fail to translocate into the ER (as indicated by the fact that they retain their signal peptide), accumulate in the cytosol, and exert a cytoprotective function in a cell-type-specific fashion (Restelli et al. 2010; Fioriti et al. 2005).

Some of these observations appear difficult to reconcile with each other. Part of the problem may stem from the fact that cyPrP is barely detectable in the absence of proteasome inhibition or PrP overexpression. Therefore, additional information is needed to conclusively evaluate the role of ER stress, the UPR, and cytoplasmic forms of PrP in prion diseases.

A recent study utilizing differential proteomics has investigated the impact of intracellular accumulation of a particular PrP mutant (the mouse homologue of the D178N–M129 mutation, which is linked to FFI) on Golgi homeostasis (Massignan et al. 2010a). Mutant PrP was found to induce changes in proteins involved in energy metabolism, redox regulation, and vesicular transport, together with a significant increase in the level of Rab GDP dissociation inhibitor alpha (GDIα), a factor known to govern vesicular trafficking by modulating the activity of Rab proteins. GDI overexpression was shown to induce selective reorganization of Rab11, a protein involved in vesicular post-Golgi trafficking, from an active, membrane-bound state to inactive, cytosol-localized form. As direct consequence of these alterations, the trafficking of GPI-anchored proteins in N2a cells expressing mutant PrP was significantly impaired. These results provide evidence for the existence of a cytotoxic feedback loop initiated by mutant PrP intracellular aggregation, which causes overexpression of GDI and accumulation of several GPI-anchored proteins (including mutant PrP) in the secretory pathway (Fig. 10.1). The possibility that a global alteration of protein trafficking to the cell membrane contributes to the pathogenesis of inherited prion diseases provides an interesting parallel with other neurodegenerative diseases linked to protein aggregation and suggests that the expression of mutant PrPs could have a broader impact on cellular homeostasis than previously thought.

10.3 Pathogenic, Noninfectious Aggregates of PrP in Mice

Although mutant PrP molecules expressed in cultured cells provide an important tool for studying PrP biogenesis, trafficking, misfolding, and aggregation, they fail to produce spontaneous signs of cytotoxicity. In contrast, when expressed in transgenic mice, some of these mutant molecules induce neurodegenerative phenotypes (Telling 2011). Several mutant PrPs have been expressed in Tg mice, for example, the mouse PrP homologues of D178N/V129 (linked to familial CJD) (Dossena et al. 2008),

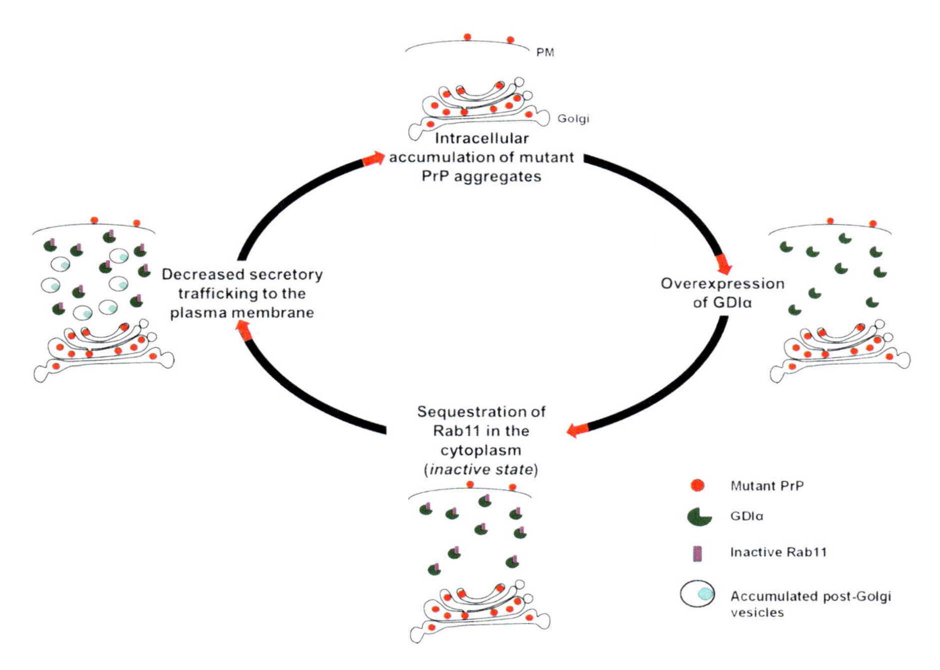

Fig. 10.1 Graphical representation of the putative neurotoxic loop activated by a mutant PrP (D177N/M128). Accumulation of mutant PrP in the Golgi is associated with overexpression of GDIα. High levels of GDIα cause the sequestration of Rab11 in the cytosol in an inactive state. The absence of active Rab11 negatively influences the post-Golgi trafficking of mutant PrP and other secreted proteins

D178N/M129 (linked to FFI) (Jackson et al. 2009), P102L (linked to GSS) (Hsiao et al. 1994; Friedman-Levi et al. 2011), and E200K (linked to CJD) (Friedman-Levi et al. 2011). Although some of these mutants induce a neurodegenerative illness, very few of them have been shown to carry infectivity. Therefore, in most cases, Tg mice represent a model of toxicity in absence of infectivity.

A clear example of pathogenicity in absence of infectivity is provided by Tg(PG14) mice. These mice express the mouse PrP homologue of a nine-octapeptide repeat insertion (referred to as PG14) that in humans is associated with an inherited form of prion dementia (Chiesa et al. 1998). Tg(PG14) mice accumulate in their brains a mutant PrP molecule that exhibits some of the major biochemical properties of PrP^{Sc}, including detergent insolubility and protease resistance (Chiesa et al. 2000). As this form accumulates in the brain, Tg(PG14) mice develop a slowly progressive neurological disorder characterized by ataxia, gliosis, PrP deposition, and massive loss of cerebellar granule cells by apoptosis. Accumulation of PG14 PrP precedes the appearance of the neuropathological changes, and its elevation in the brain correlates with nerve cell loss and the progression of the clinical symptoms. Importantly, brain homogenates from Tg(PG14) mice failed to transmit disease when inoculated intracerebrally into recipient mice (Chiesa et al. 2003).

Similarly, there is no evidence for transmissibility of inherited human prion diseases linked to the nine-octapeptide insertion mutation. Collectively, these results indicate that PG14 PrP forms PrP^{Sc}-like aggregates that are highly neurotoxic but not infectious.

Interestingly, two previous studies have reported that aggregated forms of PrP spontaneously accumulate in transgenic mice overexpressing WT PrP (Westaway et al. 1994; Chiesa et al. 2008). These mice develop an ataxic neurodegenerative syndrome characterized neuropathologically by PrP deposition and synaptic dysfunction in the molecular layer of the cerebellum. Similar to mutant PrPs, overexpressed WT PrP forms detergent-insoluble, mildly protease-resistant aggregates that react with PrP^{Sc}-directed antibodies but that are not infectious in transmission assays. These data reinforce the idea that aggregates of PrP may induce neurodegeneration in absence of infectivity.

10.3.1 Searching for the Structural Determinants of Prion Infectivity and Pathogenicity

Tg(PG14) mice provide a convenient biological model for investigating the molecular determinants of prion infectivity and toxicity. As mentioned previously, these mice do not generate spontaneous infectivity, and brain-extracted aggregates of PG14 PrP are unable to seed the misfolding of WT PrP substrate in the protein misfolding cyclic amplification assay (PMCA) (Biasini et al. 2008b). However, when inoculated with a mouse-adapted RML (Rocky Mountain Laboratory) prion strain, Tg(PG14) mice accumulate a form of PG14 PrP that is infectious upon serial passages (Chiesa et al. 2003). This RML-seeded form of the protein was referred to as PG14^{RML}, to distinguish it from the form accumulating in spontaneously ill, uninoculated mice (called PG14^{Spon}).

These two forms of PG14 PrP have been subjected to a panel of biochemical assays capable of discriminating soluble, monomeric PrP^{C} from aggregated PrP^{Sc}, including detergent insolubility, PK treatment, precipitation with sodium phosphotungstate (Na-PT), immobilized metal affinity chromatography (IMAC), and immunoprecipitation with previously described PrP^{Sc}-directed antibodies [which include a mouse monoclonal IgM antibody called 15B3 and three PrP motif-grafted monoclonal antibodies (referred to as IgG 19–33, 89–112, and 136–158)] (Biasini et al. 2008a, b; Chiesa et al. 2003). Surprisingly, no difference was detected between PG14^{Spon} and PG14^{RML} in each of these biochemical assays, indicating that the two forms, although biologically different, share fundamental conformational features. However, some notable differences between the two forms could be discerned. In particular, analyses of their aggregation state by sucrose gradient centrifugation and urea-induced dissociation revealed that PG14^{Spon} aggregates are smaller and less densely packed than PG14^{RML} aggregates (Chiesa et al. 2003).

Taken together, these experiments suggest that PG14^{Spon} and PG14^{RML} share multiple conformational similarities, but differ for their quaternary structure. Therefore, the

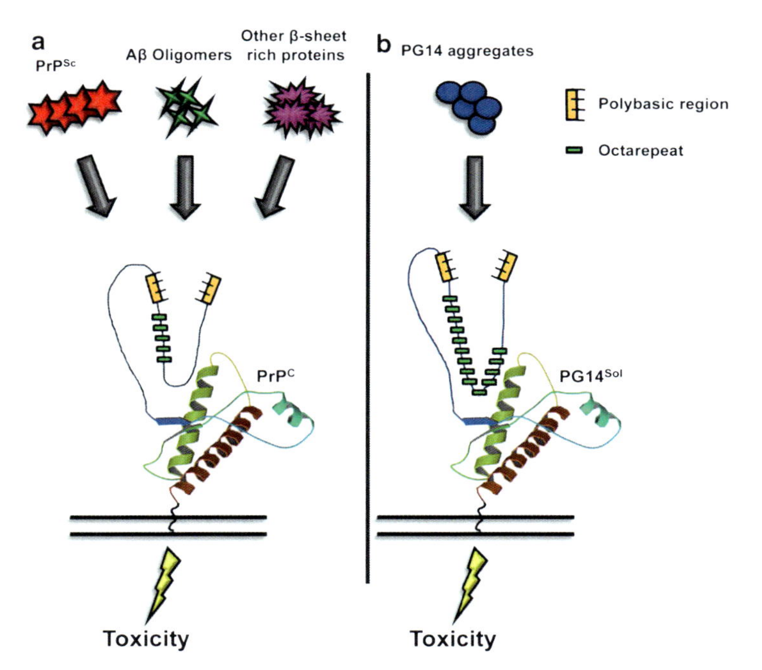

Fig. 10.2 Model of PrP^C-mediated toxicity of aggregated proteins. (**a**) PrP^C could be a receptor for infectious PrP^{Sc}, oligomers of the Aβ peptide, and possibly other misfolded proteins associated with neurodegeneration. The binding sites on PrP^C for these aggregated forms are the two clusters of positively charged amino acids in the N-terminus of the protein (residues 23–28 and 95–105). As a result of these interactions, PrP^C transduces toxic signals inside the cell. (**b**) The soluble, monomeric form of the PG14 PrP mutant (PG14^{Sol}) could act like PrP^C, binding to aggregates of PG14 (PG14^{Spon} or PG14^{RML}) and mediating their neurotoxic effects

infectivity of PG14^{RML} is likely to be encoded in the three-dimensional architecture of the aggregated particles. Interestingly, a similar conclusion regarding the molecular determinants of prion infectivity has emerged from studies of the strain phenomenon in yeast prions, where a number of genetic, biochemical, and biophysical data support the idea that prion strain variation relies on the size of the amyloid core (Tessier and Lindquist 2009). Recent success in the purification of infectious and noninfectious aggregates of PrP from the brains of Tg mice, using a protocol that includes sequential centrifugations followed by immunoprecipitation with a PrP^{Sc}-directed antibody, should open up the way to the analysis of purified PrP aggregates with high-resolution biophysical techniques (Biasini et al. 2009).

The conclusions presented above provide insights into the physical determinants of prion infectivity, but do not address the mechanisms underlying the neurotoxicity of PG14^{Spon}, PG14^{RML}, or other noninfectious aggregated states of PrP. One possible clue to this puzzle is provided by recent data demonstrating that PrP^C can act as a cell-surface transducer of toxic signals derived from PrP^{Sc}, oligomers of the Aβ peptide, or other β-sheet-rich protein assemblies (Biasini et al. 2012a) (Fig. 10.2a). Therefore, several noninfectious, misfolded states of PrP, including aggregates of

WT or mutant PrP, could be toxic by presenting specific surface conformations that would allow them to bind to PrP^{C} and unleash its cytotoxic property. This model, which has relevance for prion diseases and several other neurodegenerative disorders, implies that PrP^{C} is a protein that works at the intersection of physiology and disease (Biasini et al. 2012a).

One piece of evidence in support of this model comes from the observation that a subpopulation of PG14 molecules in both $PG14^{Spon}$ and $PG14^{RML}$ mice remain soluble (Biasini et al. 2008b). This isoform, which was named $PG14^{Sol}$, to distinguish it from the aggregated forms of PG14 PrP, shares several biological and biochemical features with WT PrP^{C}, including sensitivity to PK, reactivity with PrP^{C}-directed antibodies, and localization at the cell surface. Therefore, $PG14^{Sol}$ may act as the surface receptor that delivers toxic signals of $PG14^{Spon}$ and $PG14^{RML}$ aggregates, similar to the putative transduction activity exerted by PrP^{C} after binding to PrP^{Sc} (Fig. 10.2b). In the next chapter, we will review some of the experimental evidence supporting the idea that PrP^{C} lies at the root of the pathogenic process occurring in prion diseases and possibly other neurodegenerative disorders.

10.4 Neurotoxic Mutants Subvert the Physiological Function of PrP^{C}

The notion that the normal function of PrP^{C} can be altered or subverted to generate toxicity is supported by experiments involving topologically altered forms of PrP. In particular situations, the PrP polypeptide chain inserts its central hydrophobic domain into the lipid bilayer, resulting in two distinct topological variants called ^{Ctm}PrP and ^{Ntm}PrP (with the C or N terminus located on the extracellular side of the membrane, respectively) (Hegde et al. 1998; Stewart and Harris 2003). Mutations in the hydrophobic domain or in the N-terminal signal peptide have been shown to enhance the percentage of ^{Ctm}PrP in a cell (Stewart and Harris 2003). Interestingly, expression of ^{Ctm}PrP in Tg mice induces a spontaneous neurodegenerative phenotype that requires co-expression of WT PrP, indicating that these aberrant PrP molecules sabotage the normal function of PrP^{C} to generate toxicity (Hegde et al. 1998).

Other kinds of PrP molecules, carrying deletions in the N-terminal region of the protein, display neurotoxicity that is suppressed by wild-type PrP. These mutants are collectively referred to as ΔN PrP (Solomon et al. 2010a). When expressed in Tg mice, PrP molecules deleted for residues 32–121 or 32–134 lead to progressive degeneration of CGNs and white matter vacuolation (Shmerling et al. 1998). Surprisingly, the toxicity is even more pronounced when deletions are restricted to residues 105–125 (ΔCR) or 94–134, as these molecules induce neonatal lethal phenotype in mice (Li et al. 2007; Baumann et al. 2007). A crucial observation linking the toxicity of these artificial PrP mutants to the normal activity of PrP^{C} is the observation that co-expression of WT PrP in each of the transgenic mouse lines abrogates clinical symptoms and neuropathology. Higher doses of WT PrP are required to rescue more toxic mutations (Biasini et al. 2012a). These data imply that a functional interaction occurs between mutant and WT PrP, either via formation of a complex between the two forms or via competition for binding to a common receptor. ΔN PrP mutants share several biochemical and biological properties with WT PrP, including solubility in detergents, susceptibility to proteases, and specific localization at the cell membrane (Christensen and Harris 2009). Moreover, these molecules are not infectious and are resistant to prion-induced misfolding, as they lack the central hydrophobic domain, known to be required for the conversion of PrP^{C} into PrP^{Sc}. Therefore, ΔN PrPs represent an example of PrP-related toxicity in absence of aggregation or infectivity.

How do ΔN PrPs exert their neurotoxic effects? Important insights into this question have emerged by studying the ΔCR PrP mutant. Expression of this molecule in transfected cells induces the appearance of spontaneous ion channel activity at the cell membrane, a phenomenon that can be detected by patch-clamping techniques (Solomon et al. 2010b, 2011). Cells expressing ΔCR PrP are also hypersensitive to several cationic drugs commonly used for selection of transfected cell lines, including G418 and Zeocin (Massignan et al. 2010b). Importantly, both these interrelated activities are dose-dependently suppressed by co-expression of WT PrP, implying that they are related to a normal physiological activity of PrP^{C}. Two cellular assays based on these phenotypes have recently been used to perform structure–function studies (Solomon et al. 2010a; Massignan et al. 2011; Biasini et al. 2012b). These experiments showed that the toxicity of ΔCR PrP is cell autonomous and depends on localization at the cell membrane and on the presence of a polybasic region at the extreme N-terminus of the protein (residues 23–31) (Solomon et al. 2011; Biasini et al. 2012b). This region determines several other features of PrP^{C}, including endocytic recycling (Shyng et al. 1995), interaction with GAGs (Pan et al. 2002), and modulation of APP processing (Parkin et al. 2007). Strikingly, this region and another positively charged cluster of residues (95–105; just upstream of the CR region) also constitute the two major binding sites on PrP for Aβ oligomers (Laurén et al. 2009; Chen et al. 2010). These observations suggest unexpected connections between the function of PrP^{C}, the toxicity of ΔN mutants, and the ability of the protein to transduce neurotoxic signals delivered by several kinds of aggregated proteins.

10.5 Conclusions

The concept of protein-mediated inheritance, which lies at the heart of the original prion hypothesis, is supported by strong experimental evidence accumulated during the last 30 years. Recent advances in propagating prions in cell-free systems provide hope for the characterization of the precise mechanism of prion replication and the identification of the structural determinants of prion infectivity. Despite substantial advances in understanding prion propagation, the mechanisms underlying the pathogenicity of prions remain obscure. The idea that infectious PrP^{Sc} is also the primary pathogenic form in prion diseases has been questioned. Several alternative, noninfectious forms of PrP have been found to possess neurotoxic properties in cells and mice. There is also increasing evidence that the function of PrP^C, although still uncertain, could be involved in transducing toxic signals delivered by PrP^{Sc}, noninfectious aggregates of PrP, or even misfolded forms of proteins associated with several other neurodegenerative disorders. Clearly, a fundamental step for understanding these phenomena will be to define the physiological activity of PrP^C. It will also be crucial to develop cell-based assays capable of detecting prion toxicity, which will allow the dissection of the neurotoxic pathways activated in prion diseases. Addressing these important challenges will not only illuminate the biology of prions but also provide an unprecedented opportunity for establishing innovative therapeutic approaches for several neurodegenerative disorders.

Acknowledgments This work was supported by grants from the National Institutes of Health (NS052526, NS040975, NS065244, and NS056376) to DAH and by a grant from the Creutzfeldt–Jakob Disease Foundation to EB and DAH.

References

Ashok A, Hegde RS (2009) Selective processing and metabolism of disease-causing mutant prion proteins. PLoS Pathog 5:e1000479

Baumann F et al (2007) Lethal recessive myelin toxicity of prion protein lacking its central domain. EMBO J 26:538–547

Biasini E et al (2008a) Non-infectious aggregates of the prion protein react with several PrPSc-directed antibodies. J Neurochem 105:2190–2204

Biasini E et al (2008b) Multiple biochemical similarities between infectious and non-infectious aggregates of a prion protein carrying an octapeptide insertion. J Neurochem 104:1293–1308

Biasini E et al (2009) Immunopurification of pathological prion protein aggregates. PLoS One 4:e7816

Biasini E et al (2010) The hydrophobic core region governs mutant prion protein aggregation and intracellular retention. Biochem J 430:477–486

Biasini E et al (2012a) Prion protein at the crossroads of physiology and disease. Trends Neurosci 35(2):92–103

Biasini E et al (2012b) The toxicity of a mutant prion protein is cell-autonomous, and can be suppressed by wild-type prion protein on adjacent cells. PLoS One 7:e33472

Brandner S et al (1996) Normal host prion protein necessary for scrapie-induced neurotoxicity. Nature 379:339–343

Campana V et al (2005) The highways and byways of prion protein trafficking. Trends Cell Biol 15:102–111

Chakrabarti O, Hegde RS (2009) Functional depletion of mahogunin by cytosolically exposed prion protein contributes to neurodegeneration. Cell 137:1136–1147

Chen S et al (2010) Interaction between human prion protein and amyloid-beta (Abeta) oligomers: role of N-terminal residues. J Biol Chem 285:26377–26383

Chiesa R, Harris DA (2001) Prion diseases: what is the neurotoxic molecule? Neurobiol Dis 8:743–763

Chiesa R et al (1998) Neurological illness in transgenic mice expressing a prion protein with an insertional mutation. Neuron 21:1339–1351

Chiesa R et al (2000) Accumulation of protease-resistant prion protein (PrP) and apoptosis of cerebellar granule cells in transgenic mice expressing a PrP insertional mutation. Proc Natl Acad Sci USA 97:5574–5579

Chiesa R et al (2003) Molecular distinction between pathogenic and infectious properties of the prion protein. J Virol 77:7611–7622
Chiesa R et al (2008) Aggregated, wild-type prion protein causes neurological dysfunction and synaptic abnormalities. J Neurosci 28:13258–13267
Christensen HM, Harris DA (2009) A deleted prion protein that is neurotoxic *in vivo* is localized normally in cultured cells. J Neurochem 108:44–56
Collinge J (1993) Inherited prion diseases. Adv Neurol 61:155–165
Collinge J (2001) Prion diseases of humans and animals: their causes and molecular basis. Annu Rev Neurosci 24:519–550
Collinge J, Palmer MS (1994) Human prion diseases. Baillieres Clin Neurol 3:241–247
Collins SJ, Masters CL (1995) Transmissibility of Creutzfeldt-Jakob disease and related disorders. Sci Prog 78(Pt 3):217–227
Daude N et al (1997) Identification of intermediate steps in the conversion of a mutant prion protein to a scrapie-like form in cultured cells. J Biol Chem 272:11604–11612
Dossena S et al (2008) Mutant prion protein expression causes motor and memory deficits and abnormal sleep patterns in a transgenic mouse model. Neuron 60:598–609
Drisaldi B et al (2003) Mutant PrP is delayed in its exit from the endoplasmic reticulum, but neither wild-type nor mutant PrP undergoes retrotranslocation prior to proteasomal degradation. J Biol Chem 278:21732–21743
Fioriti L et al (2005) Cytosolic prion protein (PrP) is not toxic in N2a cells and primary neurons expressing pathogenic PrP mutations. J Biol Chem 280:11320–11328
Friedman-Levi Y et al (2011) Fatal prion disease in a mouse model of genetic E200K Creutzfeldt-Jakob disease. PLoS Pathog 7:e1002350
Frost B, Diamond MI (2010) Prion-like mechanisms in neurodegenerative diseases. Nat Rev Neurosci 11:155–159
Harris DA (1999) Cell biological studies of the prion protein. Curr Issues Mol Biol 1:65–75
Hegde RS et al (1998) A transmembrane form of the prion protein in neurodegenerative disease. Science 279:827–834
Hsiao KK et al (1994) Serial transmission in rodents of neurodegeneration from transgenic mice expressing mutant prion protein. Proc Natl Acad Sci USA 91:9126–9130
Ivanova L et al (2001) Mutant prion proteins are partially retained in the endoplasmic reticulum. J Biol Chem 276:42409–42421
Jackson WS et al (2009) Spontaneous generation of prion infectivity in fatal familial insomnia knockin mice. Neuron 63:438–450
Laurén J et al (2009) Cellular prion protein mediates impairment of synaptic plasticity by amyloid-beta oligomers. Nature 457:1128–1132
Lehmann S, Harris DA (1996) Mutant and infectious prion proteins display common biochemical properties in cultured cells. J Biol Chem 271:1633–1637
Li A et al (2007) Neonatal lethality in transgenic mice expressing prion protein with a deletion of residues 105–125. EMBO J 26:548–558
Ma J, Lindquist S (2001) Wild-type PrP and a mutant associated with prion disease are subject to retrograde transport and proteasome degradation. Proc Natl Acad Sci USA 98:14955–14960
Ma J et al (2002) Neurotoxicity and neurodegeneration when PrP accumulates in the cytosol. Science 298:1781–1785
Mallucci G et al (2003) Depleting neuronal PrP in prion infection prevents disease and reverses spongiosis. Science 302:871–874
Mallucci GR et al (2007) Targeting cellular prion protein reverses early cognitive deficits and neurophysiological dysfunction in prion-infected mice. Neuron 53:325–335
Massignan T et al (2010a) Mutant prion protein expression is associated with an alteration of the Rab GDP dissociation inhibitor alpha (GDI)/Rab11 pathway. Mol Cell Proteomics 9:611–622
Massignan T et al (2010b) A novel, drug-based, cellular assay for the activity of neurotoxic mutants of the prion protein. J Biol Chem 285:7752–7765
Massignan T et al (2011) A drug-based cellular assay (DBCA) for studying cytotoxic and cytoprotective activities of the prion protein: a practical guide. Methods 53:214–219

Nunziante M et al (2011) Proteasomal dysfunction and endoplasmic reticulum stress enhance trafficking of prion protein aggregates through the secretory pathway and increase accumulation of pathologic prion protein. J Biol Chem 286:33942–33953
Pan T et al (2002) Cell-surface prion protein interacts with glycosaminoglycans. Biochem J 368:81–90
Parkin ET et al (2007) Cellular prion protein regulates beta-secretase cleavage of the Alzheimer's amyloid precursor protein. Proc Natl Acad Sci USA 104:11062–11067
Prusiner SB (1998) Prions. Proc Natl Acad Sci USA 95:13363–13383
Quaglio E et al (2011) Expression of mutant or cytosolic PrP in transgenic mice and cells is not associated with endoplasmic reticulum stress or proteasome dysfunction. PLoS One 6:e19339
Resenberger UK et al (2011) The cellular prion protein mediates neurotoxic signalling of beta-sheet-rich conformers independent of prion replication. EMBO J 30:2057–2070
Restelli E et al (2010) Cell type-specific neuroprotective activity of untranslocated prion protein. PLoS One 5:e13725
Shmerling D et al (1998) Expression of amino-terminally truncated PrP in the mouse leading to ataxia and specific cerebellar lesions. Cell 93:203–214
Shyng SL et al (1995) Sulfated glycans stimulate endocytosis of the cellular isoform of the prion protein, PrPC, in cultured cells. J Biol Chem 270:30221–30229
Solomon IH et al (2010a) Prion neurotoxicity: insights from prion protein mutants. Curr Issues Mol Biol 12:51–61
Solomon IH et al (2010b) Neurotoxic mutants of the prion protein induce spontaneous ionic currents in cultured cells. J Biol Chem 285:26719–26726
Solomon IH et al (2011) An N-terminal polybasic domain and cell surface localization are required for mutant prion protein toxicity. J Biol Chem 286:14724–14736
Stewart RS, Harris DA (2003) Mutational analysis of topological determinants in prion protein (PrP) and measurement of transmembrane and cytosolic PrP during prion infection. J Biol Chem 278:45960–45968
Telling GC (2011) Transgenic mouse models and prion strains. Top Curr Chem 305:79–99
Tessier PM, Lindquist S (2009) Unraveling infectious structures, strain variants and species barriers for the yeast prion [PSI+]. Nat Struct Mol Biol 16:598–605
Weissmann C (2004) The state of the prion. Nat Rev Microbiol 2:861–871
Westaway D et al (1994) Degeneration of skeletal muscle, peripheral nerves, and the central nervous system in transgenic mice overexpressing wild-type prion proteins. Cell 76:117–129
Wickner RB et al (2011) Prion diseases of yeast: amyloid structure and biology. Semin Cell Dev Biol 22:469–475

Chapter 11
Cellular Mechanisms of Propagation and Clearance

Hermann M. Schatzl

Abstract Here, we focus on the implications of two cellular degradation pathways on prion replication and clearance. The first one is autophagy which can have a promoting and inhibiting role in prion infection. Lysosomal prion clearance can be enhanced in vitro and in vivo by drug-induced activation of autophagy. More recent work revealed that a certain level of autophagy is needed for establishing acute and persistent prion infection, implicating that autophagy might represent a functional equivalent for a disaggregase function. Such one was postulated for seed fragmentation in prion propagation, similar to sonication in PMCA or Hsp104 in yeast prion biology. The second pathway described here is the proteasomal one. We have challenged various cell lines by inducing ER stress or compromising proteasomal activity and analyzed the effects on PrP metabolism strictly in the secretory pathway. Both events led to enhanced detection of PrP aggregates and significant increase of PrP^{Sc} in prion-infected cells, which could be reversed by overexpression of proteins of the cellular quality control. These findings suggest a novel pathway which possibly provides additional substrate and template for prion formation when protein clearance by the proteasome is impaired and point to mechanisms which might play a role in prion de novo generation in sporadic prion diseases.

Keywords Prion propagation • Prion clearance • Autophagy • Lysosomal clearance • Endosomal recycling • Disaggregase • Cellular quality control • ER stress • Proteasomal impairment • Sporadic CJD

H.M. Schatzl, M.D. (✉)
Department of Veterinary Sciences and Department of Molecular Biology, University of Wyoming, 1000 E. University Ave, Laramie, WY 82071, USA
e-mail: hschatzl@uwyo.edu

W.-Q. Zou and P. Gambetti (eds.), *Prions and Diseases: Volume 1, Physiology and Pathophysiology*, DOI 10.1007/978-1-4614-5305-5_11,

11.1 Autophagy in Health and Disease

Degradation and recycling of organelles or cytoplasmic proteins and protein aggregates is mediated by an intracellular bulk degradation process called macroautophagy (referred here as autophagy). The name autophagy originally denotes a cellular self-digestion (self-eating) program, in its simplest form as a single cell's adaptation to starvation. During autophagy, portions of the cytosol are engulfed by a membrane sac resulting in a double-membrane vesicle, called autophagosome, which delivers cytoplasmic cargo to endosomes and lysosomes (Klionsky 2007; Mizushima 2009, 2011). After fusion with lysosomes, the protein and organelle contents of the autophagolysosome are degraded by acidic lysosomal hydrolases and recycled. Beyond its classical role in nutrient supply under starvation and turnover of organelles and proteins, autophagy greatly contributes to various physiological processes such as intracellular cleansing, differentiation, longevity, elimination of invading pathogens, antigen transport to the innate and adaptive immune system, or counteracting endoplasmic reticulum stress (Levine and Kroemer 2008; Mizushima et al. 2008; Meijer and Codogno 2004). Besides its role in physiology, autophagy is also directly implicated in pathophysiology and disease. Autophagy plays a role in cancer, in a number of infectious and inflammatory diseases, and in protein misfolding diseases (Levine and Kroemer 2008; Mizushima et al. 2008; Chu et al. 2009; Batlevi and La Spada 2011; Martinez-Vicente and Cuervo 2007). With respect to the importance of tight regulation of autophagy, perhaps the most fundamental point is that either too little or too much autophagy can be deleterious, a complex balance resulting in its dual role in survival and adaptation or cell death. However, in response to most forms of cellular stress, autophagy plays a protective role. Autophagy has long been defined as a form of non-apoptotic (type II) programmed cell death. However, a consensus is emerging that autophagy might be a cell death impostor which, in reality, functions primarily to promote cellular and organism health (Levine and Kroemer 2008).

Autophagy occurs at basal, constitutive levels in cells. In tissues where cells do not divide, such as neurons and myocytes, basal autophagy is of great relevance (Hara et al. 2006; Komatsu et al. 2006). Several studies suggest a crucial role of autophagy in neurodegenerative diseases, including Alzheimer's disease (AD), Parkinson's disease (PD), tauopathies, and polyglutamine expansion diseases like Huntington's disease (HD) (Nixon 2005; Rubinsztein et al. 2005; Rubinsztein 2006; Ventruti and Cuervo 2007). A number of in vivo studies showed that conventional autophagy knockout mice die during embryogenesis or the neonatal period. Mice with neural-tissue-specific knockouts of these genes survive the postnatal starvation period. However, these mice develop progressive motor deficits and display abnormal reflexes, and ubiquitin-positive inclusion bodies accumulate in their neurons (Hara et al. 2006; Komatsu et al. 2006). Studies showed that the CNS displays only low levels of autophagosomes under normal conditions and even after starvation, but it was also demonstrated that constitutive turnover of cytosolic contents by autophagy is indispensable, even in the absence of expression of any disease-associated mutant proteins (Nixon 2005; Mizushima et al. 2004).

The requirement for autophagy is even more evident under disease conditions, and levels of autophagosomes can be dramatically increased in injured or degenerating neurons. Available data state that autophagy has a beneficial effect of protecting against neurodegeneration. One idea is that autophagy eliminates aggregated and aggregate-prone proteins (Cherra et al. 2010; Jeong et al. 2009; Korolchuk and Rubinsztein 2011; Krainc 2010). Thus, it is reasonable to assume that autophagy could be a therapeutic target for treatment of these neurodegenerative diseases (Rubinsztein 2006). Recent studies underlined that degradation of disease-related mutant proteins is highly dependent on autophagy. It was shown that the clearance of aggregate-prone proteins, such as mutant huntingtin fragments or mutant forms of α-synuclein, respectively, can be mediated by autophagy (Martinez-Vicente and Cuervo 2007; Winslow and Rubinsztein 2008; Chew et al. 2011; Vives-Bauza et al. 2010). Animal models of HD and of other proteinopathies revealed that treatment with rapamycin, a known inducer of autophagy via the mTOR pathway, accelerates the clearance of toxic proteins. Induction of autophagy, mediated by compounds like lithium and trehalose, also has been seen to accelerate the clearance of mutant huntingtin and α-synucleins (Sarkar et al. 2007; Sarkar and Rubinsztein 2008). The beneficial effect of upregulated autophagy has also been described for other diseases associated with aggregate-prone proteins, such as AD and ALS. The possible involvement of autophagy in prion pathogenesis was first described morphologically in form of autophagic vacuoles (Boellaard et al. 1991). These were detected in neurons of prion-infected mice and hamsters and of patients with CJD and other human prion disorders (Liberski et al. 2004, 2008; Sikorska et al. 2004). The appearance of multilamellar bodies and autophagic vacuoles was observed in prion-infected cultured neuronal cells (Schatzl et al. 1997).

11.1.1 Drug-Induced Autophagy Counteracts Prion Infection In Vitro and In Vivo

Our research program over the last years attempted to decipher the molecular and cellular mechanisms which underlie prion diseases (Gilch et al. 2007, 2008; Gilch and Schatzl 2003; Krammer et al. 2009a). Such molecular understanding was then used by us to define and characterize novel molecular targets against prion diseases. One example is our finding that the clearance of prions can be intensified by drug-induced increase of the autophagic flux, both in vitro and in vivo (Aguib et al. 2009; Heiseke et al. 2009, 2010; Ertmer et al. 2004, 2007; Yun et al. 2007). As autophagosomes fuse with the endosomal–lysosomal machinery for final degradation in autophagolysosomes, PrP^{Sc} present in endosomal–lysosomal compartments can be subject to changes in the activity of autophagy (Heiseke et al. 2010). PrP^{Sc}/prions produced within cells only indirectly have access to autophagy pathways. They are neither a direct substrate for autophagy, nor is autophagy in any way specific for prions/PrP^{Sc} and their clearance.

We showed that the c-abl inhibitor imatinib (Gleevec), a drug used to treat chronic myelogenous leukemia, activates the lysosomal degradation of PrP^{Sc} and, at the same time, induces autophagy (Ertmer et al. 2004, 2007). In prion-infected mice, imatinib treatment at an early phase of peripheral infection delayed neuroinvasion and onset of clinical disease (Yun et al. 2007). Since imatinib is not effectively crossing the blood–brain barrier, there was no beneficial effect when the process of neuroinvasion was already completed (Yun et al. 2007). Follow-up work with other chemical inducers of autophagy corroborated these findings. We showed recently that lithium and trehalose enhance the clearance of PrP^{Sc} by induction of autophagy (Aguib et al. 2009; Heiseke et al. 2009). To demonstrate that indeed induction of autophagy is the underlying mechanism, we inhibited autophagy by pharmacological interference or siRNA gene silencing of essential members of the autophagy machinery. Such co-treatment impaired or antagonized the capacity of compound-induced autophagy in reducing cellular levels of PrP^{Sc}. Besides compounds inducing autophagy in an mTor-independent manner (e.g., lithium, trehalose), we studied rapamycin, a drug widely used to activate autophagy by inhibiting mTor. Rapamycin also reduced PrP^{Sc}, showing that both autophagy-inducing pathways, mTor-dependent and mTor-independent, can be involved in the degradation of PrP^{Sc} (Heiseke et al. 2009, 2010).

To test whether autophagy-inducing compounds are candidates for therapeutic approaches against prion infection, we treated intraperitoneally prion-infected mice. Oral rapamycin treatment of prion-infected mice initiated in the last third of incubation time, mimicking a preclinical therapeutic situation, showed a significant prolongation of prion incubation times as compared to mock-treated control mice (Heiseke et al. 2009). Similar findings were obtained with lithium, although less uniform (Heiseke et al. 2009). Trehalose treatment did not prolong incubation times, but showed effects on PrP^{Sc} levels in spleen, and depending on when treatment was started, the peripheral accumulation of PrP^{Sc} was delayed (Aguib et al. 2009). As was the case with imatinib (Yun et al. 2007), this reflects that the process of neuroinvasion was decelerated. Taken together, these in vivo studies strongly indicate that autophagy-inducing compounds are beneficial in prion disease scenarios and ask for further studies, including also combination of drugs (Heiseke et al. 2010).

Before doing so, we studied another compound which has a promising bioavailability profile: the selective estrogen receptor modulator tamoxifen, a widely used anticancer drug (Heel et al. 1978). In fact, tamoxifen was the most potent enhancer of PrP^{Sc} clearance in our recent studies. We assessed its mode of action and found that tamoxifen treatment leads to robust clearance of prion infection after only a few days. Attenuation of the autophagic pathway (e.g., knockdown of Beclin-1 or Atg5) antagonized these effects. Time kinetics experiments showed that induction of autophagy is also of importance after prions are taken up by the cell, restricting the accumulation of intracellular aggregated prion protein. In vivo, tamoxifen treatment was able to reduce the PrP^{Sc} load in spleens, prolonged survival in infected animals, and led to reduced microgliosis in brain tissue of treated mice (Fig. 11.1).

Taken together, our data convincingly show that autophagy is a potent modifier of the cellular clearance of prions and that chemically induced autophagy shifts the delicate equilibrium between propagation and clearance of prions towards the latter.

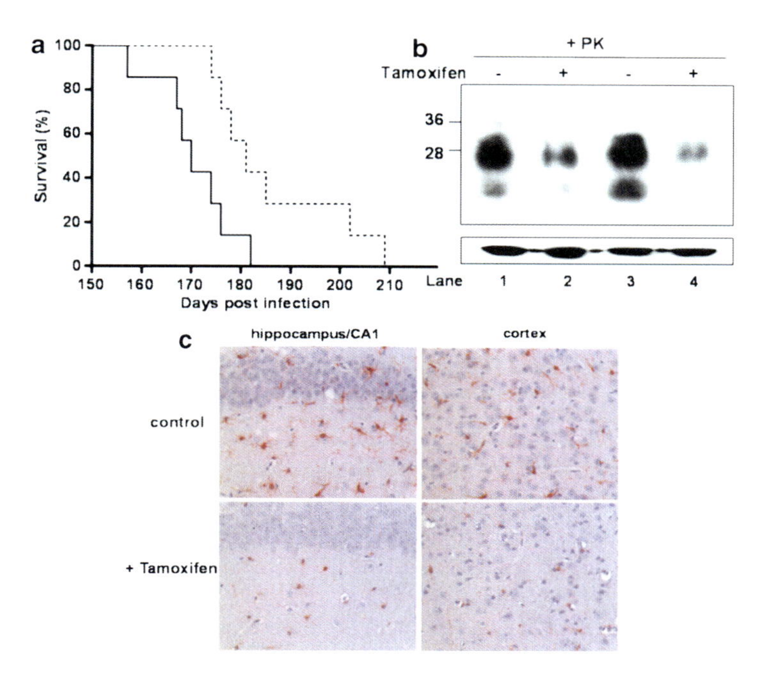

Fig. 11.1 Effect of tamoxifen in in vivo prion infection. (**a**) Prolonged survival times in tamoxifen-treated mice. Oral treatment with tamoxifen initiated at day 100 post intracerebral infection with prion strain 139A. Solid line depicts control mice (mean 170.6 ± 7.9 days) and broken tamoxifen-treated mice (mean 186.4 ± 13.6 days, $^{*}p>0.01$); $n=7$. (**b**) Reduced PrP^{Sc} level in spleens. Immunoblot detection of PrP^{Sc} in spleens of mock- (*lanes 1, 3*) and tamoxifen-treated (*lanes 2, 4*) mice at terminal time points. (**c**) Reduced microgliosis in brain tissue of tamoxifen-treated mice. Immunohistochemical analysis of disease-associated microgliosis via detection of ionized calcium binding adapter molecule 1 (Iba1) in hippocampus (*left row*) and cortex (*right row*) of treated (*lower row*) and nontreated (*upper row*) mice at 125 dpi. Representative sections are shown in a magnification of ×400

11.1.2 A Second Role for Autophagy in Prion Infection: Basal Autophagy Is Required for Establishing Prion Infection and Might Provide Disaggregase Function

Our further studies focused on a novel biological function of autophagy recently found by us: Basal (i.e., normal, non-induced) autophagy is required for cellular prion propagation. We believe that autophagy compartments function as a biological "disaggregase" providing fragmentation activity which is postulated in protein aggregation/disaggregation (Borchsenius et al. 2006; Shorter 2011). PMCA technology uses physical disintegration to accomplish this task (Castilla et al. 2005). In yeast prion biology, Hsp104, for which no mammalian homologue is known, mainly fulfills this part (Shorter 2008, 2011; Chernoff et al. 1995; Lindquist et al. 1995). The molecular characterization of this finding might be of significance also for other diseases involving prion-like mechanisms. For no neurodegenerative disease, it is understood how aggregates build up in the cell and exit and enter neighboring cells, starting the cycle there again (Brundin et al. 2010; Kaganovich et al. 2008; Ren et al. 2009; Krammer et al. 2009b).

Our initial finding which brought us to look into this direction was that autophagy is transiently induced when cells start propagating PrP^{Sc}, implicating a more general role of autophagy in prion conversion. We infected neuronal cells with different known susceptibilities to primary prion infection and analyzed whether levels of autophagy were modulated. The recipient cells harbored a prion protein tagged with an epitope for mAb 3F4 (Maas et al. 2007). Therefore, only newly synthesized PrP^{Sc} was detected and discriminated from PrP^{Sc} in inocula. Inoculated cells were lysed at various time points postinfection and analyzed both for newly converted PrP^{Sc} and LC3-II levels. Upon induction of autophagy, posttranslationally processed LC3 (LC3-I) is converted into LC3-II. An increase in the level of LC3-II is commonly used as marker for autophagy induction, as the amount of LC3-II associated with autophagosome membranes correlates with the extent of autophagosome formation (Klionsky et al. 2008). In comparison to mock-brain infection, increased amounts of LC3-II were detected in prion-susceptible cell populations upon prion inoculation. This phenomenon was observed concomitant with the ability of cells to propagate PrP^{Sc} in detectable levels. When primary prion infection manifested in cells, the increased level of LC3-II went back to levels as observed in controls. Similar results were found for the medium susceptible clone, which started propagating PrP^{Sc} at a later time point. This phenomenon was lacking in prion-unsusceptible cells, indicating that autophagosome formation is transiently induced only in cells actively propagating PrP^{Sc} and is not the result of a cellular response to PrP^{Sc} in inocula.

We then wanted to further analyze whether basal levels of autophagy indeed play a role in primary prion infection. We inoculated wild-type mouse embryonic fibroblasts (MEFwt) and autophagy-deficient MEFs ($MEFATG5^{-/-}$), originating from $ATG5^{-/-}$ transgenic mice, with prion-infected brain homogenate. Since these cells did not contain a 3F4-tagged PrP, we waited until day 20 postinfection to be sure that we do not detect PrP^{Sc} from inocula. Interestingly, $MEFATG5^{-/-}$ cells showed only very weak amounts of PrP^{Sc} at days 20 and 30 p.i., whereas MEFwt cells propagated PrP^{Sc} very efficiently. In addition, when we used siRNA-targeting Atg5 or beclin-1 genes, both genes necessary for execution of autophagy, around the time of primary prion infection of neuronal cells, we also observed a reduction in PrP^{Sc} propagation compared to mock-treated cells. This data indicated that the absence of autophagy very strongly decreased the cellular susceptibility to primary prion infection and that autophagy potentially also plays a role in maintenance of productive prion infection over time.

To rule out the possibility that the difference in prion susceptibility between $Atg5^{-/-}$ and $^{+/+}$ cells was not based on ATG5 alone and might depend on cell clone issues, we decided to reintroduce the Atg5 gene into $ATG5^{-/-}$ cells. $MEFATG5^{-/-}$ cells were stably transduced with a lentivirus construct-encoding ATG5 to restore autophagy competence (cells termed MEFATG5). As done before, autophagy-competent and autophagy-deficient cells were inoculated with 22 L prion- or mock-infected brain homogenates in parallel experiments. Reintroduction of autophagy competence in MEFATG5 cells provoked a clearly increased susceptibility to primary prion infection as compared to autophagy-deficient counterparts. Interestingly, analysis of prion-infected cells at a later time point postinfection (55 days postinfection, dpi) revealed that the reduced PrP^{Sc} level in autophagy-deficient cells is not a transient phenomenon and that lack of autophagy may even result in abrogation of cellular prion infection. In contrast, autophagy-competent cells efficiently propagated PrP^{Sc} at 55 dpi, reflecting persistent prion infection. We also quantitatively addressed this phenomenon. We stepwise increased autophagy competence by transducing $MEFATG5^{-/-}$ cells with different dilutions of ATG5-encoding virus. In contrast to autophagy-deficient cells, gradually restored autophagy competence resulted in accordingly elevated levels in prion infection and PrP^{Sc} propagation, both after 20 and 30 dpi. These results indicate that restored autophagy competence renders cells more prone to PrP^{Sc} propagation, validating a pivotal role of functional active basal autophagy in primary and persistent prion infection.

We hypothesize that autophagy plays a general role in the subcellular recycling of prions. The previous view that PrP^{Sc} is generated along the early endocytic pathway and is unidirectional transported to lysosomes for final degradation is not compatible anymore with recent findings (Beranger et al. 2002; Marijanovic et al. 2009; Gilch et al. 2009; Yamasaki et al. 2012). Without obtaining new prions steadily from the outside of the cell, such a unidirectional mechanism is rather incompatible with persistent prion propagation in terminally differentiated cells. It is likely that a fraction of PrP^{Sc} is retro-transported to the more upstream locale of prion conversion, thereby allowing continuous flow of prion generation (see Fig. 11.2). Experimental evidence for such a scenario was reported by our and the Lehmann, Zurzolo, and Horiuchi groups (Beranger et al. 2002; Marijanovic et al. 2009; Gilch et al. 2009; Yamasaki et al. 2012). We postulate that autophagic flux mechanisms play a role in this scenario; we even hypothesize that the level of autophagic flux represents a crossing point which decides whether PrP^{Sc} gets recycled or degraded. Increasing the autophagic flux counteracts the recycling pathway and thereby also affects prion conversion, taking away template for conversion and inducing its degradation (Fig. 11.2). We postulate that autophagic compartments provide disaggregase function and increase seeds as needed or are at least supportive for efficient prion propagation. When late endosomes containing PrP^{Sc} aggregates fuse with autophagosomes, the autophagic machinery and the cellular locale containing PrP^{Sc}/prions physically meet and get interconnected. How the endosomal recycling compartment (ESCRT) machinery is involved in completion of autophagy is presently subject of intensive research (Raiborg and Stenmark 2009; Rusten and Stenmark 2009; Rusten and Simonsen 2008). A general view is that the ESCRT machinery is required in fusion of autophagosomes with endosomes and lysosomes (Raiborg and Stenmark 2009; Rusten and Stenmark 2009). Two rab proteins have been found involved in this process (Rab11 and Rab7) (Rusten and Simonsen 2008; Fader and Colombo 2009). Interestingly, work from the Zurzolo laboratory has identified the ERC as a likely site of prion conversion (Marijanovic et al. 2009).

Is such a dual function of autophagy conceivable? It is known that basal autophagy and moderately enhanced autophagy help in cell survival but that both impairment

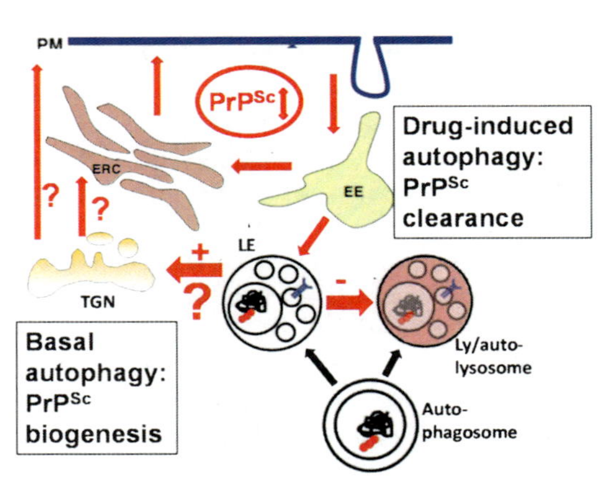

Fig. 11.2 Model for how the level of autophagy impacts prion infection. Exogenously elevated levels of autophagy (e.g., by chemical compounds) affect prion clearance and thereby prion infection. When the autophagic flux increases more autophagosomes fuse with late endosomes (LL, in some cell types also known as multivesicular bodies, MVBs) which contain PrP^{Sc}/prions to form amphisomes (not shown). This results in increased fusion to lysosomes (Ly) and steadily sequesters the template for prion conversion. At the same time, the increase in prion clearance negatively affects the role of autophagy in prion propagation (*left side*). We hypothesize that PrP^{Sc}/prions recycle from early endosomes (EE) directly or indirectly (via LE/TGN) to the ERC (endosomal recycling complex) and from there back towards the cellular compartment of prion conversion. When this goes over LE, the autophagic machinery gets connected with compartments recycling PrP^{Sc}/prions and can provide disaggregation activity to them. Based on our new findings and data as published by us and others (Beranger et al. 2002; Marijanovic et al. 2009; Gilch et al. 2009; Yamasaki et al. 2011)

of autophagy or strongly enhanced autophagy can lead to cell death (Martinez-Vicente and Cuervo 2007; Ventruti and Cuervo 2007; Ertmer et al. 2004; Cuervo et al. 2010; Wong and Cuervo 2010). We postulate that components of the autophagic flux might represent the biological equivalent for the postulated disaggregase activity in mammalian prion and prion-like biology. Whereas PMCA uses physical disintegration, for yeast prions this is done by Hsp104. Interestingly, Hsp104 also has a dual role, and depending on the level of activity, it is involved in aggregation and disaggregation (Shorter and Lindquist 2004). Having a system in hand which provides PrP^{Sc} originating from cells with normal and impaired autophagy, we can study now how this affects molecular, biophysical, and infectivity features of prions.

11.2 Effect of Proteasome Dysfunction and ER Stress on PrP^{Sc} Biogenesis

We recently found that proteasomal impairment and ER stress can have a direct impact on the level and quality of PrP^{c} in the secretory pathway and its "fitness" for substrate in prion conversion. This finding is different from previous findings which focused on PrP moieties in the cytosol (Ma et al. 2002; Ma and Lindquist 2001, 2002; Kristiansen et al. 2005, 2007; Cohen and Taraboulos 2003). Our data suggest a novel pathway which contributes to "conventional" prion propagation. Such conversion favoring or disfavoring cellular conditions might also be of relevance for the pathogenesis of sporadic CJD, where initial conversion might take place without a bona fide PrP^{Sc} template. Improving protein quality in ER and post-ER compartments in trans in order to generate PrP^{c} populations which have a more stable conformation and/or are less efficiently converted into PrP^{Sc} might provide translational potential (Nunziante et al. 2011).

Proteasomal dysfunction and ER stress enhance trafficking of prion protein aggregates through the secretory pathway and increase PrP^{Sc}.

In infectious forms of prion diseases, a direct interaction between PrP^{Sc} template and PrP^{c} substrate underlies the conformational change of PrP^{c} into PrP^{Sc} (Prusiner 1998). It is assumed that a preceding plasma membrane localization of PrP^{c} is mandatory for conversion into PrP^{Sc} (Prusiner 2001; Caughey and Raymond 1991; Caughey et al. 1998; Borchelt et al. 1992). Much less is known in this respect about events occurring in ER and in early secretory compartments. The cellular mechanisms underlying sporadic prion diseases are mostly unknown and are difficult to assess in experimental systems. Various models propose the existence of a PrP isoform which is more prone to conversion into PrP^{Sc} (Billeter et al. 1997; Glockshuber 2001; Hornemann and Glockshuber 1998). The fundamental role of the ER environment and of the ER-associated degradation pathway (ERAD) in metabolism and turnover of wild-type and some mutant prion proteins has been highlighted in the past with regard to implications for prion diseases (Rogers et al. 1990; Yedidia et al. 2001; Lorenz et al. 2002; Drisaldi et al. 2003). Whereas work done by other groups mainly focused on aberrant PrP moieties in the cytosol or in aggresomes and its possible impact in execution of neurodegeneration (Ma et al. 2002; Ma and Lindquist 2002; Kristiansen et al. 2005, 2007; Yedidia et al. 2001), the aim of our study was to investigate how perturbations of ER homeostasis or proteasomal impairment affect PrP^{c} metabolism in the secretory pathway and thereby directly PrP^{Sc} biogenesis.

We found that induction of ER stress resulted in a general attenuation of PrP^{c} level (Nunziante et al. 2011). In addition, we found aggregated PrP species that localized mainly in secretory compartments and at the cell surface. Inhibition of proteasomal function led to a significant increase of the total PrP^{c} level and to accumulation of detergent soluble and insoluble PrP^{c} isoforms. PrP species detected under these conditions were fully glycosylated, were properly processed through the secretory pathway, and localized at the outer leaflet of the plasma membrane. This was the case in cells with endogenous PrP expression, in primary neurons as well as in PrP-transfected cells. The majority of studies conducted on proteasomal degradation of PrP describe cytosolic accumulation of toxic PrP aggregates upon inhibition of this pathway. Although not extensively investigated for PrP metabolism before, it was assumed for other proteins that ER and quality control compartments are connected to the secretory pathway. In our hands, experimental manipulation of both pathways led to accumulation of insoluble PrP species in the secretory pathway, but the events underlying their formation seemed to be different, as were the effects on PrP^{c} localization and expression.

Strikingly, inhibition of proteasomal activity amplified PrP^{Sc} levels in persistently prion-infected cells. The direct correlation between proteasome and PrP^{Sc} accumulation within cells represents a new aspect in prion metabolism. Previous studies reported formation of cytoplasmic PrP^{Sc} aggregates which associated with aggresomes and led to apoptotic death in prion-infected neurons, but only after mild inhibition of the proteasome (Kristiansen et al. 2005, 2007). In addition, purified PrP^{Sc} preparations were seen to inhibit the proteolytic activity of the proteasome (Kristiansen et al. 2007). These data support the view of a cytosolic localization for portions of PrP^{Sc} either by retro-translocation or by endosomal–lysosomal membrane destabilization (Laszlo et al. 1992). In our study, upon proteasomal inhibition, PrP^{c} and detergent-insoluble aggregates were extensively transported to the cell surface, one of the putative sites for prion formation. Such PrP molecules might represent additional substrate binding to existing PrP^{Sc} seeds and leading to the increased formation of PrP^{Sc} as detected in our study.

We further underlined the fundamental role of the early secretory pathway in folding and transport of PrP^{c} with respect to prion formation by overexpressing molecules known to promote cellular quality. Overexpression of EDEM-3 or ERGIC-53 significantly reduced PrP aggregates and PrP^{Sc} in infected cells. EDEM proteins are ER-resident lectins which recognize *N*-linked glycans on aberrantly folded proteins, accelerate their release from the calnexin/calreticulin cycle, and sort them for ERAD-degradation (Molinari et al. 2003; Oda et al. 2003; Ruddock and Molinari 2006). It is therefore plausible that by enhancing ERAD-degradation of PrP aggregates, EDEM-3 subtracts the substrate necessary for prion conversion. A similar explanation for reduction of PrP^{Sc} could apply to ERGIC-53, which selectively transports functionally folded proteins from ER to ERGIC vesicles and also operates in the quality control of glycoproteins (Appenzeller et al. 1999). ERGIC-53 might therefore promote proper folding of PrP^{c} and selectively transport this cargo to the cell surface. This PrP^{c} population would have a more stable conformation and/or be less efficiently converted into PrP^{Sc}.

Taken together, our data support the notion that ER and cellular quality control mechanisms tightly modulate PrP maturation and PrP^{Sc} formation. We show that proteasomal degradation and ERAD play a physiological role for endogenous PrP^{c} in the secretory pathway. Impairments in this pathway as well as disturbances in ER homeostasis cause accumulation of PrP aggregates which are increasingly recycled through the secretory pathway, resulting in enhanced PrP^{Sc} replication. Of note, such conversion favoring or disfavoring cellular conditions might also be of relevance for the pathogenesis of sporadic CJD, where initial conversion might take place without a bona fide PrP^{Sc} template.

References

Aguib Y, Heiseke A, Gilch S, Riemer C, Baier M, Schatzl HM, Ertmer A (2009) Autophagy induction by trehalose counteracts cellular prion infection. Autophagy 5:361–369

Appenzeller C, Andersson H, Kappeler F, Hauri HP (1999) The lectin ERGIC-53 is a cargo transport receptor for glycoproteins. Nat Cell Biol 1:330–334

Batlevi Y, La Spada AR (2011) Mitochondrial autophagy in neural function, neurodegenerative disease, neuron cell death, and aging. Neurobiol Dis 43:46–51
Beranger F, Mange A, Goud B, Lehmann S (2002) Stimulation of PrP(C) retrograde transport toward the endoplasmic reticulum increases accumulation of PrP(Sc) in prion-infected cells. J Biol Chem 277:38972–38977
Billeter M, Riek R, Wider G, Hornemann S, Glockshuber R, Wuthrich K (1997) Prion protein NMR structure and species barrier for prion diseases. Proc Natl Acad Sci USA 94:7281–7285
Boellaard JW, Kao M, Schlote W, Diringer H (1991) Neuronal autophagy in experimental scrapie. Acta Neuropathol 82:225–228
Borchelt DR, Taraboulos A, Prusiner SB (1992) Evidence for synthesis of scrapie prion proteins in the endocytic pathway. J Biol Chem 267:16188–16199
Borchsenius AS, Muller S, Newnam GP, Inge-Vechtomov SG, Chernoff YO (2006) Prion variant maintained only at high levels of the Hsp104 disaggregase. Curr Genet 49:21–29
Brundin P, Melki R, Kopito R (2010) Prion-like transmission of protein aggregates in neurodegenerative diseases. Nat Rev Mol Cell Biol 11:301–307
Castilla J, Saa P, Hetz C, Soto C (2005) In vitro generation of infectious scrapie prions. Cell 121:195–206
Caughey B, Raymond GJ (1991) The scrapie-associated form of PrP is made from a cell surface precursor that is both protease- and phospholipase-sensitive. J Biol Chem 266:18217–18223
Caughey B, Race RE, Ernst D, Buchmeier MJ, Chesebro B (1998) Prion protein biosynthesis in scrapie-infected and uninfected neuroblastoma cells. J Virol 63:175–181
Chernoff YO, Lindquist SL, Ono B, Inge-Vechtomov SG, Liebman SW (1995) Role of the chaperone protein Hsp104 in propagation of the yeast prion-like factor [psi+]. Science 268:880–884
Cherra SJ III, Dagda RK, Chu CT (2010) Review: autophagy and neurodegeneration: survival at a cost? Neuropathol Appl Neurobiol 36:125–132
Chew KC, Ang ET, Tai YK, Tsang F, Lo SQ, Ong E, Ong WY, Shen HM, Lim KL, Dawson VL, Dawson TM, Soong TW (2011) Enhanced autophagy from chronic toxicity of iron and mutant A53T alpha-synuclein: implications for neuronal cell death in Parkinson disease. J Biol Chem 286:33380–33389
Chu CT, Plowey ED, Dagda RK, Hickey RW, Cherra SJ III, Clark RS (2009) Autophagy in neurite injury and neurodegeneration: in vitro and in vivo models. Methods Enzymol 453:217–249
Cohen E, Taraboulos A (2003) Scrapie-like prion protein accumulates in aggresomes of cyclosporin A-treated cells. EMBO J 22:404–417
Cuervo AM, Wong ES, Martinez-Vicente M (2010) Protein degradation, aggregation, and misfolding. Mov Disord 25(Suppl 1):S49–S54
Drisaldi B, Stewart RS, Adles C, Stewart LR, Quaglio E, Biasini E, Fioriti L, Chiesa R, Harris DA (2003) Mutant PrP is delayed in its exit from the endoplasmic reticulum, but neither wild-type nor mutant PrP undergoes retrotranslocation prior to proteasomal degradation. J Biol Chem 278:21732–21743
Ertmer A, Gilch S, Yun SW, Flechsig E, Klebl B, Stein-Gerlach M, Klein MA, Schatzl HM (2004) The tyrosine kinase inhibitor STI571 induces cellular clearance of PrPSc in prion-infected cells. J Biol Chem 279:41918–41927
Ertmer A, Huber V, Gilch S, Yoshimori T, Erfle V, Duyster J, Elsasser HP, Schatzl HM (2007) The anticancer drug imatinib induces cellular autophagy. Leukemia 21:936–942
Fader CM, Colombo MI (2009) Autophagy and multivesicular bodies: two closely related partners. Cell Death Differ 16:70–78
Gilch S, Schatzl HM (2003) Promising developments bringing prion diseases closer to therapy and prophylaxis. Trends Mol Med 9:367–369
Gilch S, Nunziante M, Ertmer A, Schatzl HM (2007) Strategies for eliminating PrP(c) as substrate for prion conversion and for enhancing PrP(Sc) degradation. Vet Microbiol 123:377–386
Gilch S, Krammer C, Schatzl HM (2008) Targeting prion proteins in neurodegenerative disease. Expert Opin Biol Ther 8:923–940
Gilch S, Bach C, Lutzny G, Vorberg I, Schatzl HM (2009) Inhibition of cholesterol recycling impairs cellular PrP(Sc) propagation. Cell Mol Life Sci 66:3979–3991

Glockshuber R (2001) Folding dynamics and energetics of recombinant prion proteins. Adv Protein Chem 57:83–105
Hara T, Nakamura K, Matsui M, Yamamoto A, Nakahara Y, Suzuki-Migishima R, Yokoyama M, Mishima K, Saito I, Okano H, Mizushima N (2006) Suppression of basal autophagy in neural cells causes neurodegenerative disease in mice. Nature 441:885–889
Heel RC, Brogden RN, Speight TM, Avery GS (1978) Tamoxifen - review of its pharmacological properties and therapeutic use in treatment of breast-cancer. Drugs 16:1–24
Heiseke A, Aguib Y, Riemer C, Baier M, Schatzl HM (2009) Lithium induces clearance of protease resistant prion protein in prion-infected cells by induction of autophagy. J Neurochem 109:25–34
Heiseke A, Aguib Y, Schatzl HM (2010) Autophagy, prion infection and their mutual interactions. Curr Issues Mol Biol 12:87–98
Hornemann S, Glockshuber R (1998) A scrapie-like unfolding intermediate of the prion protein domain PrP(121–231) induced by acidic pH. Proc Natl Acad Sci USA 95:6010–6014
Jeong H, Then F, Melia TJ, Mazzulli JR, Cui L, Savas JN, Voisine C, Paganetti P, Tanese N, Hart AC, Yamamoto A, Krainc D (2009) Acetylation targets mutant huntingtin to autophagosomes for degradation. Cell 137:60–72
Kaganovich D, Kopito R, Frydman J (2008) Misfolded proteins partition between two distinct quality control compartments. Nature 454:1088–1095
Klionsky DJ (2007) Autophagy: from phenomenology to molecular understanding in less than a decade. Nat Rev Mol Cell Biol 8:931–937
Klionsky DJ et al (2008) Guidelines for the use and interpretation of assays for monitoring autophagy in higher eukaryotes. Autophagy 4:151–175
Komatsu M, Waguri S, Chiba T, Murata S, Iwata J, Tanida I, Ueno T, Koike M, Uchiyama Y, Kominami E, Tanaka K (2006) Loss of autophagy in the central nervous system causes neurodegeneration in mice. Nature 441:880–884
Korolchuk VI, Rubinsztein DC (2011) Regulation of autophagy by lysosomal positioning. Autophagy 7:927–928
Krainc D (2010) Clearance of mutant proteins as a therapeutic target in neurodegenerative diseases. Arch Neurol 67:388–392
Krammer C, Vorberg I, Schatzl HM, Gilch S (2009a) Therapy in prion diseases: from molecular and cellular biology to therapeutic targets. Infect Disord Drug Targets 9:3–14
Krammer C, Kryndushkin D, Suhre MH, Kremmer E, Hofmann A, Pfeifer A, Scheibel T, Wickner RB, Schatzl HM, Vorberg I (2009b) The yeast Sup35NM domain propagates as a prion in mammalian cells. Proc Natl Acad Sci USA 106:462–467
Kristiansen M, Messenger MJ, Klohn PC, Brandner S, Wadsworth JD, Collinge J, Tabrizi SJ (2005) Disease-related prion protein forms aggresomes in neuronal cells leading to caspase activation and apoptosis. J Biol Chem 280:38851–38861
Kristiansen M, Deriziotis P, Dimcheff DE, Jackson GS, Ovaa H, Naumann H, Clarke AR, van Leeuwen FW, Menendez-Benito V, Dantuma NP, Portis JL, Collinge J, Tabrizi SJ (2007) Disease-associated prion protein oligomers inhibit the 26S proteasome. Mol Cell 26:175–188
Laszlo L, Lowe J, Self T, Kenward N, Landon M, McBride T, Farquhar C, McConnell I, Brown J, Hope J (1992) Lysosomes as key organelles in the pathogenesis of prion encephalopathies. J Pathol 166:333–341
Levine B, Kroemer G (2008) Autophagy in the pathogenesis of disease. Cell 132:27–42
Liberski PP, Sikorska B, Bratosiewicz-Wasik J, Gajdusek DC, Brown P (2004) Neuronal cell death in transmissible spongiform encephalopathies (prion diseases) revisited: from apoptosis to autophagy. Int J Biochem Cell Biol 36:2473–2490
Liberski PP, Brown DR, Sikorska B, Caughey B, Brown P (2008) Cell death and autophagy in prion diseases (transmissible spongiform encephalopathies). Folia Neuropathol 46:1–25
Lindquist S, Patino MM, Chernoff YO, Kowal AS, Singer MA, Liebman SW, Lee KH, Blake T (1995) The role of Hsp104 in stress tolerance and [PSI+] propagation in Saccharomyces cerevisiae. Cold Spring Harb Symp Quant Biol 60:451–460

Lorenz H, Windl O, Kretzschmar HA (2002) Cellular phenotyping of secretory and nuclear prion proteins associated with inherited prion diseases. J Biol Chem 277:8508–8516
Ma J, Lindquist S (2001) Wild-type PrP and a mutant associated with prion disease are subject to retrograde transport and proteasome degradation. Proc Natl Acad Sci USA 98: 14955–14960
Ma J, Lindquist S (2002) Conversion of PrP to a self-perpetuating PrPSc-like conformation in the cytosol. Science 298:1785–1788
Ma J, Wollmann R, Lindquist S (2002) Neurotoxicity and neurodegeneration when PrP accumulates in the cytosol. Science 298:1781–1785
Maas E, Geissen M, Groschup MH, Rost R, Onodera T, Schatzl H, Vorberg IM (2007) Scrapie infection of prion protein-deficient cell line upon ectopic expression of mutant prion proteins. J Biol Chem 282:18702–18710
Marijanovic Z, Caputo A, Campana V, Zurzolo C (2009) Identification of an intracellular site of prion conversion. PLoS Pathog 5:e1000426
Martinez-Vicente M, Cuervo AM (2007) Autophagy and neurodegeneration: when the cleaning crew goes on strike. Lancet Neurol 6:352–361
Meijer AJ, Codogno P (2004) Regulation and role of autophagy in mammalian cells. Int J Biochem Cell Biol 36:2445–2462
Mizushima N (2009) Regulation of autophagosome formation in mammalian cells. Autophagy 5:898–899
Mizushima N (2011) Autophagy in protein and organelle turnover. Cold Spring Harb Symp Quant Biol 76:397–402
Mizushima N, Yamamoto A, Matsui M, Yoshimori T, Ohsumi Y (2004) In vivo analysis of autophagy in response to nutrient starvation using transgenic mice expressing a fluorescent autophagosome marker. Mol Biol Cell 15:1101–1111
Mizushima N, Levine B, Cuervo AM, Klionsky DJ (2008) Autophagy fights disease through cellular self-digestion. Nature 451:1069–1075
Molinari M, Calanca V, Galli C, Lucca P, Paganetti P (2003) Role of EDEM in the release of misfolded glycoproteins from the calnexin cycle. Science 299:1397–1400
Nixon RA (2005) Endosome function and dysfunction in Alzheimer's disease and other neurodegenerative diseases. Neurobiol Aging 26:373–382
Nunziante M, Ackermann K, Dietrich K, Wolf H, Gadtke L, Gilch S, Vorberg I, Groschup M, Schatzl HM (2011) Proteasomal dysfunction and endoplasmic reticulum stress enhance trafficking of prion protein aggregates through the secretory pathway and increase accumulation of pathologic prion protein. J Biol Chem 286:33942–33953
Oda Y, Hosokawa N, Wada I, Nagata K (2003) EDEM as an acceptor of terminally misfolded glycoproteins released from calnexin. Science 299:1394–1397
Prusiner SB (1998) Prions. Proc Natl Acad Sci USA 95:13363–13383
Prusiner SB (2001) Shattuck lecture–neurodegenerative diseases and prions. N Engl J Med 344:1516–1526
Raiborg C, Stenmark H (2009) The ESCRT machinery in endosomal sorting of ubiquitylated membrane proteins. Nature 458:445–452
Ren PH, Lauckner JE, Kachirskaia I, Heuser JE, Melki R, Kopito RR (2009) Cytoplasmic penetration and persistent infection of mammalian cells by polyglutamine aggregates. Nat Cell Biol 11:219–225
Rogers M, Taraboulos A, Scott M, Groth D, Prusiner SB (1990) Intracellular accumulation of the cellular prion protein after mutagenesis of its Asn-linked glycosylation sites. Glycobiology 1:101–109
Rubinsztein DC (2006) The roles of intracellular protein-degradation pathways in neurodegeneration. Nature 443:780–786
Rubinsztein DC, DiFiglia M, Heintz N, Nixon RA, Qin ZH, Ravikumar B, Stefanis L, Tolkovsky A (2005) Autophagy and its possible roles in nervous system diseases, damage and repair. Autophagy 1:11–22

Ruddock LW, Molinari M (2006) N-glycan processing in ER quality control. J Cell Sci 119:4373–4380
Rusten TE, Simonsen A (2008) ESCRT functions in autophagy and associated disease. Cell Cycle 7:1166–1172
Rusten TE, Stenmark H (2009) How do ESCRT proteins control autophagy? J Cell Sci 122:2179–2183
Sarkar S, Rubinsztein DC (2008) Small molecule enhancers of autophagy for neurodegenerative diseases. Mol Biosyst 4:895–901
Sarkar S, Davies JE, Huang ZB, Tunnacliffe A, Rubinsztein DC (2007) Trehalose, a novel mTOR-independent autophagy enhancer, accelerates the clearance of mutant huntingtin and alpha-synuclein. J Biol Chem 282:5641–5652
Schatzl HM, Laszlo L, Holtzman DM, Tatzelt J, Dearmond SJ, Weiner RI, Mobley WC, Prusiner SB (1997) A hypothalamic neuronal cell line persistently infected with scrapie prions exhibits apoptosis. J Virol 71:8821–8831
Shorter J (2008) Hsp104: a weapon to combat diverse neurodegenerative disorders. Neurosignals 16:63–74
Shorter J (2011) The mammalian disaggregase machinery: hsp110 synergizes with hsp70 and hsp40 to catalyze protein disaggregation and reactivation in a cell-free system. PLoS One 6:e26319
Shorter J, Lindquist S (2004) Hsp104 catalyzes formation and elimination of self-replicating Sup35 prion conformers. Science 304:1793–1797
Sikorska B, Liberski PP, Giraud P, Kopp N, Brown P (2004) Autophagy is a part of ultrastructural synaptic pathology in Creutzfeldt–Jakob disease: a brain biopsy study. Int J Biochem Cell Biol 36:2563–2573
Ventruti A, Cuervo AM (2007) Autophagy and neurodegeneration. Curr Neurol Neurosci Rep 7:443–451
Vives-Bauza C, Zhou C, Huang Y, Cui M, de Vries RL, Kim J, May J, Tocilescu MA, Liu W, Ko HS, Magrane J, Moore DJ, Dawson VL, Grailhe R, Dawson TM, Li C, Tieu K, Przedborski S (2010) PINK1-dependent recruitment of Parkin to mitochondria in mitophagy. Proc Natl Acad Sci USA 107:378–383
Winslow AR, Rubinsztein DC (2008) Autophagy in neurodegeneration and development. Biochim Biophys Acta 1782:723–729
Wong E, Cuervo AM (2010) Autophagy gone awry in neurodegenerative diseases. Nat Neurosci 13:805–811
Yamasaki T, Suzuki A, Shimizu T, Watarai M, Hasebe R, Horiuchi M (2012) Characterization of intracellular localization of PrPSc in prion-infected cells using monoclonal antibody that recognizes the region consisting of amino acids 119–127 of mouse PrP. J Gen Virol 93(Pt 3): 668–80
Yedidia Y, Horonchik L, Tzaban S, Yanai A, Taraboulos A (2001) Proteasomes and ubiquitin are involved in the turnover of the wild-type prion protein. EMBO J 20:5383–5391
Yun SW, Ertmer A, Flechsig E, Gilch S, Riederer P, Gerlach M, Schatzl HM, Klein MA (2007) The tyrosine kinase inhibitor imatinib mesylate delays prion neuroinvasion by inhibiting prion propagation in the periphery. J Neurovirol 13:328–337

Chapter 12
Molecular Mechanisms Encoding Quantitative and Qualitative Traits of Prion Strains

Jiri G. Safar

Abstract Yeast, fungal, and mammalian prions determine heritable as well as infectious traits. In mammals, prions cause a group of fatal and rapidly progressive neurodegenerative diseases, originally described as transmissible spongiform encephalopathies (TSEs). Variations in prions, which cause different disease phenotypes, are referred to as strains. Mammalian prion strains are differentiated by qualitative characteristics such as clinical symptoms, brain pathology, targeted brain anatomical areas and cells, or Western blot patterns of glycosylated or deglycosylated pathogenic prion protein (PrPSc). Quantitative prion traits are determined by incubation time, prion dose response, proteolytic sensitivity, and conformational stability of PrPSc. The high degree of fidelity with which prion strains replicate requires a precise molecular mechanism that can account for all these characteristics. Remarkable progress in the past decade produced many lines of evidence arguing that prion traits are encoded in the self-replicating conformation of PrPSc that is unique for each strain. Thus, prions behave like proteinaceous genes. The determination of the full spectrum of human and animal prion strains and the conformational features in the pathogenic human prion protein that govern replication of prion strains is essential for the development of diagnostic as well as therapeutic strategies.

Keywords Prion strains • Conformation of prion protein • Protein misfolding cyclic amplification (PMCA) • Conformation-dependent immunoassay (CDI) • Neurodegeneration

---

J.G. Safar, M.D. (✉)
Department of Pathology and Department of Neurology, National Prion Disease Surveillance Center, School of Medicine, Case Western Reserve University, 2085 Adelbert Rd, Cleveland, OH 44106, USA
e-mail: jiri.safar@case.edu

W.-Q. Zou and P. Gambetti (eds.), *Prions and Diseases: Volume 1, Physiology and Pathophysiology*, DOI 10.1007/978-1-4614-5305-5_12,

Abbreviations

ALS	Amyotrophic lateral sclerosis
CDI	Conformation-dependent immunoassay
CHO	N-linked complex glycosylation chains
CJD	Creutzfeldt–Jakob disease
CPA	Cell panel assay
ER	Endoplasmic reticulum
FFI	Fatal familial insomnia
GSS	Gerstmann–Sträussler–Scheinker syndrome
PMCA	Protein misfolding cyclic amplification
PRNP	Prion protein gene
PrP	Prion protein
PrP^{C}	Normal or cellular prion protein
PrP^{Sc}	Pathogenic prion protein
$rPrP^{Sc}$	Protease-resistant conformers of pathogenic prion protein (PrP 27-30)
sCJD	Sporadic Creutzfeldt–Jakob disease
SFI	Sporadic fatal insomnia
$sPrP^{Sc}$	Protease-sensitive conformers of pathogenic prion protein
SSCA	Standard scrapie cell assay
TSE	Transmissible spongiform encephalopathy
VPSPr	Variable protease-sensitive prionopathy
WB	Western blot

12.1 Prion Diversity

Unique characteristics of mammalian prion isolates, which cause distinctive disease phenotypes, are referred to as strains. Prion strains were initially isolated based on distinctive clinical symptoms in goats with scrapie (Pattison and Millson 1961). Subsequently, strains were isolated in rodents based on divergent incubation times and neuropathologic profiles (Fraser and Dickinson 1973; Dickinson and Fraser 1977). New strains have been produced upon passage from one species to another (Kimberlin et al. 1987), from nontransgenic (Tg) mice to mice expressing a foreign or artificial PrP transgene (Scott et al. 1997), or most recently in vitro from recombinant prion protein (Legname et al. 2006; Wang et al. 2010).

For several decades, the existence of several prion strains was offered as an argument for the existence of a scrapie-specific nucleic acid (Bruce and Dickinson 1987; Dickinson and Outram 1988). However, despite numerous attempts to find such a nucleic acid using several approaches and despite mounting evidence against the existence of a strain-coding polynucleotide (Meyer et al. 1991; Kellings et al. 1992, 1994; Safar et al. 2005a), an explanation for prion strains remained a conundrum and a major challenge to basic principles of molecular biology (Safar et al. 2005a; Prusiner 1998a; Weissmann 2004). Moreover, the discovery that different strains of prions can be propagated indefinitely with high fidelity in inbred mouse lines expressing only a single PrP sequence and the finding that prion strains were selective with regard to the cells in which they can replicate raised fundamental questions (a) How many mammalian prion strains exist? (b) How can cells distinguish different prion strains, as reflected in the cells' ability to propagate them? (c) How are strain-specific characteristics encoded if the prion is composed solely of PrP with the same sequence?

12.2 Distinct Phenotypes of Prion Strains in Bioassay

An important milestone in the history of research on prion strains was the experimental transmission of scrapie from sheep to mice ~18 months after intracerebral inoculation of brain extracts (Chandler 1961). On second passage, the incubation periods shortened to 4–5 months and remained constant on subsequent passages. The demonstration that scrapie could be transmitted to a small laboratory rodent made possible many new experimental studies that were previously impracticable in sheep or goats and helped to identify and characterize the first prion isolates by distinct clinical symptoms, incubation time, and brain pathology (Fraser and Dickinson 1973; Dickinson et al. 1972). A second milestone occurred with the development of an incubation time bioassay in Syrian hamsters, which reduced the time required to measure prions in samples with high titers by a factor of nearly 6; only 70 days were required instead of the 360 days previously needed. Equally important, four animals could be used instead of the 60 mice that were required for endpoint titrations, and this made possible a large number of parallel experiments (Prusiner et al. 1982, 1999a). However, there were disadvantages to using hamsters instead of mice: (1) the number of inbred hamster strains was small, (2) they we susceptible to only some prion strains, and (3) there were no procedures for transfer and ablation of genes in the hamster. Thus, the third milestone became the production of transgenic (Tg) mice overexpressing prion protein homologous to the original prion host, for example, mouse (Mo), Syrian hamster, or human (Hu) PrP. In contrast to nontransgenic hosts, Tg mouse models of prion diseases produced the original species of prions, and overexpression of the PRNP gene led to significantly shorter incubation times (Carlson et al. 1994a; Scott et al. 1989). Most importantly, the transmission experiments established stable laboratory strains of prions with defined biological characteristics that became standard experimental tools in prion research (Prusiner et al. 1999a, 2004a, b; Scott et al. 2004).

Because of the wealth of data accumulated in animal experiments, the parameters distinguishing distinct mammalian prion isolates fell into qualitative or quantitative categories:

1. Qualitative traits:
   (a) Clinical symptoms of the host (Pattison and Millson 1961)
   (b) Anatomical distribution and characteristics of brain lesions (Fraser and Dickinson 1973; Dickinson and Fraser 1977)

   (c) Anatomical distribution of pathogenic PrP^{Sc} in the brain (Gambetti et al. 2003; Taraboulos et al. 1992)
   (d) Mass of unglycosylated or deglycosylated $rPrP^{Sc}$ on Western blots (WBs) (Parchi et al. 1996; Bessen and Marsh 1994; Telling et al. 1996)
   (e) Glycoform pattern of $rPrP^{Sc}$ on WBs (Collinge et al. 1996)
   (f) Conformational characteristics of PrP^{Sc} in conformation-dependent immunoassay (CDI) (Safar et al. 1998)
2. Quantitative traits:
   (a) Incubation time (Pattison and Millson 1961)
   (b) Dose–response curve in endpoint titration (Kimberlin and Walker 1978)
   (c) Susceptibility of pathogenic PrP^{Sc} to proteases (Safar et al. 1998)
   (d) Conformational stability of PrP^{Sc} (Safar et al. 1998, 2011; Peretz et al. 2001)

12.3 Prion Species

A prion species is defined by the amino acid sequence of the donor's (host's) PrP. Transmission of prions between different animal species frequently results in low transmission rates and long incubation times, which shorten upon repeated transmission to the recipient species (Scott et al. 2004; Safar et al. 2011; Bruce and Dickinson 1979). This so-called species barrier is attributed to differences in the PrP sequences between prion donor and new host that hinder the response of host PrP^C to the incoming $rPrP^{Sc}$ seed (Scott et al. 2004; Collinge and Clarke 2007). A "species barrier" may also exist within the same animal species; for example, there are two distinct polymorphic PrP alleles in different mouse lines, the Prnpa (108L, 189T) and the Prnpb allele (108F, 189V), and transfer of prions between mice with divergent PrP alleles is subject to a barrier similar to that observed in the transfer between different animal species (Prusiner et al. 2004a; Carlson et al. 1994b; Tremblay et al. 2004).

In the case of interspecies prion transfer to mice, the barrier may be overcome by replacing the murine PrP genes with their counterpart from the donor, for example, Syrian hamster (Prusiner et al. 1990), cattle (Scott et al. 1999), human (Telling et al. 1994), or cervids (Browning et al. 2004). Importantly, in PrP-deficient ($Prnp^{0/0}$) mice, neither prion disease nor prion replication has been found (Büeler et al. 1993). But replacement of the murine PrP gene with its homologs from another species does not recreate the physiology of the donor species, and genes other than PrP may play a role in susceptibility to prions, thereby resulting in different incubation times (Tamguney et al. 2008; Stephenson et al. 2000; Prusiner et al. 1999b). From these experiments and those in vitro, several authors have proposed an auxiliary role for an as yet hypothetical host-derived cofactor in prion replication, which could be a polynucleotide, glycosaminoglycan, lipid, or chaperone facilitating conversion (Kaneko et al. 1997; Kim et al. 2010; Deleault et al. 2010, 2012; Piro and Supattapone 2011; Geoghegan et al. 2007).

Cumulatively, the expression of foreign, mutant, or chimeric PrP transgenes in mice has created a wealth of knowledge about prions that was previously unattainable. Most importantly, this knowledge helped to separate the phenomena generated by "species barrier" from true strain characteristics encoded in the prion itself (Scott et al. 2004, 2005; Collinge and Clarke 2007). It has also helped to define the central domain (residues 96–167) in the PrP amino acid sequence determining "species barrier" (Scott et al. 2004), demonstrated an inverse relationship between the level of PrP^{C} expression and the incubation time (Scott et al. 1989), and allowed differentiation of the natural prion isolates from de novo prions generated with mutant and recombinant PrP (Legname et al. 2006; Wang et al. 2010; Tremblay et al. 2004; Safar et al. 2000).

12.4 Cell Tropism of Prion Strains

A few traits, such as clinical symptoms, pathology, and CNS distribution of pathogenic PrP^{Sc}, probably indicate distinct susceptibility of different cells to prions (Mahal et al. 2007). Different prion strains are evident in different locations of lesions and PrP^{Sc} deposition in the brain and may exhibit different tropism for cell lines (Mahal et al. 2007). Because the uptake of PrP^{Sc} by cultured cells appears to be a nonspecific process, the distinct susceptibility of various cells to different prion strains probably reflects the capacity of the cell to replicate prions at a rate exceeding natural clearance (Bergstrom et al. 2006; Mishra et al. 2004).

Some authors studying Western blot patterns of PrP 27-30 proposed that the observed differences in glycosylation specify prion strains (Collinge et al. 1996). However, this proposal is difficult to reconcile with the addition of high mannose oligosaccharides to Asn-linked consensus sites on PrP in the ER and subsequent remodeling of the sugar chains in the Golgi (Endo et al. 1989). Modification of the complex CHOs attached to PrP^{C} is clearly completed prior to the PrP^{C} trafficking to the cell surface (Borchelt et al. 1990; Caughey and Raymond 1991), which indicates that the Asn-linked CHOs of PrP^{Sc} do not instruct the addition of such complex-type sugars to PrP^{C}. Mutagenesis of the complex-type sugar attachment sites seemed to increase PrP^{Sc} formation in cultured cells (Taraboulos et al. 1990) but resulted in prolonged incubation times in Tg mice and differences in the patterns of PrP^{C} distribution and PrP^{Sc} deposition in mice expressing mutant PrPs (DeArmond et al. 1997; Tuzi et al. 2008). Finally, the idea that strain recognition is mediated by the nature of the glycans carried by PrP^{Sc} is not supported by the finding that two distinct prion strains could be propagated by PMCA using unglycosylated PrPC (Piro et al. 2009). Cumulatively these studies indicate that Asn-linked glycosylation might alter the stability and susceptibility of PrP^{C} to conversion, thereby resulting in distinctive patterns of PrP^{Sc} deposition and glycosylation on WBs.

An important contribution to the understanding of cellular phenomena related to prion strains came from the cell panel assay (CPA) developed by Charles Weissmann and colleagues. Conventionally the distinction between mouse-adapted prion strains requires determination of incubation times in at least two mouse lines extending over 6–10 months. The CPA, which can distinguish between various murine prion strains in less than 2 weeks (Mahal et al. 2007), is based on the standard scrapie cell assay (SSCA), a method for the rapid and sensitive quantification of prions in vitro. The CPA carried out on a set of four cell lines, PK1, R33, CAD5, and LD9, showed different responses to various prions (Mahal et al. 2007; Karapetyan et al. 2009) and allowed reliable distinction of RML, 22L, 301C, and Me7 mouse prion strains. Additionally, when transferred from brain to cultured cells, "cell-adapted" prions outcompeted their "brain-adapted" counterparts, but the opposite occurred when prions were returned from cells to brain. Thus, the authors concluded that prions, although lacking a nucleic acid genome, are subject to mutation and selective amplification (Li et al. 2010).

However, the mechanisms underlying specificity for brain areas and for cultured cell lines in vitro are likely to be somewhat different. Persistent infection requires that the rate of PrP^{Sc} synthesis be at least equal to the rate of PrP^{Sc} depletion (Weissmann 2004). In cell culture, depletion of PrP^{Sc} is caused by degradation, secretion, and cell division, whereas in brain, where PrP^{Sc} accumulates predominantly in neurons, depletion does not occur by cell division. Thus, slowing cell division of cultured cells not only increases the accumulation of PrP^{Sc} but may also allow cells to become chronically infected by strains to which they are resistant under normal growth conditions (Ghaemmaghami et al. 2007). The fact that many drugs that "cure" chronically infected cell lines are largely ineffective in abrogating prion disease in vivo reflects at least in part the fact that in the brain PrP^{Sc} depletion does not occur by cell division (Ghaemmaghami et al. 2007; Collinge et al. 2009; Trevitt and Collinge 2006).

12.5 Conformational Mechanism of Prion Strain Propagation

Most researchers now accept the model according to which the infectious pathogen responsible for TSEs is pathogenic PrP^{Sc} (Prusiner 1982). This protein is a misfolded, β-sheet-rich isoform of the normal cellular prion protein, PrP^{C}, which is predominantly α-helical (Collinge and Clarke 2007; Prusiner 1998b, 2004; Caughey et al. 2009; Cobb and Surewicz 2009; Morales et al. 2007). The discovery that proteins may be infectious represents a new paradigm of molecular biology and medicine. Although originally deemed heretical, this protein-only model is now supported by a wealth of biochemical, genetic, and animal studies (Collinge and Clarke 2007; Prusiner 1998b, 2004; Caughey et al. 2009; Cobb and Surewicz 2009; Morales et al. 2007), including recent success in generating infectious prions in vitro (Wang et al. 2010; Kim et al. 2010; Legname et al. 2004; Castilla et al. 2005; Barria et al. 2009; Deleault et al. 2007; Geoghegan et al. 2009). The PrP^{Sc} conformer is believed to self-replicate by a mechanism which remains poorly understood, but which involves binding to PrP^{C} and causing this protein to convert to the PrP^{Sc} state (Fig. 12.1) (Kocisko et al. 1994; Prusiner 1997).

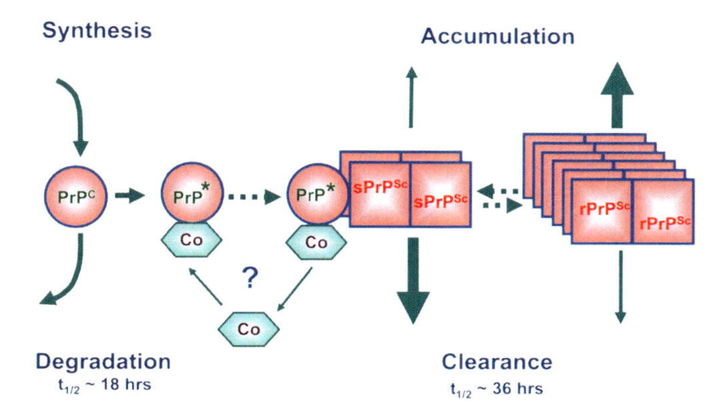

Fig. 12.1 Schematic reaction coordinates of sPrP^{Sc} and rPrP^{Sc} formation and accumulation. Different conversion and clearance rates of PrP^{Sc} dictate the speed of accumulation and thus incubation time in particular prion isolates

The first suggestion that properties of PrP^{Sc} might be distinct in various strains of prions arose from an analysis of two prion isolates from mink that had been passaged in Syrian hamsters and labeled drowsy (DY) and hyper (HY) according dominant clinical symptoms (Bessen and Marsh 1992, 1994). The more pronounced resistance of HY PrP^{Sc} to limited proteinase K digestion and distinct sedimentation velocity suggested dissimilar physical properties of PrP^{Sc}, but the results did not correlate with other isolates that produced similar incubation times and indistinguishable patterns of PrP^{Sc} on WBs (Scott et al. 1997). Only when prion strains generated de novo in humans with inherited prion diseases were passaged in Tg(MHu2M) mice could an argument be made for the distinctive conformation or ligands of PrP^{Sc} present in different prion strains (Telling et al. 1996; Prusiner 1997). These studies were fortuitous in the sense that fCJD(E200K) and fatal familial insomnia (FFI) produced different sizes of rPrP^{Sc} fragments after limited proteinase K digestion on WBs.

The WB-based studies of PrP^{Sc} were limited to the most protease-resistant fraction of PrP^{Sc}. It has also been difficult to analyze low levels of PrP^{Sc} in the presence of high levels of PrP^{C}. Moreover, the limited digestion by proteinase K resulting in either 19- or 21-kDa bands after deglycosylation of PrP 27-30 could not explain the broad biological diversity observed in more than 30 rodent-adapted prion strains in bioassays. In response to these problems, we developed a rapid, specific, and highly sensitive method for the detection and conformational characterization of PrP^{Sc} designated as conformation-dependent immunoassay (CDI) (Safar et al. 1998). After assay calibration with recombinant PrP that has refolded into different conformations, we could distinguish α-helical, β-sheet, and random coil conformations of PrP, either alone or in a mixture. Thus, the assay enabled us to directly measure the amount of PrP^{Sc} in brain homogenates without prior digestion with proteinase K to eliminate PrP^{C}. The assay is conformation sensitive, and with selective precipitation of PrP^{Sc} before differential immunoassay, PrP^{Sc} could be measured in a sandwich format in the presence of ~10,000-fold excess of PrP^C with a sensitivity similar to that of bioassays (Safar et al. 1998, 2002, 2005b, 2008; Kim et al. 2011).

The CDI led to the discovery of a variable fraction of pathogenic prion protein that is actually protease sensitive ($sPrP^{Sc}$) (Fig. 12.1) and allowed us to differentiate all eight strains examined by differently exposed epitopes, response to limited digestion with proteinase K, and stability in a chaotrope guanidine hydrochloride (Gdn HCl) (Safar et al. 1998). Thus, our data provided compelling evidence that eight different strains passaged in the same host (Syrian hamsters) possess at least eight distinct conformations. The differences in conformation of PrP^{Sc} detected by CDI in different prion strains in brain homogenates suggested two markedly distinct conformational mechanisms responsible for propagation of different prion characteristics. Under one possibility, each strain would be encoded by the PrP^{Sc} molecules in a definite number of conformations, and a specific mixture (ratio) of the same building blocks would replicate itself in the next passage. The second possibility is that each strain characteristic is encoded in a unique conformer of PrP^{Sc}, which then replicates with a high degree of fidelity and thus reproduces the strain characteristics.

Thus, in addition to a structure for PrP^C that is distinct from PrP^{Sc}, our data on prion strains in Syrian hamsters suggested that there may be several PrP^{Sc} conformers with distinct stabilities (energies) (Fig. 12.4) (Shirley 1995). This hypothesis represents an obvious departure from earlier work demonstrating that most proteins had a single folded structure that was uniquely encoded in the sequence (Anfinsen 1973). What is the structural basis of these alternative PrP^{Sc} conformers? Work on diphtheria toxin identified distinct crystal forms that displayed different tertiary and quaternary structures for a single polypeptide sequence (Bennett et al. 1995). To describe this observation, the notion of domain swapping was introduced whereby a region of one monomer displaced the corresponding region in another monomer to create an interlocking molecular handshake (Cohen and Prusiner 1998). This phenomenon has now been observed in a variety of other protein structures with the swapped elements as small as an isolated α-helix or β-strand and as large as an entire folded domain. We suspect that a similar phenomenon may be responsible for prion strains. The early experimental data obtained with infrared spectroscopy or with mass spectroscopy after hydrogen/deuterium exchange (H/X MS) confirm the conformational plasticity of PrP^{Sc} (Cobb and Surewicz 2009; Jones and Surewicz 2005; Caughey et al. 1998). In fact, conformational polymorphism (i.e., ability to form different strains) appears to be a general feature of amyloids and was observed, for instance, in fibrils formed by Aβ peptide associated with Alzheimer's disease (Paravastu et al. 2008; Petkova et al. 2002).

The data also argue that PrP^{Sc} must act as a template in the replication of nascent PrP^{Sc} molecules. It seems likely that the binding of PrP^C or a metastable intermediate PrP* (Figs. 12.1 and 12.4) (Safar et al. 1994) constitutes the initial step in PrP^{Sc} formation and that this is also the rate-limiting step in prion replication (Safar et al. 1998; Kaneko et al. 1997; Cohen and Prusiner 1998; Prusiner et al. 1998). The finding that the rate of PrP^{Sc} amplification by PMCA varies considerably for different murine strains supports the view that PrP^{Sc} structure is likely rate determining also in vivo (Karapetyan et al. 2009). However, the rate of PrP^{Sc} synthesis must also reflect the activation energy required for the conversion process and thus is likely a function of both the conformation of the PrP^{Sc} multimer, which is believed to be strain dependent, and of the conformation of the PrP^{C} serving as substrate (Fig. 12.4). The conformational stability of PrP^{C} may depend on posttranslational modifications of PrP such as glycosylation or on association with cellular components which, by favoring certain PrP conformations, could promote preferential propagation of particular strains in different cells. The remarkable affinity of PrP^{C} for nucleic acids (King et al. 2007) and the requirement for polyanions in the PMCA reaction using purified PrP^{C} as substrate (Deleault et al. 2005) together support the view that cell components other than PrP^{C} may play an auxiliary role in prion strain replication (Geoghegan et al. 2007). Thus, the optimal conversion process of different prion strains might require different cofactors, and it is likely that the cofactor content or structure in a particular cell type may contribute to its capacity for propagating a particular strain (Fig. 12.1).

12.6 Human Prion Strains

Although remarkable progress has been made in understanding the pathology, biochemistry, and structure of rodent-adapted prion strains (Prusiner et al. 2004b; Caughey et al. 2009; Cobb and Surewicz 2009; Morales et al. 2007; Watts and Westaway 2007; Telling 2008), understanding of the molecular basis of human prion diseases has lagged behind. The human prion diseases are more complex, and a single pathologic process may present as a sporadic, genetic, or infectious illness (Prusiner 2004). The most common human prion disease is sporadic Creutzfeldt–Jakob disease (sCJD), accounting for ~85% of cases. Although sCJD was shown to be transmissible to nonhuman primates 40 years ago (Gibbs et al. 1968; Brown et al. 1994), the origin, pathogenesis, and the number of human prion strains causing the disease remain unknown.

Lack of progress in the area of human prions stems from three barriers. First, these diseases present greater variability on complex genetic background; second, experiments with human material are prohibitive; and finally, relatively few investigators focus on human prion diseases. Nevertheless, researchers today generally agree that the genotype at codon 129 of the chromosomal gene PRNP underly the susceptibility to prions and to some degree the phenotypes of diseases (Gambetti et al. 2003; Bishop et al. 2010; Giles et al. 2010). In contrast to the experiments with laboratory rodent prion strains, in which the digestion of brain PrP^{Sc} with proteolytic enzyme proteinase K (PK) consistently results in a single protease-resistant domain with mass ~19 kDa, the outcome in sCJD is more complex. Distinctive glycosylation patterns and up to four PK-resistant fragments of the pathogenic prion protein ($rPrP^{Sc}$) found in sCJD brains are easily distinguishable on Western blot (WB) (Gambetti et al. 2003; Telling et al. 1996; Collinge et al. 1996; Parchi et al. 1997; Wadsworth et al. 1999; Zou et al. 2003) (Fig. 12.2).

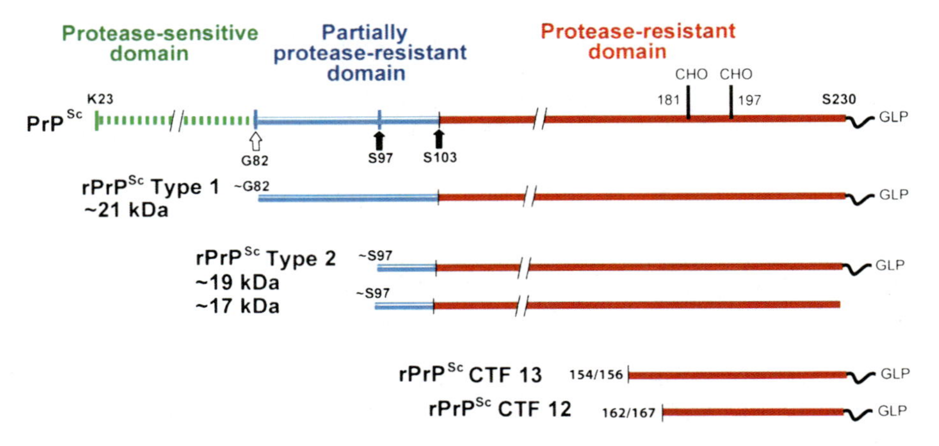

Fig. 12.2 Schematic representation of sCJD PrP^{Sc} and outline of classification of WB fragments of $rPrP^{Sc}$ (PrP 27-30) (Gambetti et al. 2003; Parchi et al. 1997, 1999; Zou et al. 2003). Major cleavage sites by PK are indicated by *arrows*; *GLP* glycolipid; *CHO* N-linked complex glycosylation chains

Although the disease phenotypes of patients with sCJD are remarkably heterogeneous, the WB findings together with human PRNP gene polymorphism led Parchi, Gambetti, and colleagues to posit a clinicopathologic classification of sCJD into five or six subtypes. Importantly, it has been shown that the WB characteristics of PrP^{Sc} breed true upon transmission to susceptible transgenic mice and guinea pigs (*Cavia porcellus*) (Gambetti et al. 2003; Telling et al. 1996; Safar et al. 2011; Parchi et al. 1997) (Fig. 12.2). Subsequently, Collinge and collaborators (Collinge et al. 1996; Collinge and Clarke 2007; Wadsworth et al. 1999; Hill et al. 1997) introduced an alternative classification of the PrP^{Sc} types and their pairing with CJD phenotypes that differed from the previous one in two aspects (a) it recognized three different electrophoretic mobilities of PrP^{Sc} and (b) differentiated distinct glycoform ratios in PrP^{Sc} (Collinge and Clarke 2007).

Because the disease duration and phenotypes associated with 21-kDa fragments of unglycosylated PrP^{Sc} (type 1) frequently differ from the 19-kDa fragments of PrP^{Sc} (type 2) (Fig. 12.3) (Gambetti et al. 2003; Telling et al. 1996; Parchi et al. 1997; Monari et al. 1994), these findings argue that the PrP^{Sc} type may represent another modifier of the phenotype in human prion diseases. Consequently, WB-based clinicopathologic classifications became useful tool in studies of prion pathogenesis in transgenic mice models of human prion diseases and in human brains (Telling et al. 1996; Collinge and Clarke 2007). Because two distinct PK cleavage sites in PrP^{Sc} types 1 and 2 most likely originate from different conformations, some investigators contend that PrP^{Sc} types 1 and 2 code distinct prion strains (Parchi et al. 1996; Telling et al. 1996; Collinge et al. 1996; Monari et al. 1994). However, the findings of the co-occurrence of PrP^{Sc} types 1 and 2 in 40% or more of sCJD cases suggested that the originally observed differences were quantitative rather than

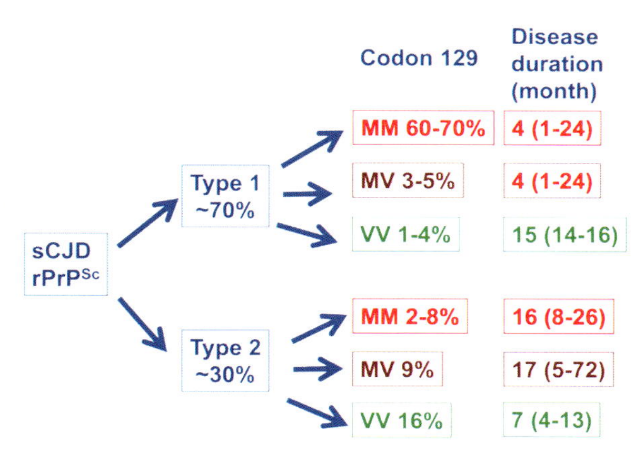

Fig. 12.3 Impact of the polymorphism in codon 129 of the PRNP gene and WB pattern of PrP^{Sc} on the susceptibility to sCJD and duration of the disease. The *first number* indicates the mean duration of the disease; the range is in the *parenthesis* (Gambetti et al. 2003; Parchi et al. 1997, 1999; Zou et al. 2003)

qualitative (Puoti et al. 1999; Kovacs et al. 2002; Head et al. 2004; Lewis et al. 2005; Schoch et al. 2006; Cali et al. 2009). Additionally, the extensive phenotypic heterogeneity of sCJD, along with a growing number of studies including bioassays, all suggests that the range of prions causing sCJD exceeds the number of categories recognized within the original WB-based clinicopathologic schemes (Safar et al. 2005b; Uro-Coste et al. 2008; Polymenidou et al. 2005). Finally, up to 90% of PrP^{Sc} is protease sensitive (s), and the conformation and the role of this fraction in the pathogenesis of the disease are unknown and remain a subject of speculation (Safar et al. 2005b, c; Cronier et al. 2008) because it is destroyed by proteinase K treatment, which is necessary to eliminate PrP^{C} (Safar et al. 2005b). Cumulatively, no direct structural data are available for sCJD brain PrP^{Sc} beyond the evidence that it is variably resistant to proteolytic digestion.

To determine the conformational range and strain-dependent structural characteristics of sCJD PrP^{Sc} in patients who were homozygous for codon 129 of the PRNP gene and thus advance our understanding of the molecular pathogenesis of human prion diseases, we introduced the conformation-dependent immunoassay (CDI) (Safar et al. 1998, 2002, 2005b). The conformational stability of the protein in a denaturant such as Gdn HCl (Shirley 1995) is reflecting the original conformation of the protein. If the protein has the same amino acid sequence, the difference in stability indicates the difference in conformation. Thus, even relatively minute variations in a protein structure can be determined. Using this concept, we developed conformational stability assay in which PrP^{Sc} is first exposed to denaturant Gdn HCl and then to europium-labeled mAb against the epitopes hidden in the native conformation (Safar et al. 1998). With sequentially increasing concentration of Gdn HCl, PrP^{Sc} dissociates and unfolds from native β-sheet-structured aggregates and more epitopes become available to antibody binding. Because PrP^{Sc} is insoluble oligomer and denaturation of this protein is irreversible in vitro, the Gibbs free energy change (ΔG) of PrP^{Sc} cannot be calculated (Safar et al. 1994). Therefore, we introduced instead the Gdn HCl value found at the half-maximal denaturation ($[GdnHCl]_{1/2}$) as a measure of the relative conformational stability of PrP^{Sc}.

The differences in $[\text{GdnHCl}]_{1/2}$ reveal evidence of distinct conformations of PrP^{Sc} (Safar et al. 1994, 1998; Shirley 1995).

The process of disaggregation and unfolding of PrP^{Sc} in the presence of increasing concentration of Gdn HCl has been described as follows:

$$\left[\text{Pr P}^{\text{Sc}}\right]_{\text{n}} \rightarrow \left[\text{sPrP}^{\text{Sc}}\right]_{\text{n}} \rightarrow \text{iPrP} \rightarrow \text{uPrP}$$

where $[PrP^{Sc}]_n$ are native aggregates of PrP^{Sc}, $[sPrP^{Sc}]_n$ are soluble protease-sensitive oligomers of PrP^{Sc}, iPrP is an intermediate, and uPrP is completely unfolded (denatured) PrP (Safar et al. 1993, 1994, 2011; Tzaban et al. 2002; Safar 2012). Since CDI is not dependent on protease treatment, it allowed us to address fundamental questions concerning the concentration and conformation of different isoforms of sCJD PrP^{Sc}, including protease-sensitive (s) and protease-resistant (r) PrP^{Sc} (Kim et al. 2011; Safar 2012). Consequently, the CDI monitors the global transition from native aggregates to fully denatured monomers of PrP^{Sc}. In contrast, the WB-based techniques monitor either the partial solubilization of PrP^{Sc} (Pirisinu et al. 2011) or conversion of $rPrP^{Sc}$ to protease-sensitive conformers (Peretz et al. 2001) after exposure to denaturant. Therefore, stability data on protease-sensitive oligomers and intermediates of PrP^{Sc} cannot be obtained with WB and may lead to some markedly different values (Choi et al. 2011).

We found with CDI a remarkable heterogeneity of PrP^{Sc} conformations within sCJD patients homozygous for codon 129 polymorphism of the PRNP gene and a range corresponded to that of stabilities found in ~30 distinct strains of natural and de novo laboratory rodent prions that were examined so far (Safar et al. 1998; Peretz et al. 2001; Kim et al. 2011; Colby et al. 2010). The unexpected differential effect of PK treatment with increasing stability of type 1 and decreasing stability of type 2 PrP^{Sc}(129M) suggests that in contrast to type 1, the protease-resistant core of type 2 is less stable (Fig. 12.4). The increased frequency of exposed epitopes and decreased stability in type 2 PrP^{Sc} after PK treatment (Kim et al. 2011) are counterintuitive and may indicate one of three possibilities: that the PK sensitivity is not an obligatory measure of protein stability and rPrPSc may be in some prion strains less stable than sPrPSc, that removal of the N-terminus from PrP^{Sc} resulted in less stable conformation with more exposed 108–112 epitopes, or that the ligand protecting the 108–112 epitopes and stabilizing the PrPSc was removed by PK. Whether the epitopes' hindrance in undigested PrP^{Sc} is the result of lipid, glycosaminoglycan, nucleic acid, or protein binding to the conformers unique to the MM2 sCJD PrP^{Sc} remains to be established. Since sCJD cases with type 2 PrPSc(129M) have generally extended disease durations, the molecular mechanism underlying this effect calls for detailed investigation. Cumulatively, our findings indicate that sCJD PrP^{Sc} exhibits extensive conformational heterogeneity and suggest that a wide spectrum of sCJD prions cause the disease (Safar 2012). Whether this heterogeneity originates in a stochastic misfolding process that generates many distinct self-replicating conformations (Collinge and Clarke 2007; Prusiner 2001) or in a complex process of evolutionary selection during development of the disease (Li et al. 2010) remains to be established (Kim et al. 2011; Safar 2012).

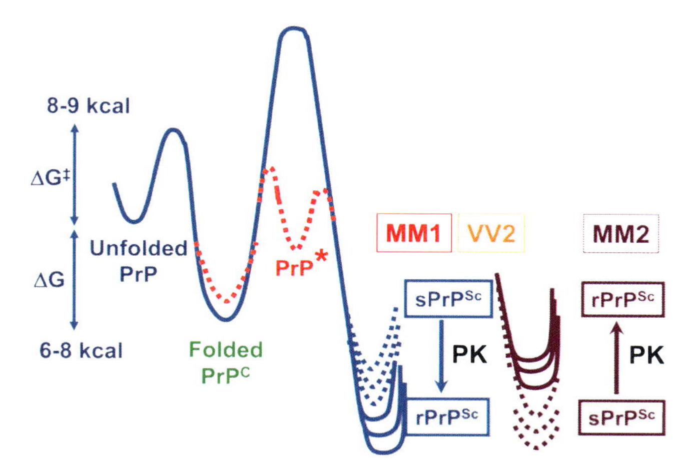

Fig. 12.4 Schematic representation of the energy landscape of different PrP^{Sc} conformers in sCJD and the impact of protease treatment. Distinct conformers within the same WB type are depicted with *multiple lines*

Despite the inevitable influence of the potential difficulties in evaluating initial symptoms and variable genetic background, our recent data indicate that the levels as well as stability of sPrP^{Sc} are a good predictor of the progression rate in sCJD (Kim et al. 2011). The disease progression rate and incubation time jointly represent replication rate, propagation, and clearance of prions from the brain (Prusiner et al. 2004a; Safar et al. 2005c). Therefore, the correlations among the levels of sPrP^{Sc}, the stability of sPrP^{Sc}, and the duration of the disease all indicate that sPrP^{Sc} conformers play an important role in the pathogenesis. When sPrP^{Sc} is less stable than rPrP^{Sc}, the difference in stability correlates with less accumulated sPrP^{Sc} and shorter duration of the disease. An opposite effect is observed when sPrP conformers are more stable than rPrP^{Sc}—more accumulated sPrP^{Sc} and extended disease duration (Fig. 12.4) (Kim et al. 2011). These observations parallel the experiments on yeast prions and suggest that the stability of misfolded protein is inversely related to the replication rate (Kim et al. 2011; Tanaka et al. 2006). Thus, the data from both yeast and human prions lead to the hypothesis that the less stable prions replicate faster by exposing more available sites for growth of the aggregates. Although the modulating effect of prion clearance in the mammalian brains is likely (Safar et al. 2005c), faster prion replication leads to shorter incubation time and faster progression of the disease.

12.7 Outlook

The continuing mystery surrounding replication of the PrP^{Sc} conformer poses a fundamental challenge in modern biology, and important questions regarding prion strains remain to be answered. For example, is each strain composed of a unique conformer or of a spectrum of conformations, which may shift by selection or conformational evolution? Additionally, the conformational concept of prion strain replication raises the question of which conformational features of PrP^{Sc} are important for replication and which determine clearance. Although there is now convincing evidence that the PrP^{Sc} conformation of distinct strains is different, it is not known to what extent the conformation or replication rate of different conformers might depend on factors other than conformation of the PrP, for example, the nature of the glycans or additional cell-derived ligands (cofactors). An attractive experiment would be to obtain large quantities of highly purified PrP^{Sc} from a single cell line, infected separately with several different prion strains; determine the glycans carried by each strain-associated PrP^{Sc}; and search for associated molecules, such as small RNAs or other cell components. Finally, the deepest insight will be gained once the three-dimensional structure of PrP^{Sc} can be determined at high resolution, currently a still formidable task.

Acknowledgments This work was supported by grants from NIA (AG-14359), NINDS (NS074317), CDC (UR8/CCU515004), and the Charles S. Britton Fund.

References

Anfinsen CB (1973) Principles that govern the folding of protein chains. Science 181:223–230

Barria MA, Mukherjee A, Gonzalez-Romero D, Morales R, Soto C (2009) De novo generation of infectious prions in vitro produces a new disease phenotype. PLoS Pathog 5:e1000421

Bennett MJ, Schlunegger MP, Eisenberg D (1995) 3D domain swapping: a mechanism for oligomer assembly. Protein Sci 4:2455–2468

Bergstrom AL, Jensen TK, Heegaard PM, Cordes H, Hansen VB, Laursen H et al (2006) Short-term study of the uptake of PrP(Sc) by the Peyer's patches in hamsters after oral exposure to scrapie. J Comp Pathol 134:126–133

Bessen RA, Marsh RF (1992) Biochemical and physical properties of the prion protein from two strains of the transmissible mink encephalopathy agent. J Virol 66:2096–2101

Bessen RA, Marsh RF (1994) Distinct PrP properties suggest the molecular basis of strain variation in transmissible mink encephalopathy. J Virol 68:7859–7868

Bishop MT, Will RG, Manson JC (2010) Defining sporadic Creutzfeldt-Jakob disease strains and their transmission properties. Proc Natl Acad Sci USA 107:12005–12010

Borchelt DR, Scott M, Taraboulos A, Stahl N, Prusiner SB (1990) Scrapie and cellular prion proteins differ in their kinetics of synthesis and topology in cultured cells. J Cell Biol 110:743–752

Brown P, Gibbs CJ Jr, Rodgers-Johnson P, Asher DM, Sulima MP, Bacote A et al (1994) Human spongiform encephalopathy: the National Institutes of Health series of 300 cases of experimentally transmitted disease. Ann Neurol 35:513–529

Browning SR, Mason GL, Seward T, Green M, Eliason GA, Mathiason C et al (2004) Transmission of prions from mule deer and elk with chronic wasting disease to transgenic mice expressing cervid PrP. J Virol 78:13345–13350

Bruce ME, Dickinson AG (1979) Biological stability of different classes of scrapie agent. In: Prusiner SB, Hadlow WJ (eds) Slow transmissible diseases of the nervous system, vol 2. Academic, New York, pp 71–86

Bruce ME, Dickinson AG (1987) Biological evidence that the scrapie agent has an independent genome. J Gen Virol 68:79–89

Büeler H, Aguzzi A, Sailer A, Greiner R-A, Autenried P, Aguet M et al (1993) Mice devoid of PrP are resistant to scrapie. Cell 73:1339–1347
Cali I, Castellani R, Alshekhlee A, Cohen Y, Blevins J, Yuan J et al (2009) Co-existence of scrapie prion protein types 1 and 2 in sporadic Creutzfeldt-Jakob disease: its effect on the phenotype and prion-type characteristics. Brain 132:2643–2658
Carlson GA, Ebeling C, Yang S-L, Telling G, Torchia M, Groth D et al (1994a) Prion isolate specified allotypic interactions between the cellular and scrapie prion proteins in congenic and transgenic mice. Proc Natl Acad Sci USA 91:5690–5694
Carlson GA, DeArmond SJ, Torchia M, Westaway D, Prusiner SB (1994b) Genetics of prion diseases and prion diversity in mice. Philos Trans R Soc Lond B Biol Sci 343:363–369
Castilla J, Saa P, Hetz C, Soto C (2005) *In vitro* generation of infectious scrapie prions. Cell 121:195–206
Caughey B, Raymond GJ (1991) The scrapie-associated form of PrP is made from a cell surface precursor that is both protease- and phospholipase-sensitive. J Biol Chem 266:18217–18223
Caughey B, Raymond GJ, Bessen RA (1998) Strain-dependent differences in b-sheet conformations of abnormal prion protein. J Biol Chem 273:32230–32235
Caughey B, Baron GS, Chesebro B, Jeffrey M (2009) Getting a grip on prions: oligomers, amyloids, and pathological membrane interactions. Annu Rev Biochem 78:177–204
Chandler RL (1961) Encephalopathy in mice produced by inoculation with scrapie brain material. Lancet 277:1378–1379
Choi YP, Peden AH, Groner A, Ironside JW, Head MW (2011) Distinct stability states of disease-associated human prion protein identified by conformation-dependent immunoassay. J Virol 84:12030–12038
Cobb NJ, Surewicz WK (2009) Prion diseases and their biochemical mechanisms. Biochemistry 48:2574–2585
Cohen FE, Prusiner SB (1998) Pathologic conformations of prion proteins. Annu Rev Biochem 67:793–819
Colby DW, Wain R, Baskakov IV, Legname G, Palmer CG, Nguyen HO et al (2010) Protease-sensitive synthetic prions. PLoS Pathog 6:e1000736
Collinge J, Clarke AR (2007) A general model of prion strains and their pathogenicity. Science 318:930–936
Collinge J, Sidle KCL, Meads J, Ironside J, Hill AF (1996) Molecular analysis of prion strain variation and the aetiology of "new variant" CJD. Nature 383:685–690
Collinge J, Gorham M, Hudson F, Kennedy A, Keogh G, Pal S et al (2009) Safety and efficacy of quinacrine in human prion disease (PRION-1 study): a patient-preference trial. Lancet Neurol 8:334–344
Cronier S, Gros N, Tattum MH, Jackson GS, Clarke AR, Collinge J et al (2008) Detection and characterization of proteinase K-sensitive disease-related prion protein with thermolysin. Biochem J 416:297–305
DeArmond SJ, Sánchez H, Yehiely F, Qiu Y, Ninchak-Casey A, Daggett V et al (1997) Selective neuronal targeting in prion disease. Neuron 19:1337–1348
Deleault NR, Geoghegan JC, Nishina K, Kascsak R, Williamson RA, Supattapone S (2005) Protease-resistant prion protein amplification reconstituted with partially purified substrates and synthetic polyanions. J Biol Chem 280:26873–26879
Deleault NR, Harris BT, Rees JR, Supattapone S (2007) Formation of native prions from minimal components in vitro. Proc Natl Acad Sci USA 104:9741–9746
Deleault NR, Kascsak R, Geoghegan JC, Supattapone S (2010) Species-dependent differences in cofactor utilization for formation of the protease-resistant prion protein in vitro. Biochemistry 49(18):3928–3934
Deleault NR, Piro JR, Walsh DJ, Wang F, Ma J, Geoghegan JC et al (2012) Isolation of phosphatidylethanolamine as a solitary cofactor for prion formation in the absence of nucleic acids. Proc Natl Acad Sci USA 109:8546–8551
Dickinson AG, Fraser HG (1977) Scrapie: pathogenesis in inbred mice: an assessment of host control and response involving many strains of agent. In: ter Meulen V, Katz M (eds) Slow virus infections of the central nervous system. Springer, New York, pp 3–14

Dickinson AG, Outram GW (1988) Genetic aspects of unconventional virus infections: the basis of the virino hypothesis. In: Bock G, Marsh J (eds) Novel infectious agents and the central nervous system. CIBA Foundation Symposium 135. Wiley, Chichester, pp 63–83
Dickinson AG, Fraser H, Meikle VMH, Outram GW (1972) Competition between different scrapie agents in mice. Nat New Biol 237:244–245
Endo T, Groth D, Prusiner SB, Kobata A (1989) Diversity of oligosaccharide structures linked to asparagines of the scrapie prion protein. Biochemistry 28:8380–8388
Fraser H, Dickinson AG (1973) Scrapie in mice. Agent-strain differences in the distribution and intensity of grey matter vacuolation. J Comp Pathol 83:29–40
Gambetti P, Kong Q, Zou W, Parchi P, Chen SG (2003) Sporadic and familial CJD: classification and characterisation. Br Med Bull 66:213–239
Geoghegan JC, Valdes PA, Orem NR, Deleault NR, Williamson RA, Harris BT et al (2007) Selective incorporation of polyanionic molecules into hamster prions. J Biol Chem 282:36341–36353
Geoghegan JC, Miller MB, Kwak AH, Harris BT, Supattapone S (2009) Trans-dominant inhibition of prion propagation in vitro is not mediated by an accessory cofactor. PLoS Pathog 5:e1000535
Ghaemmaghami S, Phuan PW, Perkins B, Ullman J, May BC, Cohen FE et al (2007) Cell division modulates prion accumulation in cultured cells. Proc Natl Acad Sci USA 104:17971–17976
Gibbs CJ Jr, Gajdusek DC, Asher DM, Alpers MP, Beck E, Daniel PM et al (1968) Creutzfeldt-Jakob disease (spongiform encephalopathy): transmission to the chimpanzee. Science 161:388–389
Giles K, Glidden DV, Patel S, Korth C, Groth D, Lemus A et al (2010) Human prion strain selection in transgenic mice. Ann Neurol 68:151–161
Head MW, Bunn TJ, Bishop MT, McLoughlin V, Lowrie S, McKimmie CS et al (2004) Prion protein heterogeneity in sporadic but not variant Creutzfeldt-Jakob disease: UK cases 1991–2002. Ann Neurol 55:851–859
Hill AF, Desbruslais M, Joiner S, Sidle KCL, Gowland I, Collinge J et al (1997) The same prion strain causes vCJD and BSE. Nature 389:448–450
Jones EM, Surewicz WK (2005) Fibril conformation as the basis of species- and strain-dependent seeding specificity of mammalian prion amyloids. Cell 121:63–72
Kaneko K, Zulianello L, Scott M, Cooper CM, Wallace AC, James TL et al (1997) Evidence for protein X binding to a discontinuous epitope on the cellular prion protein during scrapie prion propagation. Proc Natl Acad Sci USA 94:10069–10074
Karapetyan YE, Saa P, Mahal SP, Sferrazza GF, Sherman A, Sales N et al (2009) Prion strain discrimination based on rapid in vivo amplification and analysis by the cell panel assay. PLoS One 4:e5730
Kellings K, Meyer N, Mirenda C, Prusiner SB, Riesner D (1992) Further analysis of nucleic acids in purified scrapie prion preparations by improved return refocussing gel electrophoresis (RRGE). J Gen Virol 73:1025–1029
Kellings K, Prusiner SB, Riesner D (1994) Nucleic acids in prion preparations: unspecific background or essential component? Philos Trans R Soc Lond B Biol Sci 343:425–430
Kim JI, Cali I, Surewicz K, Kong Q, Raymond GJ, Atarashi R et al (2010) Mammalian prions generated from bacterially expressed prion protein in the absence of any mammalian cofactors. J Biol Chem 285:14083–14087
Kim C, Haldiman T, Cohen Y, Chen W, Blevins J, Sy MS et al (2011) Protease-sensitive conformers in broad spectrum of distinct PrP structures in sporadic Creutzfeldt-Jakob disease are indicator of progression rate. PLoS Pathog 7:e1002242
Kimberlin RH, Walker CA (1978) Pathogenesis of mouse scrapie: effect of route of inoculation on infectivity titres and dose–response curves. J Comp Pathol 88:39–47
Kimberlin RH, Cole S, Walker CA (1987) Temporary and permanent modifications to a single strain of mouse scrapie on transmission to rats and hamsters. J Gen Virol 68:1875–1881
King DJ, Safar JG, Legname G, Prusiner SB (2007) Thioaptamer interactions with prion proteins: sequence-specific and non-specific binding sites. J Mol Biol 369:1001–1014

Kocisko DA, Come JH, Priola SA, Chesebro B, Raymond GJ, Lansbury PT Jr et al (1994) Cell-free formation of protease-resistant prion protein. Nature 370:471–474
Kovacs GG, Head MW, Hegyi I, Bunn TJ, Flicker H, Hainfellner JA et al (2002) Immunohistochemistry for the prion protein: comparison of different monoclonal antibodies in human prion disease subtypes. Brain Pathol 12:1–11
Legname G, Baskakov IV, Nguyen H-OB, Riesner D, Cohen FE, DeArmond SJ et al (2004) Synthetic mammalian prions. Science 305:673–676
Legname G, Nguyen H-OB, Peretz D, Cohen FE, DeArmond SJ, Prusiner SB (2006) Continuum of prion protein structures enciphers a multitude of prion isolate-specified phenotypes. Proc Natl Acad Sci USA 103:19105–19110
Lewis V, Hill AF, Klug GM, Boyd A, Masters CL, Collins SJ (2005) Australian sporadic CJD analysis supports endogenous determinants of molecular-clinical profiles. Neurology 65:113–118
Li J, Browning S, Mahal SP, Oelschlegel AM, Weissmann C (2010) Darwinian evolution of prions in cell culture. Science 327:869–872
Mahal SP, Baker CA, Demczyk CA, Smith EW, Julius C, Weissmann C (2007) Prion strain discrimination in cell culture: the cell panel assay. Proc Natl Acad Sci USA 104:20908–20913
Meyer N, Rosenbaum V, Schmidt B, Gilles K, Mirenda C, Groth D et al (1991) Search for a putative scrapie genome in purified prion fractions reveals a paucity of nucleic acids. J Gen Virol 72:37–49
Mishra RS, Basu S, Gu Y, Luo X, Zou WQ, Mishra R et al (2004) Protease-resistant human prion protein and ferritin are cotransported across Caco-2 epithelial cells: implications for species barrier in prion uptake from the intestine. J Neurosci 24:11280–11290
Monari L, Chen SG, Brown P, Parchi P, Petersen RB, Mikol J et al (1994) Fatal familial insomnia and familial Creutzfeldt-Jakob disease: different prion proteins determined by a DNA polymorphism. Proc Natl Acad Sci USA 91:2839–2842
Morales R, Abid K, Soto C (2007) The prion strain phenomenon: molecular basis and unprecedented features. Biochim Biophys Acta 1772:681–691
Paravastu AK, Leapman RD, Yau WM, Tycko R (2008) Molecular structural basis for polymorphism in Alzheimer's beta-amyloid fibrils. Proc Natl Acad Sci USA 105:18349–18354
Parchi P, Castellani R, Capellari S, Ghetti B, Young K, Chen SG et al (1996) Molecular basis of phenotypic variability in sporadic Creutzfeldt-Jakob disease. Ann Neurol 39:767–778
Parchi P, Capellari S, Chen SG, Petersen RB, Gambetti P, Kopp P et al (1997) Typing prion isoforms. Nature 386:232–233
Parchi P, Giese A, Capellari S, Brown P, Schulz-Schaeffer W, Windl O et al (1999) Classification of sporadic Creutzfeldt-Jakob disease based on molecular and phenotypic analysis of 300 subjects. Ann Neurol 46:224–233
Pattison IH, Millson GC (1961) Scrapie produced experimentally in goats with special reference to the clinical syndrome. J Comp Pathol 71:101–108
Peretz D, Scott M, Groth D, Williamson A, Burton D, Cohen FE et al (2001) Strain-specified relative conformational stability of the scrapie prion protein. Protein Sci 10:854–863
Petkova AT, Ishii Y, Balbach JJ, Antzutkin ON, Leapman RD, Delaglio F et al (2002) A structural model for Alzheimer's beta-amyloid fibrils based on experimental constraints from solid state NMR. Proc Natl Acad Sci USA 99:16742–16747
Pirisinu L, Di Bari M, Marcon S, Vaccari G, D'Agostino C, Fazzi P et al (2011) A new method for the characterization of strain-specific conformational stability of protease-sensitive and protease-resistant PrP. PLoS One 5:e12723
Piro JR, Supattapone S (2011) Photodegradation illuminates the role of polyanions in prion infectivity. Prion 5:49–51
Piro JR, Harris BT, Nishina K, Soto C, Morales R, Rees JR et al (2009) Prion protein glycosylation is not required for strain-specific neurotropism. J Virol 83:5321–5328
Polymenidou M, Stoeck K, Glatzel M, Vey M, Bellon A, Aguzzi A (2005) Coexistence of multiple PrPSc types in individuals with Creutzfeldt-Jakob disease. Lancet Neurol 4:805–814

Prusiner SB (1982) Novel proteinaceous infectious particles cause scrapie. Science 216:136–144
Prusiner SB (1997) Prion diseases and the BSE crisis. Science 278:245–251
Prusiner SB (1998a) Prions (Les Prix Nobel Lecture). In: Frängsmyr T (ed) Les Prix Nobel. Almqvist & Wiksell International, Stockholm, pp 268–323
Prusiner SB (1998b) Prions. Proc Natl Acad Sci USA 95:13363–13383
Prusiner SB (2001) Shattuck lecture—neurodegenerative diseases and prions. N Engl J Med 344:1516–1526
Prusiner SB (ed) (2004) Prion biology and diseases. Cold Spring Harbor Laboratory Press, Cold Spring Harbor
Prusiner SB, Cochran SP, Groth DF, Downey DE, Bowman KA, Martinez HM (1982) Measurement of the scrapie agent using an incubation time interval assay. Ann Neurol 11:353–358
Prusiner SB, Scott M, Foster D, Pan K-M, Groth D, Mirenda C et al (1990) Transgenetic studies implicate interactions between homologous PrP isoforms in scrapie prion replication. Cell 63:673–686
Prusiner SB, Scott MR, DeArmond SJ, Cohen FE (1998) Prion protein biology. Cell 93:337–348
Prusiner SB, Tremblay P, Safar J, Torchia M, DeArmond SJ (1999a) Bioassays of prions. In: Prusiner SB (ed) Prion biology and diseases. Cold Spring Harbor Laboratory Press, Cold Spring Harbor, pp 113–145
Prusiner SB, Scott MR, DeArmond SJ, Carlson G (1999b) Transmission and replication of prions. In: Prusiner SB (ed) Prion biology and diseases. Cold Spring Harbor Laboratory Press, Cold Spring Harbor, pp 147–190
Prusiner SB, Scott MR, DeArmond SJ, Carlson G (2004a) Transmission and replication of prions. In: Prusiner SB (ed) Prion biology and diseases. Cold Spring Harbor Laboratory Press, Cold Spring Harbor, pp 187–242
Prusiner SB, Legname G, DeArmond SJ, Cohen FE, Safar J, Riesner D et al (2004b) Some strategies and methods for the study of prions. In: Prusiner SB (ed) Prion biology and diseases. Cold Spring Harbor Laboratory Press, Cold Spring Harbor, pp 857–920
Puoti G, Giaccone G, Rossi G, Canciani B, Bugiani O, Tagliavini F (1999) Sporadic Creutzfeldt-Jakob disease: co-occurrence of different types of PrP(Sc) in the same brain. Neurology 53:2173–2176
Safar JG (2012) Molecular pathogenesis of sporadic prion diseases in man. Prion 6:108–115
Safar J, Roller PP, Gajdusek DC, Gibbs CJ Jr (1993) Conformational transitions, dissociation, and unfolding of scrapie amyloid (prion) protein. J Biol Chem 268:20276–20284
Safar J, Roller PP, Gajdusek DC, Gibbs CJ Jr (1994) Scrapie amyloid (prion) protein has the conformational characteristics of an aggregated molten globule folding intermediate. Biochemistry 33:8375–8383
Safar J, Wille H, Itri V, Groth D, Serban H, Torchia M et al (1998) Eight prion strains have PrP^{Sc} molecules with different conformations. Nat Med 4:1157–1165
Safar J, Cohen FE, Prusiner SB (2000) Quantitative traits of prion strains are enciphered in the conformation of the prion protein. Arch Virol Suppl 2000:227–235
Safar JG, Scott M, Monaghan J, Deering C, Didorenko S, Vergara J et al (2002) Measuring prions causing bovine spongiform encephalopathy or chronic wasting disease by immunoassays and transgenic mice. Nat Biotechnol 20:1147–1150
Safar JG, Kellings K, Serban A, Groth D, Cleaver JE, Prusiner SB et al (2005a) Search for a prion-specific nucleic acid. J Virol 79:10796–10806
Safar JG, Geschwind MD, Deering C, Didorenko S, Sattavat M, Sanchez H et al (2005b) Diagnosis of human prion disease. Proc Natl Acad Sci USA 102:3501–3506
Safar JG, DeArmond SJ, Kociuba K, Deering C, Didorenko S, Bouzamondo-Bernstein E et al (2005c) Prion clearance in bigenic mice. J Gen Virol 86:2913–2923
Safar JG, Lessard P, Tamguney G, Freyman Y, Deering C, Letessier F et al (2008) Transmission and detection of prions in feces. J Infect Dis 198:81–89
Safar JG, Giles K, Lessard P, Letessier F, Patel S, Serban A (2011) et al. Conserved properties of human and bovine prion strains on transmission to guinea pigs. Lab Invest
Schoch G, Seeger H, Bogousslavsky J, Tolnay M, Janzer RC, Aguzzi A et al (2006) Analysis of prion strains by PrPSc profiling in sporadic Creutzfeldt-Jakob disease. PLoS Med 3:e14

Scott M, Foster D, Mirenda C, Serban D, Coufal F, Wälchli M et al (1989) Transgenic mice expressing hamster prion protein produce species-specific scrapie infectivity and amyloid plaques. Cell 59:847–857
Scott MR, Groth D, Tatzelt J, Torchia M, Tremblay P, DeArmond SJ et al (1997) Propagation of prion strains through specific conformers of the prion protein. J Virol 71:9032–9044
Scott MR, Will R, Ironside J, Nguyen H-OB, Tremblay P, DeArmond SJ et al (1999) Compelling transgenetic evidence for transmission of bovine spongiform encephalopathy prions to humans. Proc Natl Acad Sci USA 96:15137–15142
Scott M, Peretz D, Ridley RM, Baker HF, DeArmond SJ, Prusiner SB (2004) Transgenetic investigations of the species barrier and prion strains. In: Prusiner SB (ed) Prion biology and diseases. Cold Spring Harbor Laboratory Press, Cold Spring Harbor, pp 435–482
Scott MR, Peretz D, Nguyen H-OB, DeArmond SJ, Prusiner SB (2005) Transmission barriers for bovine, ovine, and human prions in transgenic mice. J Virol 79:5259–5271
Shirley BA (ed) (1995) Protein stability and folding: theory and practice. Humana, Totowa, NJ
Stephenson DA, Chiotti K, Ebeling C, Groth D, DeArmond SJ, Prusiner SB et al (2000) Quantitative trait loci affecting prion incubation time in mice. Genomics 69:47–53
Tamguney G, Giles K, Glidden DV, Lessard P, Wille H, Tremblay P et al (2008) Genes contributing to prion pathogenesis. J Gen Virol 89:1777–1788
Tanaka M, Collins SR, Toyama BH, Weissman JS (2006) The physical basis of how prion conformations determine strain phenotypes. Nature 442:585–589
Taraboulos A, Rogers M, Borchelt DR, McKinley MP, Scott M, Serban D et al (1990) Acquisition of protease resistance by prion proteins in scrapie-infected cells does not require asparagine-linked glycosylation. Proc Natl Acad Sci USA 87:8262–8266
Taraboulos A, Jendroska K, Serban D, Yang S-L, DeArmond SJ, Prusiner SB (1992) Regional mapping of prion proteins in brains. Proc Natl Acad Sci USA 89:7620–7624
Telling GC (2008) Transgenic mouse models of prion diseases. Methods Mol Biol 459:249–263
Telling GC, Scott M, Hsiao KK, Foster D, Yang S-L, Torchia M et al (1994) Transmission of Creutzfeldt-Jakob disease from humans to transgenic mice expressing chimeric human-mouse prion protein. Proc Natl Acad Sci USA 91:9936–9940
Telling GC, Parchi P, DeArmond SJ, Cortelli P, Montagna P, Gabizon R et al (1996) Evidence for the conformation of the pathologic isoform of the prion protein enciphering and propagating prion diversity. Science 274:2079–2082
Tremblay P, Ball HL, Kaneko K, Groth D, Hegde RS, Cohen FE et al (2004) Mutant PrP^{Sc} conformers induced by a synthetic peptide and several prion strains. J Virol 78:2088–2099
Trevitt CR, Collinge J (2006) A systematic review of prion therapeutics in experimental models. Brain 129:2241–2265
Tuzi NL, Cancellotti E, Baybutt H, Blackford L, Bradford B, Plinston C et al (2008) Host PrP glycosylation: a major factor determining the outcome of prion infection. PLoS Biol 6:e100
Tzaban S, Friedlander G, Schonberger O, Horonchik L, Yedidia Y, Shaked G et al (2002) Protease-sensitive scrapie prion protein in aggregates of heterogeneous sizes. Biochemistry 41:12868–12875
Uro-Coste E, Cassard H, Simon S, Lugan S, Bilheude JM, Perret-Liaudet A et al (2008) Beyond PrP9res) type 1/type 2 dichotomy in Creutzfeldt-Jakob disease. PLoS Pathog 4:e1000029
Wadsworth JDF, Hill AF, Joiner S, Jackson GS, Clarke AR, Collinge J (1999) Strain-specific prion-protein conformation determined by metal ions. Nat Cell Biol 1:55–59
Wang F, Wang X, Yuan CG, Ma J (2010) Generating a prion with bacterially expressed recombinant prion protein. Science 327:1132–1135
Watts JC, Westaway D (2007) The prion protein family: diversity, rivalry, and dysfunction. Biochim Biophys Acta 1772:654–672
Weissmann C (2004) The state of the prion. Nat Rev Microbiol 2:861–871
Zou WQ, Capellari S, Parchi P, Sy MS, Gambetti P, Chen SG (2003) Identification of novel proteinase K-resistant C-terminal fragments of PrP in Creutzfeldt-Jakob disease. J Biol Chem 278:40429–40436

Chapter 13
Modeling the Cell Biology of Prions

Richard Rubenstein and Robert B. Petersen

Abstract Cell models have been useful for elucidating the function of proteins and/or their role in pathogenesis. Even before the discovery that the prion protein was a normal cellular protein (Oesch et al. Cell 40 (4):735–746, 1985), cell models were developed to investigate prion infection (Rubenstein et al. J Gen Virol 65 (Pt 12):2191–2198, 1984). Subsequently, with the discovery of familial forms of human prion diseases (Hsiao et al. Nature 338 (6213):342–345, 1989), cell models were developed to investigate the effect of mutations on the metabolism of the prion protein and, in parallel, the normal synthesis and processing of the cellular prion protein. In this chapter, we review the progress made in these two areas to date.

Keywords Cell models • Prion protein • Prions • TSE agent replication • Cellular cultures • Pathogenic mutations

13.1 Cellular Cultures Supporting TSE Agent Replication

Cell cultures represent relevant and useful experimental models to study transmissible spongiform encephalopathies (TSEs) or prion diseases. Our current understanding of the cell biology of both the normal prion protein (PrP^C) and the pathogenic isoform (PrP^{Sc}) has utilized infected cell culture models. Cell culture models have also been useful in the development and validation of anti-prion drugs as well as

R. Rubenstein, Ph.D.
Department of Neurology and Physiology/Pharmacology, SUNY Downstate Medical Center, Brooklyn, NY, USA

R.B. Petersen, Ph.D. (✉)
Department of Pathology, Neuroscience, and Neurology, Case Western Reserve University, 2103 Cornell Road, 5-126 Wolstein Research Building, Cleveland, OH 44106, USA
e-mail: rbp@case.edu

W.-Q. Zou and P. Gambetti (eds.), *Prions and Diseases: Volume 1, Physiology and Pathophysiology*, DOI 10.1007/978-1-4614-5305-5_13,

offering an alternative approach to the transmission/infectivity assays historically performed in animal models. Cell culture models have also been used to study prion-induced cytopathological changes, which might help to explain the prion disease-associated neuropathogenesis observed in vivo.

Several cell culture models permissive to prion replication are available and some of them allow subpassaging to monitor stable and persistent replication of the infectious agent. The target cell type presumed to be most useful and informative would be cells of neuronal origin from the central nervous system (CNS), since the majority of prion infectivity is found in the CNS and the associated pathology is predominantly neurodegeneration. These include uncloned and cloned mouse neuroblastoma cell lines (N2a, C-1300, N1E-115) (Race et al. 1987; Nishida et al. 2000; Butler et al. 1988; Markovits et al. 1983; Ostlund et al. 2001) and murine GT1 hypothalamic neural cells (Schatzl et al. 1997). The GT1 cells are differentiated gonadotrophin-releasing hormone neurons, and in contrast to some of the neuroblastoma cell lines, they are susceptible to the 139A and 22L mouse-adapted scrapie strains, as well as prions from familial GSS and sporadic CJD. GT1 cells are particularly useful for studying prion infection-associated cytopathic effects, since they become stably infected in contrast to N2a cells and therefore do not require periodic subcloning to maintain an infected culture (Nishida et al. 2000; Schatzl et al. 1997).

In addition, neuronal stem cells isolated from conventional or transgenic mice propagate mouse-adapted prions (Giri et al. 2006; Milhavet et al. 2006). Recently, hippocampal-derived HpL3-4 cells obtained from a PrP^{C} knockout mouse and transfected with mouse PrP^{C} were shown to be permissive to the mouse-adapted 22L scrapie strain (Maas et al. 2007). Finally, cells from the peripheral nervous system, such as MSC80, murine Schwann-like cells, replicate low levels of the Rocky Mountain Laboratories (RML) mouse scrapie strain (Follet et al. 2002). Non-neuronal cell lines can also efficiently propagate prions. Common fibroblast cell lines (Vorberg et al. 2004), a microglial cell line (MG20) established from transgenic mice overexpressing PrP (Iwamaru et al. 2007), and PC12 rat pheochromocytoma cells (Rubenstein et al. 1984, 1991) are susceptible to various murine prion strains. Notably, the mouse-adapted bovine spongiform encephalopathy (BSE) agent was successfully propagated in MG20 cells. Infection of a skeletal myoblast cell line (C2C12) was recently described (Dlakic et al. 2007), and could be used to investigate the mechanism underlying the prion infection of muscles observed in sheep and cervids.

One well-established feature of the animal bioassay is the species-specificity relationship between the source of the infectious agent and the recipient animal, which dictates both efficiency of infection and latency. In the cell system, both homologous (i.e., species matched) and heterologous (i.e., species mismatched) cell culture model systems have been successfully used. The rationale for using homologous cies-barrier phenomenon observed in animal bioassays in which the efficiency of infection is reduced if there are dissimilar primary amino acid sequences in the PrP of the species from which the prion agent and the host cells were derived. However, cell culture studies have demonstrated that this is not as straightforward as it seems. For example, only a limited number of mouse-adapted scrapie strains can replicate in murine-derived host cell lines, and rat-derived PC12 cells can only be infected with selected mouse (but not rat)-adapted scrapie strains.

Although the source of the infectious agent is typically homogenized brain tissue originating from infected animals, partially purified preparations of scrapie-associated fibrils or PrP^{Sc} have also been used to achieve a higher-titer inoculum (Race et al. 1987). Cultures are either maintained in a nondividing, neuronal state, or passaged several times, and continually monitored for the disappearance (i.e., dilution) of the initial inoculum and appearance of de novo agent replication. To monitor propagation, cells are harvested at different times after exposure to the source of agent and cell lysates are used in animal bioassays. Alternatively, once it had been demonstrated that there is a close association between PrP^{Sc} and agent replication, the appearance and increase of the proteinase K (PK)-resistant PrP^{Sc} isoforms can be monitored by immunodetection and used as a biomarker of prion agent replication. Cell blotting techniques have been successfully used to detect PrP^{Sc} when only 1% of the cells are infected (Bosque and Prusiner 2000) and a filter retention assay for PrP^{Sc} (Winklhofer et al. 2001), which measures both PK resistance and presence of a detergent-insoluble aggregated state, has also been used. In addition, Vilette et al. have used a post-embedding method able to detect single infected cells (Vilette et al. 2001). This method has the advantage of evaluating the percentage of infected cells present in a particular culture. This is important when one considers the reports that for N2a cells only 1% of the cells were actually infected (Race 1991), although more efficient cell-culture models (Bosque and Prusiner 2000; Nishida et al. 2000) seem to have up to 30% of cells actually accumulating PrP^{Sc}.

The amount of infectivity present in the culture is also an important issue. Recent data on permissible cell lines revealed that cultures have the potential to accumulate as many infectious units per milligram of protein as brain from affected animals (Vilette et al. 2001).

Studying prion propagation in cell culture originally used animal-derived infected cells in which infected cultures were obtained from infected animals. The SMB cell line was established from the brain of a mouse clinically affected by the Chandler scrapie strain (Clarke and Haig 1970a, b). The majority of the initial studies on infecting cells in vitro used murine neuroblastoma cell lines (Race et al. 1987; Butler et al. 1988; Nishida et al. 2000; Markovits et al. 1983; Ostlund et al. 2001; Borchelt et al. 1990). Several investigators have described various biochemical and, at best, only subtle phenotypic differences in scrapie-infected cells. In addition, both increases and decreases in the rates of cell proliferation have been reported in infected cell lines. Unfortunately, it is not clear that the changes described were necessarily only due to the scrapie agent as opposed to clonal differences or to other factors present in the inoculum used to infect the cells. In addition, since the concentration of PrP has been shown to influence infectability, replication, and transmissibility of the prion agent in vivo, using an overexpressing cell line, such as the murine N2a neuroblastoma, allows these cells to be readily infected by the three mouse-adapted scrapie strains, Chandler, 139A, and 22L (Nishida et al. 2000).

A common feature of susceptible cell lines is that they only support the propagation of TSE strains that have been experimentally adapted to rodents. Recently, Vilette et al. developed a new heterologous model for naturally occurring sheep scrapie.

This model was obtained by stable expression of the ovine PrP gene in a rabbit epithelial cell line (RK13) (Vilette et al. 2001). The authors showed that the expression of heterologous PrP in an otherwise refractory system, such as the rabbit system, is sufficient tocross the species barrier ex vivo.

Infected cell culture models have provided some valuable insights into the biogenesis of PrP^{Sc} in terms of conversion, subcellular localizations, physiopathological consequences, and species-barrier determinants. They have also contributed to the screening and the study of possible therapeutic compounds and to the development of new strategies for the investigation of TSE-specific biomarkers. Studies with infected cell cultures have shown that PrP^{C} and PrP^{Sc} are associated with the cell surface differently since only the former can be released by phosphatidylinositol-specific phospholipase C (PIPLC) treatment of intact infected cells (Caughey et al. 1990; Lehmann and Harris 1996). Analysis of several types of infected cells, including N2a, GT1, and HaB (Schatzl et al. 1997; Taraboulos et al. 1990), made it clear that PrP^{Sc} resides within the cell and accumulates in late endosomes and/or lysosomes (McKinley et al. 1991; Pimpinelli et al. 2005) where amino terminal trimming of PrP^{Sc} may occur (Caughey et al. 1991).

Furthermore, although PrP^{C} is rapidly synthesized and degraded, while the abnormal PrP^{Sc} isoform is relatively stable (Borchelt et al. 1990; Caughey et al. 1989; Nunziante et al. 2003), the infected cells do have the capacity, processing functions, and proteases to degrade PrP^{Sc} (Beringue et al. 2004; Enari et al. 2001; Feraudet et al. 2005; Peretz et al. 2001; Perrier et al. 2004).

The information obtained from the use of infected cell cultures to study events associated with neurodegeneration have been limited. Replication of the prion agent in cultured cells can result in specific alterations in cellular metabolism, some of which can affect cell survival. For instance, infection with several murine prion strains impairs the cellular response of GT1 and N2a cells to oxidative stress (Milhavet et al. 2000), presumably through a decrease in superoxide dismutase activity. It is interesting to note that prion-infected cell lines accumulating infectious titers similar to those in brain tissue do not show any obvious cytopathic effect, with the possible exception of RML-infected GT1 cells that undergo apoptosis inconsistently. The use of primary cultures may lead to a better understanding of the effect of prion agent replication on neuronal death. For example, infection of primary cultures of neurons and astrocytes by a sheep scrapie agent resulted only in neuronal apoptosis involving JNK-c-Jun signaling (Cronier et al. 2004).

Numerous compounds have been used successfully to inhibit PrP^{Sc} formation in vitro, but the results in vivo have been disappointing. An example is provided by cellular heparan sulfates, which are sulfated linear polysaccharides typically linked to proteins to form heparan sulfate proteoglycans located at the cell surface (Turnbull et al. 2001). A number of studies suggest that heparan sulfates are involved in the biogenesis of PrP^{Sc} possibly by bringing together components involved in the conversion process such as PrP^{C}, PrP^{Sc}, and other possible cofactors. A variety of sulfated glycans, including pentosan polysulfate (Birkett et al. 2001; Caughey and Raymond 1993), dextran sulfate 500 (Barret et al. 2003; Beringue et al. 2004; Caughey and Raymond 1993), and heparin (Gabizon et al. 1993), are potent inhibitors of PrP^{Sc} accumulation in several cell lines infected with murine prions presumably by competitive inhibition of cellular heparan sulfates for the binding to PrP^{C} (Gabizon et al. 1993).

The use of cell culture models to determine the therapeutic value of compounds in vivo has been disappointing. A large number of compounds have been found to inhibit PrP^{Sc} accumulation in prion-infected cultures, mainly in N2a cells (Kocisko et al. 2003); however, most of them showed no or very limited effects when subsequently tested in infected animals (Trevitt and Collinge 2006). This does not necessarily mean that infected cell models are not adequate to screen for anti-prion drugs, but rather indicates that prion propagation in organisms is a complex biological process. In addition to drugs, passive immunization with anti-PrP antibodies (Abs) has been tested in cell culture models (Enari et al. 2001; Peretz et al. 2001; Perrier et al. 2004; Gilch et al. 2003). These Abs significantly reduced prion agent replication in cell culture by preventing the conversion of PrP^{C} into PrP^{Sc} through blockage of PrP^{C}–PrP^{Sc} binding and/or by stabilizing the PrP^{C} on the cell surface. Although infected animals injected with antibodies did, under certain circumstances, show a modest increase in survival times, it did not reflect the extent demonstrated in cell culture (Sigurdsson et al. 2003; White et al. 2003).

The utilization of a cell culture system as a replacement for the expensive and time-consuming animal bioassay has been explored. However, this has been hampered because of low sensitivity due, in part, to the small percentage of cells actually infected (Race et al. 1987). The isolation of N2a subclones with higher permissiveness (Bosque and Prusiner 2000; Enari et al. 2001), along with improved detection of PrP^{Sc}, allowed the development of a quantitative, highly sensitive scrapie cell-based infectivity assay (SCA) for the RML murine prion strain (Klohn et al. 2003). Although the SCA is almost as sensitive as the mouse bioassay while being much less expensive and ten times faster, it is limited in that N2a cells are not permissive to natural strains of the infectious agents.

Much research is still needed for the development of better cell culture models. These models will be important tools to dissect the properties of the prion agents including their molecular composition, the basis of cell permissiveness, and the identification of the biochemical and molecular mechanisms causing neuronal death. Some interesting studies along these lines have been reported. Weissmann's group recently demonstrated that the composition of the glycan can affect infection efficiency (Browning et al. 2011). In another study, sialyation of the glycosylphosphatidyl inositol anchor was shown to play a significant role in PrP aggregation, which is associated with neurodegeneration (Bate and Williams 2012).

13.2 Cell Models of Pathogenic Mutations in the Prion Protein

Following the discovery that PrP was a normal cellular protein, pathogenic mutations associated with familial prion diseases were discovered (Hsiao et al. 1989). This provided the opportunity to study the metabolism of the mutant protein in cell culture

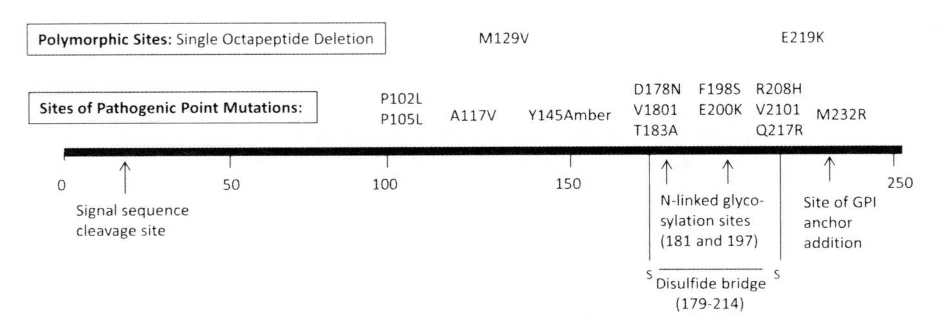

Fig. 13.1 Linear map of the prion protein showing a limited number of the pathogenic mutations in the human PrP as well as the sites of major post-translational modifications

models with the hope that this might shed light on the conditions that lead to pathogenic conversion of the prion protein. Cell culture models had been used to study the metabolism of PrP in infected cells, which included a detailed study of the synthesis and processing of PrP^C (Caughey et al. 1989). PrP^C is modified in the endoplasmic reticulum by the addition of a glycosylphospatidyl inositol (GPI) anchor and the nonobligatory addition of *N*-linked glycans; there are two N-linked glycosylation sites in the prion protein (Robakis et al. 1986; Locht et al. 1986). The N-linked glycans serve as a major source of heterogeneity in the prion protein (Rudd et al. 1999). As mentioned above, the glycans appear to influence infection (Browning et al. 2011) and may provide the basis for strain determination, as previously conjectured (Rudd et al. 2001).

To study the effect of point mutations on the synthesis and metabolism of PrP^C, a variety of cell models have been established (The insert mutation in the octapeptide repeat is not included based on the complexity of the clinical phenotype. See Solomon et al. (2010) for a review.). Some of the models use heterologous pairings of cells and homologues of the pathogenic human mutations (Lehmann and Harris 1995, 1996; Ma and Lindquist 2001). An alternative model, in which the human PrP-coding region was placed under the control of a strong promoter in an episomal vector (Petersen et al. 1996) and transfected into a human neuroblastoma cell line (M17), is the only cell model that has been validated by comparison with human tissue. The use of an episomal vector eliminates problems of copy number and integration site effects. In addition, the instantaneous rate of protein synthesis is the same for all constructs so that modifications that may be concentration dependent, i.e., glycosylation, GPI anchoring, etc., are unaffected. In general, however, similar results have been obtained in all systems.

Detailed studies examining the metabolic defects associated with expression of mutant PrP, PrP^M, suggest that the effects of the mutations fall into two general categories (see Fig. 13.1). The first category includes mutations around the normal amino terminal cleavage site at residue 111/112, which includes those at codons 102, 105, and 117. In the human neuroblastoma cell models, these mutations do not appear to affect overall metabolism, but seem to cause an altered cleavage of PrP (Mishra et al. 2002) (RBP, unpublished). The truncated fragments are generally associated with Gerstmann–Straussler–Scheinker syndrome, a familiar form of prion disease that presents with prion plaques. Novel PrP fragments have been found in cells expressing the F198S, E200K, and Q217R mutations, indicating an alteration in protein processing (Zaidi et al. 2005; Capellari et al. 2000a; Singh et al. 1997).

The second category of mutations is clustered in and around the site of post-translational modifications that include nonobligatory addition of two *N*-linked glycans and the formation of a disulfide bridge. These mutations include those at codons D178N (Petersen et al. 1996), T183A (Capellari et al. 2000b), F198S (Zaidi et al. 2005), E200K (Capellari et al. 2000a), and Q217R (Singh et al. 1997). In this category, the mutations shared four common alterations in metabolism or processing: (1) PrP^{M} is unstable and degraded. This is particularly evident in the unglycosylated form of the PrP^{M}, which is virtually absent in some of these mutants. Also, treatment with tunicamycin, which inhibits glycosylation, results in the rapid degradation of PrP^{M} compared to PrP^{C} in the secretory pathway. These results support the role of *N*-linked glycans in facilitating protein folding. An early study suggested that loss of the first glycosylation site blocks transport of the mutant protein (Lehmann and Harris 1997); however, it was later established that the T183A mutation results in a structural change in the protein (Capellari et al. 2000b). (2) As a result of the decreased stability of the PrP^{M}, less PrP^{M} is found on the surface of cells expressing PrP^{M} and the ratio of glycoforms found at the cell surface parallels that found in the cells with marked underrepresentation of the unglycosylated form. Approximately 90% of PrP^{C} reaches the cell membrane, indicating that 10% of the normal protein fails to fold properly (Cohen and Taraboulos 2003). (3) A greater proportion of PrP^{M} partitions in a detergent insoluble fraction, indicating that PrP^{M} is aggregated in cells. (4) Most of these defects can be alleviated, in part, by incubating the cells at reduced temperature, 24 °C, suggesting that the processing defects arise due to misfolding of the PrP^{M}; misfolding has been shown to be partially corrected by reduced temperature (Singh et al. 1997). It is interesting to note that biophysical studies using recombinant PrP indicated that, with the exception of mutations at codons 183 and 198, these mutations do not appear to affect the physical properties of PrP^{M} versus PrP^{C} (Liemann and Glockshuber 1999).

In addition to the common changes resulting from the mutations, some of the mutants exhibited their own specific alterations. First, the Q217R mutation results in the production of a 32-kDa PrP lacking the GPI anchor, which attaches PrP to the cell surface (Singh et al. 1997). The F198S mutation results in the most profound reduction in the unglycosylated form of PrP^{M} (Zaidi et al. 2005). This arises for two reasons. First, the unglycosylated form is unstable, and second, the mutation replaces the phenylalanine residue, which is in the middle of the second glycosylation site, with a serine residue that is known to produce a more efficient glycosylation site. The F198S mutation also established that while the protein can achieve a normal conformation when expressed in a cell, after denaturation it fails to refold into a native conformation (Zaidi et al. 2005). The E200K mutation, which is just beyond the second N-linked glycosylation site, results in the delayed maturation of PrP^{M}

and the production of an abnormally modified glycan that is observed by its abnormal migration in SDS gels (Capellari et al. 2000a). Finally, comparison of the D178N/129M mutation (FFI) and D178N/129V mutation (CJD^{178}) did not exhibit specific differences, although the reduction of the unglycosylated form was more pronounced in the FFI expressing cells (Petersen et al. 1996). It is interesting to note that the prion disease referred to as sporadic fatal insomnia is also linked to codon 129 methionine, indicating that methionine may be required for the initiation of the disease process in the thalamus (Parchi et al. 1999).

Although some of the mutations studied resulted in general and specific changes in the metabolism of PrP^{M}, such as aggregation, none of the transfected human neuroblastoma lines produced bona fide protease resistant PrP (PrP^{Sc}), as assessed by the gel migration pattern or infectivity. In fact, using an antibody that recognizes the carboxyl terminal region of PrP, wild-type PrP is as resistant to proteinase K treatment as PrP^{M} (Capellari et al. 2000a). The carboxyl terminal region of PrP is inherently resistant to protease digestion, and mutations in the region of post-translational modification appear to extend the tertiary structure through residues 90–112 that are typically unstructured. Thus, the weak protease resistance that has been associated with PrP^{M} expressed in cell culture probably reflects a conformational change of the protein. In retrospect, it is not surprising that the cell culture models expressing the mutant PrP failed to produce PrP^{Sc}. The inherited human prion diseases are age-related diseases, so while the mutations may be necessary for the development of disease, they are clearly not sufficient. This suggests that some age-related deficit in the cellular repair/defense mechanisms is required to enable the initiation of the disease process.

While the cells expressing the mutant PrP grow normally, inhibition of the proteasomal degradation has been observed to result in neuronal cytotoxicity. The first observation of PrP^{M} accumulation was in cells expressing the nonsense mutation at codon 145, Y145Stop (Zanusso et al. 1999), in which the mutant protein accumulated in the cell after inhibition of the proteasome with lactacystin. Subsequent studies demonstrated that the codon 177 murine homologue of the human D178N mutation also accumulated in cells, even in the absence of proteasome inhibition (Ma and Lindquist 2001), and that this resulted in neurotoxicity (Ma et al. 2002).

13.3 Conclusion

Cell models have been invaluable for studying the infectious process at a cellular level. The infected cell model is currently facilitating studies that will help clarify the origin of prion strains. The cell models of the pathogenic prion mutations indicate that the mutations are not sufficient to produce the disease-associated form of PrP, but show the potential for chronic stress in the secretory pathway that may facilitate the disease process.

References

Barret A, Tagliavini F, Forloni G, Bate C, Salmona M, Colombo L, De Luigi A, Limido L, Suardi S, Rossi G, Auvre F, Adjou KT, Sales N, Williams A, Lasmezas C, Deslys JP (2003) Evaluation of quinacrine treatment for prion diseases. J Virol 77(15):8462–8469

Bate C, Williams A (2012) Neurodegeneration induced by clustering of sialylated glycosylphosphatidylinositols of prion proteins. J Biol Chem 287(11):7935–7944. doi:10.1074/jbc.M111.275743

Beringue V, Vilette D, Mallinson G, Archer F, Kaisar M, Tayebi M, Jackson GS, Clarke AR, Laude H, Collinge J, Hawke S (2004) PrPSc binding antibodies are potent inhibitors of prion replication in cell lines. J Biol Chem 279(38):39671–39676. doi:10.1074/jbc.M402270200

Birkett CR, Hennion RM, Bembridge DA, Clarke MC, Chree A, Bruce ME, Bostock CJ (2001) Scrapie strains maintain biological phenotypes on propagation in a cell line in culture. EMBO J 20(13):3351–3358. doi:10.1093/emboj/20.13.3351

Borchelt DR, Scott M, Taraboulos A, Stahl N, Prusiner SB (1990) Scrapie and cellular prion proteins differ in their kinetics of synthesis and topology in cultured cells. J Cell Biol 110(3):743–752

Bosque PJ, Prusiner SB (2000) Cultured cell sublines highly susceptible to prion infection. J Virol 74(9):4377–4386

Browning S, Baker CA, Smith E, Mahal SP, Herva ME, Demczyk CA, Li J, Weissmann C (2011) Abrogation of complex glycosylation by swainsonine results in strain- and cell-specific inhibition of prion replication. J Biol Chem 286(47):40962–40973. doi:10.1074/jbc.M111.283978

Butler DA, Scott MR, Bockman JM, Borchelt DR, Taraboulos A, Hsiao KK, Kingsbury DT, Prusiner SB (1988) Scrapie-infected murine neuroblastoma cells produce protease-resistant prion proteins. J Virol 62(5):1558–1564

Capellari S, Parchi P, Russo CM, Sanford J, Sy MS, Gambetti P, Petersen RB (2000a) Effect of the E200K mutation on prion protein metabolism. Comparative study of a cell model and human brain. Am J Pathol 157(2):613–622

Capellari S, Zaidi SI, Long AC, Kwon EE, Petersen RB (2000b) The Thr183Ala mutation, not the loss of the first glycosylation site, alters the physical properties of the prion protein. J Alzheimers Dis 2(1):27–35

Caughey B, Raymond GJ (1993) Sulfated polyanion inhibition of scrapie-associated PrP accumulation in cultured cells. J Virol 67(2):643–650

Caughey B, Race RE, Ernst D, Buchmeier MJ, Chesebro B (1989) Prion protein biosynthesis in scrapie-infected and uninfected neuroblastoma cells. J Virol 63(1):175–181

Caughey B, Neary K, Buller R, Ernst D, Perry LL, Chesebro B, Race RE (1990) Normal and scrapie-associated forms of prion protein differ in their sensitivities to phospholipase and proteases in intact neuroblastoma cells. J Virol 64(3):1093–1101

Caughey B, Raymond GJ, Ernst D, Race RE (1991) N-terminal truncation of the scrapie-associated form of PrP by lysosomal protease(s): implications regarding the site of conversion of PrP to the protease-resistant state. J Virol 65(12):6597–6603

Clarke MC, Haig DA (1970a) Evidence for the multiplication of scrapie agent in cell culture. Nature 225(5227):100–101

Clarke MC, Haig DA (1970b) Multiplication of scrapie agent in cell culture. Res Vet Sci 11(5):500–501

Cohen E, Taraboulos A (2003) Scrapie-like prion protein accumulates in aggresomes of cyclosporin A-treated cells. EMBO J 22(3):404–417. doi:10.1093/emboj/cdg045

Cronier S, Laude H, Peyrin JM (2004) Prions can infect primary cultured neurons and astrocytes and promote neuronal cell death. Proc Natl Acad Sci USA 101(33):12271–12276. doi:10.1073/pnas.0402725101

Dlakic WM, Grigg E, Bessen RA (2007) Prion infection of muscle cells in vitro. J Virol 81(9):4615–4624. doi:10.1128/JVI.02628-06

Enari M, Flechsig E, Weissmann C (2001) Scrapie prion protein accumulation by scrapie-infected neuroblastoma cells abrogated by exposure to a prion protein antibody. Proc Natl Acad Sci USA 98(16):9295–9299. doi:10.1073/pnas.151242598

Feraudet C, Morel N, Simon S, Volland H, Frobert Y, Creminon C, Vilette D, Lehmann S, Grassi J (2005) Screening of 145 anti-PrP monoclonal antibodies for their capacity to inhibit PrPSc replication in infected cells. J Biol Chem 280(12):11247–11258. doi:10.1074/jbc.M407006200

Follet J, Lemaire-Vieille C, Blanquet-Grossard F, Podevin-Dimster V, Lehmann S, Chauvin JP, Decavel JP, Varea R, Grassi J, Fontes M, Cesbron JY (2002) PrP expression and replication by Schwann cells: implications in prion spreading. J Virol 76(5):2434–2439

Gabizon R, Meiner Z, Halimi M, Ben-Sasson SA (1993) Heparin-like molecules bind differentially to prion-proteins and change their intracellular metabolic fate. J Cell Physiol 157(2):319–325. doi:10.1002/jcp.1041570215

Gilch S, Wopfner F, Renner-Muller I, Kremmer E, Bauer C, Wolf E, Brem G, Groschup MH, Schatzl HM (2003) Polyclonal anti-PrP auto-antibodies induced with dimeric PrP interfere efficiently with PrPSc propagation in prion-infected cells. J Biol Chem 278(20):18524–18531. doi:10.1074/jbc.M210723200

Giri RK, Young R, Pitstick R, DeArmond SJ, Prusiner SB, Carlson GA (2006) Prion infection of mouse neurospheres. Proc Natl Acad Sci USA 103(10):3875–3880. doi:10.1073/pnas.0510902103

Hsiao K, Baker HF, Crow TJ, Poulter M, Owen F, Terwilliger JD, Westaway D, Ott J, Prusiner SB (1989) Linkage of a prion protein missense variant to Gerstmann-Straussler syndrome. Nature 338(6213):342–345. doi:10.1038/338342a0

Iwamaru Y, Takenouchi T, Ogihara K, Hoshino M, Takata M, Imamura M, Tagawa Y, Hayashi-Kato H, Ushiki-Kaku Y, Shimizu Y, Okada H, Shinagawa M, Kitani H, Yokoyama T (2007) Microglial cell line established from prion protein-overexpressing mice is susceptible to various murine prion strains. J Virol 81(3):1524–1527. doi:10.1128/JVI.01379-06

Klohn PC, Stoltze L, Flechsig E, Enari M, Weissmann C (2003) A quantitative, highly sensitive cell-based infectivity assay for mouse scrapie prions. Proc Natl Acad Sci USA 100(20):11666–11671. doi:10.1073/pnas.1834432100

Kocisko DA, Baron GS, Rubenstein R, Chen J, Kuizon S, Caughey B (2003) New inhibitors of scrapie-associated prion protein formation in a library of 2000 drugs and natural products. J Virol 77(19):10288–10294

Lehmann S, Harris DA (1995) A mutant prion protein displays an aberrant membrane association when expressed in cultured cells. J Biol Chem 270(41):24589–24597

Lehmann S, Harris DA (1996) Mutant and infectious prion proteins display common biochemical properties in cultured cells. J Biol Chem 271(3):1633–1637

Lehmann S, Harris DA (1997) Blockade of glycosylation promotes acquisition of scrapie-like properties by the prion protein in cultured cells. J Biol Chem 272(34):21479–21487

Liemann S, Glockshuber R (1999) Influence of amino acid substitutions related to inherited human prion diseases on the thermodynamic stability of the cellular prion protein. Biochemistry (Mosc) 38(11):3258–3267

Locht C, Chesebro B, Race R, Keith JM (1986) Molecular cloning and complete sequence of prion protein cDNA from mouse brain infected with the scrapie agent. Proc Natl Acad Sci USA 83(17):6372–6376

Ma J, Lindquist S (2001) Wild-type PrP and a mutant associated with prion disease are subject to retrograde transport and proteasome degradation. Proc Natl Acad Sci USA 98(26):14955–14960. doi:10.1073/pnas.011578098

Ma J, Wollmann R, Lindquist S (2002) Neurotoxicity and neurodegeneration when PrP accumulates in the cytosol. Science 298(5599):1781–1785. doi:10.1126/science.1073725

Maas E, Geissen M, Groschup MH, Rost R, Onodera T, Schatzl H, Vorberg IM (2007) Scrapie infection of prion protein-deficient cell line upon ectopic expression of mutant prion proteins. J Biol Chem 282(26):18702–18710. doi:10.1074/jbc.M701309200

Markovits P, Dautheville C, Dormont D, Dianoux L, Latarjet R (1983) In vitro propagation of the scrapie agent. I. Transformation of mouse glia and neuroblastoma cells after infection with the mouse-adapted scrapie strain c-506. Acta Neuropathol 60(1–2):75–80

McKinley MP, Taraboulos A, Kenaga L, Serban D, Stieber A, DeArmond SJ, Prusiner SB, Gonatas N (1991) Ultrastructural localization of scrapie prion proteins in cytoplasmic vesicles of infected cultured cells. Lab Invest 65(6):622–630

Milhavet O, McMahon HE, Rachidi W, Nishida N, Katamine S, Mange A, Arlotto M, Casanova D, Riondel J, Favier A, Lehmann S (2000) Prion infection impairs the cellular response to oxidative stress. Proc Natl Acad Sci USA 97(25):13937–13942. doi:10.1073/pnas.250289197

Milhavet O, Casanova D, Chevallier N, McKay RD, Lehmann S (2006) Neural stem cell model for prion propagation. Stem Cells 24(10):2284–2291. doi:10.1634/stemcells.2006-0088

Mishra RS, Gu Y, Bose S, Verghese S, Kalepu S, Singh N (2002) Cell surface accumulation of a truncated transmembrane prion protein in Gerstmann-Straussler-Scheinker disease P102L. J Biol Chem 277(27):24554–24561

Nishida N, Harris DA, Vilette D, Laude H, Frobert Y, Grassi J, Casanova D, Milhavet O, Lehmann S (2000) Successful transmission of three mouse-adapted scrapie strains to murine neuroblastoma cell lines overexpressing wild-type mouse prion protein. J Virol 74(1):320–325

Nunziante M, Gilch S, Schatzl HM (2003) Essential role of the prion protein N terminus in subcellular trafficking and half-life of cellular prion protein. J Biol Chem 278(6):3726–3734. doi:10.1074/jbc.M206313200

Oesch B, Westaway D, Walchli M, McKinley MP, Kent SB, Aebersold R, Barry RA, Tempst P, Teplow DB, Hood LE et al (1985) A cellular gene encodes scrapie PrP 27-30 protein. Cell 40(4):735–746

Ostlund P, Lindegren H, Pettersson C, Bedecs K (2001) Altered insulin receptor processing and function in scrapie-infected neuroblastoma cell lines. Brain Res Mol Brain Res 97(2): 161–170

Parchi P, Capellari S, Chin S, Schwarz HB, Schecter NP, Butts JD, Hudkins P, Burns DK, Powers JM, Gambetti P (1999) A subtype of sporadic prion disease mimicking fatal familial insomnia. Neurology 52(9):1757–1763

Peretz D, Williamson RA, Kaneko K, Vergara J, Leclerc E, Schmitt-Ulms G, Mehlhorn IR, Legname G, Wormald MR, Rudd PM, Dwek RA, Burton DR, Prusiner SB (2001) Antibodies inhibit prion propagation and clear cell cultures of prion infectivity. Nature 412(6848):739–743. doi:10.1038/35089090

Perrier V, Solassol J, Crozet C, Frobert Y, Mourton-Gilles C, Grassi J, Lehmann S (2004) Anti-PrP antibodies block PrPSc replication in prion-infected cell cultures by accelerating PrPC degradation. J Neurochem 89(2):454–463. doi:10.1111/j.1471-4159.2004.02356.x

Petersen RB, Parchi P, Richardson SL, Urig CB, Gambetti P (1996) Effect of the D178N mutation and the codon 129 polymorphism on the metabolism of the prion protein. J Biol Chem 271(21):12661–12668

Pimpinelli F, Lehmann S, Maridonneau-Parini I (2005) The scrapie prion protein is present in flotillin-1-positive vesicles in central- but not peripheral-derived neuronal cell lines. Eur J Neurosci 21(8):2063–2072. doi:10.1111/j.1460-9568.2005.04049.x

Race R (1991) The scrapie agent in vitro. Curr Top Microbiol Immunol 172:181–193

Race RE, Fadness LH, Chesebro B (1987) Characterization of scrapie infection in mouse neuroblastoma cells. J Gen Virol 68(Pt 5):1391–1399

Robakis NK, Devine-Gage EA, Jenkins EC, Kascsak RJ, Brown WT, Krawczun MS, Silverman WP (1986) Localization of a human gene homologous to the PrP gene on the p arm of chromosome 20 and detection of PrP-related antigens in normal human brain. Biochem Biophys Res Commun 140(2):758–765

Rubenstein R, Carp RI, Callahan SM (1984) In vitro replication of scrapie agent in a neuronal model: infection of PC12 cells. J Gen Virol 65(Pt 12):2191–2198

Rubenstein R, Deng H, Scalici CL, Papini MC (1991) Alterations in neurotransmitter-related enzyme activity in scrapie-infected PC12 cells. J Gen Virol 72(Pt 6):1279–1285

Rudd PM, Endo T, Colominas C, Groth D, Wheeler SF, Harvey DJ, Wormald MR, Serban H, Prusiner SB, Kobata A, Dwek RA (1999) Glycosylation differences between the normal and pathogenic prion protein isoforms. Proc Natl Acad Sci USA 96(23):13044–13049

Rudd PM, Wormald MR, Wing DR, Prusiner SB, Dwek RA (2001) Prion glycoprotein: structure, dynamics, and roles for the sugars. Biochemistry (Mosc) 40(13):3759–3766

Schatzl HM, Laszlo L, Holtzman DM, Tatzelt J, DeArmond SJ, Weiner RI, Mobley WC, Prusiner SB (1997) A hypothalamic neuronal cell line persistently infected with scrapie prions exhibits apoptosis. J Virol 71(11):8821–8831
Sigurdsson EM, Sy MS, Li R, Scholtzova H, Kascsak RJ, Kascsak R, Carp R, Meeker HC, Frangione B, Wisniewski T (2003) Anti-prion antibodies for prophylaxis following prion exposure in mice. Neurosci Lett 336(3):185–187
Singh N, Zanusso G, Chen SG, Fujioka H, Richardson S, Gambetti P, Petersen RB (1997) Prion protein aggregation reverted by low temperature in transfected cells carrying a prion protein gene mutation. J Biol Chem 272(45):28461–28470
Solomon IH, Schepker JA, Harris DA (2010) Prion neurotoxicity: insights from prion protein mutants. Curr Issues Mol Biol 12(2):51–61
Taraboulos A, Serban D, Prusiner SB (1990) Scrapie prion proteins accumulate in the cytoplasm of persistently infected cultured cells. J Cell Biol 110(6):2117–2132
Trevitt CR, Collinge J (2006) A systematic review of prion therapeutics in experimental models. Brain 129(Pt 9):2241–2265. doi:10.1093/brain/awl150
Turnbull J, Powell A, Guimond S (2001) Heparan sulfate: decoding a dynamic multifunctional cell regulator. Trends Cell Biol 11(2):75–82
Vilette D, Andreoletti O, Archer F, Madelaine MF, Vilotte JL, Lehmann S, Laude H (2001) Ex vivo propagation of infectious sheep scrapie agent in heterologous epithelial cells expressing ovine prion protein. Proc Natl Acad Sci USA 98(7):4055–4059. doi:10.1073/pnas.061337998
Vorberg I, Raines A, Story B, Priola SA (2004) Susceptibility of common fibroblast cell lines to transmissible spongiform encephalopathy agents. J Infect Dis 189(3):431–439. doi:10.1086/381166
White AR, Enever P, Tayebi M, Mushens R, Linehan J, Brandner S, Anstee D, Collinge J, Hawke S (2003) Monoclonal antibodies inhibit prion replication and delay the development of prion disease. Nature 422(6927):80–83. doi:10.1038/nature01457
Winklhofer KF, Hartl FU, Tatzelt J (2001) A sensitive filter retention assay for the detection of PrP(Sc) and the screening of anti-prion compounds. FEBS Lett 503(1):41–45
Zaidi SI, Richardson SL, Capellari S, Song L, Smith MA, Ghetti B, Sy MS, Gambetti P, Petersen RB (2005) Characterization of the F198S prion protein mutation: enhanced glycosylation and defective refolding. J Alzheimers Dis 7(2):159–171, discussion 173–180
Zanusso G, Petersen RB, Jin T, Jing Y, Kanoush R, Ferrari S, Gambetti P, Singh N (1999) Proteasomal degradation and N-terminal protease resistance of the codon 145 mutant prion protein. J Biol Chem 274(33):23396–23404

Chapter 14
Prion Strain Interference

Charles R. Schutt, Ronald A. Shikiya, and Jason C. Bartz

Abstract Prions are transmissible agents that comprised of a misfolded protein PrP^{Sc} that is posttranslationally derived from the normal isoform PrP^{C}. Prion strains are operationally defined by differences in the distribution and intensity of spongiform degeneration and distribution of PrP^{Sc} in the CNS. The mechanism by which prion strains are encoded is not known; however, current evidence suggests that the conformation of PrP^{Sc} encodes prion strain diversity. In natural prion disease, more than one prion strain can exist in an individual. Prion strains, when present in the same host, can interfere with each other, a process that may be important during prion adaptation following interspecies transmission. While the parameters that influence prion strain interference are beginning to be described, the mechanism responsible for strain interference is not known.

Keywords Prion diseases • Adaptation • Strains • Interference

14.1 Introduction

Prions are transmissible agents comprised of a misfolded protein PrP^{Sc} that is post-translationally derived from the normal isoform PrP^{C}. PrP^{C} is a cell-surface protein that is attached to the cellular membrane via a glycosylphosphatidylinositol anchor that is expressed in numerous cell types, but is most abundant in the central nervous system (CNS) (Basler et al. 1986; Prusiner 1991). Prion replication is initiated at the cell surface by the binding of PrP^{Sc} to PrP^{C} where these molecules are subsequently

C.R. Schutt, Ph.D • R.A. Shikiya, Ph.D. • J.C. Bartz, Ph.D. (✉)
Department of Medical Microbiology and Immunology, School of Medicine,
Creighton University, Omaha, NE 68178, USA
e-mail: jbartz@creighton.edu

W.-Q. Zou and P. Gambetti (eds.), *Prions and Diseases: Volume 1, Physiology and Pathophysiology*, DOI 10.1007/978-1-4614-5305-5_14,

endocytosed and the conversion of PrP^C to PrP^{Sc} occurs at the cell surface and/or in the endosomal/lysosomal system (Caughey and Raymond 1991). This conversion has been replicated in several cell-free systems, resulting in PrP molecules with biochemical and infectious properties of PrP^{Sc} (Castilla et al. 2005; Colby et al. 2007; Kocisko et al. 1994).

Prion diseases are neurodegenerative diseases of animals including humans. Animal prion diseases include scrapie of sheep and goats, transmissible mink encephalopathy of ranch-raised mink, bovine spongiform encephalopathy, and chronic wasting disease of captive and free-ranging deer, elk, and moose. Human prion diseases comprise kuru of the Fore people of Papua New Guinea, Creutzfeldt–Jacob disease (CJD), Gerstmann–Straussler–Scheinker syndrome, and fatal familial insomnia. Prion diseases are unique in biology because they have infectious, familial, and sporadic etiologies (Parchi and Gambetti 1995). Infectious prions can be detected in patients from all disease etiologies, suggesting de novo formation of prion infectivity which is consistent with recent studies where infectious prions were experimentally generated from noninfectious components (Deleault et al. 2007; Sigurdson et al. 2009). Prion diseases are zoonotic and the emergence of variant CJD (vCJD) is caused by the transmission of BSE to humans by an unknown route of infection (Bruce et al. 1997; Lasmezas et al. 1996).

Prion strains are operationally defined by differences in the distribution and intensity of spongiform degeneration and distribution of PrP^{Sc} in the CNS. Prion strain can differ in incubation period, clinical signs, agent distribution in the host, and host range (Bartz et al. 2005; Dickinson and Prusiner 1979; Kimberlin et al. 1987, 1989). These phenotypic parameters of prion strains are maintained during experimental passage. The mechanism by which prion strains are encoded is not known; however, evidence suggests that the conformation of PrP^{Sc} may be involved (Bessen and Marsh 1992a, 1994, 1995; Caughey et al. 1998; Telling et al. 1996b). The mechanisms by which changes in PrP^{Sc} conformation result in the strain-specific differences in the phenotype of disease are unknown.

In natural prion disease, more than one prion strain can exist in an individual. Transmission of field isolates of prion disease to rodents can result in the emergence of several distinct prion strains, suggesting that more than one strain is present in the field isolate (Dickinson 1976; Kimberlin and Walker 1978). Alternatively, interspecies transmission may result in the generation of new strains that have increased fitness for the new host species (Bartz et al. 2000; Dickinson and Prusiner 1979). Truncated forms of PrP^{Sc} with different molecular weights have been identified in individual humans infected with CJD (Polymenidou et al. 2005). These data strongly suggest that more than one prion strain can coexist in an individual human affected with prion disease.

Prion strains, when present in the same host, can interfere with each other. Prion strain interference was first described by Alan Dickinson where he demonstrated that inoculation of the 22C agent (the blocking strain) in mice prior to superinfection with the 22A agent (the superinfecting strain) could lead to an extension of the incubation period of 22A (Dickinson et al. 1972).

14.2 Parameters Governing Prion Strain Interference

14.2.1 Overview

The general parameters that govern prion strain interference are beginning to be understood. The blocking strain must be able to replicate to interfere with the superinfecting strain, and that increasing the titer of the blocking strain or increasing the interval between blocking strain infection and superinfection increases the interference effect. Prion strain interference has been demonstrated using numerous prion strains in two rodent animal models by multiple routes of infection, including oral infection, suggesting that prion strain interference is a more generalized phenomenon (Table 13.1) (Bartz et al. 2004, 2007; Dickinson et al. 1972, 1975; Kimberlin and Walker 1985; Manuelidis 1998; Nilsson et al. 2010; Nishida et al. 2005; Schutt and Bartz 2008; Shikiya et al. 2010). The relative onset of agent replication may be a more important parameter than if the strains are coinfected or superinfected. Finally, for strain interference to occur in the CNS, the two strains must infect the same neuroanatomical pathways. The evidence supporting these generalizations is outlined below.

14.2.2 The Interval Between Prion Strains Inoculation Influences Interference

Prion strain interference was first demonstrated when Dickinson et al. (1972) intracerebrally (i.c.) inoculated VM mice (*sinc* genotype *p7p7*) with the long incubation period mouse-adapted scrapie strain 22C nine, five, or one week prior to i.c. inoculation of the shorter incubation period strain 22A. Even though the 22A agent was able to cause disease in all mice based on the lesions profile, there was a significant extension of incubation period in the mice with a five- or nine-week interval between inoculations. Prions strain interference has also been demonstrated using the two mouse-adapted human strains SY (a long incubation period strain isolated from a sCJD patient) and FU (a short incubation period strain isolated from a Gerstmann–Straussler–Scheinker patient). When SY was i.c. inoculated 80 or 92 days prior to inoculation with FU, FU protease K-resistant PrP, pathology or clinical signs were not detected, demonstrating that FU has been blocked (Manuelidis and Lu 2003; Manuelidis 1998). Interference has also been demonstrated using i.v. inoculation. When SY was inoculated 80 days prior to FU, there was a significant increase in the incubation time of the superinfected FU (Manuelidis and Lu 2003).

Similar interfering effects have been demonstrated by the sciatic nerve (i.sc.) route of infection in hamsters using strains isolated from transmissible mink encephalopathy (TME). When the DY TME agent was inoculated into the sciatic nerve 30 or 60 days prior to the HY TME agent, there was no evidence of prion

Table 14.1 Summary of prion interference studies in animals

Blocking strain	Superinfecting strain	Route of inoculation	Host species	Interference effect	Reference
22C scrapie	22A scrapie	i.c.	VM (*Sinc*$^{p7/p7}$) mice	N,I,B[a]	Dickinson et al. (1972)
22A scrapie	22C scrapie	i.p.	RIII (*Sinc*$^{s7/s7}$) mice		Dickinson et al. (1975)
22A scrapie	22C scrapie	i.p.	CW (*Sinc*$^{s7/s7}$) mice		Kimberlin and Walker (1985)
22A scrapie	22C scrapie	i.p.	RIII (*Sinc*$^{s7/s7}$) mice	B	Taylor et al. (1986)
SY CJD	FU-1 GSS	i.c.	CD-1 (*Sinc*$^{s7/s7}$) mice		Manuelidis and Lu (2003), Manuelidis (1998), Manuelidis and Yun Lu (2000)
SY CJD	FU-1 GSS	i.v.	CD-1 (*Sinc*$^{s7/s7}$) mice		Manuelidis and Lu (2003)
SY CJD	FU-1 GSS	i.c.	RAG-1$^{-/-}$(*Sinc*$^{s7/s7}$) mice	N	Manuelidis and Lu (2003)
SY CJD	FU-1 GSS	i.c.	C57BIL/6 (Sinc$^{s7/s7}$) mice	I	Manuelidis and Lu (2003)
DY TME	HY TME	i.sc.	Syrian hamster	N,I,B	Bartz et al. (2007)
DY TME	263 K	i.sc.	Syrian hamster	B	Schutt and Bartz (2008)
DY TME	HaCWD	i.sc.	Syrian hamster	B	Schutt and Bartz (2008)
DY TME	HY TME	i.p.	Syrian hamster	N,I	Bartz et al. (2004, 2007)
DY TME	HY TME	Per os	Syrian hamster	N,I	Schutt and Bartz (2008)
DY TME	HY TME	i.c.	Syrian hamster	N,I,B	Bartz et al. (2000)
DY TME	HY TME	i.p.	Syrian hamster	N	Bessen and Marsh (1992b)
TME	22A scrapie	i.p.	VM (*Sinc*$^{p7/p7}$) mice	N	Taylor et al. (1986)
TME	22C scrapie	i.p.	VL (*Sinc*$^{s7/s7}$) mice	N	Taylor et al. (1986)
TME	79A scrapie	i.p.	VL (*Sinc*$^{s7/s7}$) mice	N	Taylor et al. (1986)
TME	79V scrapie	i.p.	VM (*Sinc*$^{p7/p7}$) mice	N	Taylor et al. (1986)
TME	79V scrapie	i.c., i.p.	BRVR (*Sinc*$^{s7/s7}$) mice	N	Taylor et al. (1986)
TME	87A scrapie	i.c., i.p.	BALB (*Sinc*$^{s7/s7}$) mice	N	Taylor et al. (1986)
TME	139A scrapie	i.p.	VL (*Sinc*$^{s7/s7}$) mice	N	Taylor et al. (1986)
TME	ME7H scrapie	i.c., i.p.	BRVR (*Sinc*$^{s7/s7}$) mice	N	Taylor et al. (1986)
Ts-1 scrapie	Ts-2 scrapie	i.c.	CD-1 (*Sinc*$^{s7/s7}$) mice	N	Hirogari et al. (2003)
Ts-1 scrapie	Ts-1 scrapie	i.c.	CD-1 (*Sinc*$^{s7/s7}$) mice	N	Hirogari et al. (2003)

[a]*N* no interference, *I* strain interference, *B* complete blocking

strain interference based on clinical signs, Western blot migration, and incubation period of disease (Bartz et al. 2004, 2007). When the interval was extended to 90 days between i.sc. inoculations, the DY TME agent was able to extend the incubation period of the HY TME agent by 12 days (Bartz et al. 2007). A 120-day interval between i.sc. inoculations resulted in the DY TME agent completely blocking the HY TME agent (or the 263 K and HaCWD agents) from causing disease based on these same three criteria (Bartz et al. 2007; Schutt and Bartz 2008). The ability of DY TME to interfere with or block HY TME corresponded with the detection of DY PrP^{Sc} in the lumbar spinal cord, consistent with the hypothesis that replication of the blocking strain is required for prion strain interference (Bartz et al. 2007; Shikiya et al. 2010). The DY TME agent can interfere with HY TME when both strains are inoculated per os. Per os infection of hamsters 120 days prior to per os superinfection with the HY TME agent results in an increase in the incubation period of HY TME by 9 days compared to the control group inoculated with HY TME agent alone (Schutt and Bartz 2008). These experiments demonstrate that the interval between inoculations is an important parameter for prion strain interference and that greater intervals between inoculation of the blocking and superinfecting strains allow for higher levels of blocking strain replication, increasing the interference effect. Overall, the relative onset of replication of the blocking and superinfecting strain is a critical parameter in strain interference, not whether the prion strains are inoculated at the same time or separately.

14.2.3 The Relative Titer of the Blocking and Superinfecting Strains Can Influence Interference

Dickinson first demonstrated that as the titer of the blocking strain was increased, there was a corresponding increase in the interfering or blocking effect; however, details of these experiments were not provided (Dickinson and Prusiner 1979). Expanding upon this observation, the titer of DY TME can determine if and when the HY TME strain emerges from a mixture (Bartz et al. 2000; Shikiya et al. 2010). Inoculation of hamsters with a mixture of a 10^{-2} dilution of DY TME brain and a 10^{-6} dilution of HY TME brain resulted in all of the animals succumbing to HY TME. When a tenfold higher relative dose of DY TME was used, nearly all of the hamsters succumbed to DY TME infection based on clinical signs and the strain-specific migration of PrP^{Sc} on Western blot. However, when brain homogenate from these hamsters was i.c. passaged a second time in hamsters, all of the animals succumbed to the HY TME agent. When a hundred-fold increase in the relative dose of DY TME agent to HY TME agent was used, the animals succumbed to DY TME agent, which was maintained upon second serial passage in hamsters.

The effect of DY TME agent dose on the emergence of the HY TME agent in vivo has been recapitulated in vitro using protein misfolding cyclic amplification (PMCA). In these studies, the ratio of the DY and HY TME agents was an important parameter that determined if and when HY TME would emerge. Similar to in vivo studies, higher ratios of DY TME to HY TME agent resulted in an increase in the ability of DY TME to interfere or completely block HY TME from emerging in PCMA (Shikiya et al. 2010). In animal studies where DY TME is able to completely block HY TME from causing disease, small amounts of HY TME can be detected using PMCA, suggesting that strain interference may not be complete (Shikiya et al. 2010). Due to limitations of the life span of the host, higher ratios of DY TME agent to HY TME agent are not feasible. PMCA strain interference can overcome this limitation and has been able to examine a wider range of ratios of DY to HY TME agent. PMCA strain interference has identified conditions where complete blockage of HY TME agent replication has occurred (Shikiya et al. 2010). Since PMCA replicates HY TME agent with similar efficiency as in animals, these in vitro strain interference studies suggest that complete blockage of agent replication in animals is feasible (Shikiya et al. 2010; Shikiya and Bartz 2011).

14.2.4 Blocking Strain Replication Is Required for Strain Interference

Transmission of the TME agent to mice does not cause disease (Marsh et al. 1969; Taylor et al. 1986). Inoculation of mice with TME prior to superinfection with the mouse strains 22A, 22C, 79A, 79V, 87A, 139A, and ME7 indicated that prion strain interference did not occur. When the 22A agent was inoculated prior to the 22C agent, incubation period for 22C was extended by over 300 days compared to controls (Taylor et al. 1986). If a prion agent is inactivated by chemical treatment, there is no extension of incubation period. Kimberlin and Walker demonstrated that in Compton white mice (*Sinc* genotype *s7s7*), the 22A agent was able to extend the incubation period of the 22C agent (Kimberlin and Walker 1985). When the 22A agent was incubated in boiling water for 15 min, 12 M urea or 5 Mrad of ionizing radiation, the interference effect on 22C was eliminated. Conversely, treatment with 1% β-propionolactone (BPL) or 1% sodium dodecyl sulfate (SDS) did not affect the ability of 22A to interfere with 22C (Kimberlin and Walker 1985).

14.2.5 Infection of Common Neuroanatomical Pathways Is Required for Interference

As described above, the DY TME agent was able to extend the incubation period or block the clinical signs of HY TME agent when both agents are inoculated into the same sciatic nerve (Bartz et al. 2007; Schutt and Bartz 2008). Since sciatic nerve inoculation targets both HY and DY PrP^{Sc} to ventral motor neurons (VMNs) in the lumbar spinal cord, this suggests that these cells are where interference occurs in this system (Shikiya et al. 2010). The only neuropathological change in DY TME-infected VMNs is the deposition of PrP^{Sc}, indicating that cellular damage to VMNs is not contributing to the interference effect (Shikiya et al. 2010). Associated with VMNs, HY and DY PrP^{Sc} are both deposited on the cell membrane, suggesting this is where strain interference occurs. Consistent with these observations, superinfection of the sciatic nerve contralateral to the sciatic nerve inoculated with the DY TME agent directs the HY TME agent to VMNs that are not infected with the DY TME agent, and the animals subsequently develop HY TME with an incubation period similar to animals inoculated with the HY TME agent alone (Bartz et al. 2007).

14.3 Prion Strain Interference and the Replication Site Hypothesis

14.3.1 The Replication Site Hypothesis

The replication site hypothesis was developed, in part, in an attempt to explain prion strain interference. The ability of blocking strains to extend the incubation period of superinfecting strains was attributed to the blocking strain occupying replication sites that were common to both the blocking and superinfecting strains (Dickinson and Prusiner 1979). Occupation of the replication sites by the blocking strain would prevent the superinfecting strain from gaining access to them, resulting in a decrease in superinfecting strain replication. The increase in incubation period or the complete blockage of the superinfecting strain would be controlled by the degree to which the blocking strain occupied the replication sites. Evidence supporting this hypothesis is the observation that splenectomy, which removes extraneural replication sites, increased the ability of the blocking strain to interfere with the superinfecting strain following intraperitoneal inoculation (Dickinson and Prusiner 1979). The replication site, as proposed by Dickinson, is hypothesized to be the gene product of *Sinc*, which is now known to be congruent with PrP^{C} (Hunter et al. 1987). This would suggest that prion strain interference is due to blocking strain PrP^{Sc} preventing superinfecting strain PrP^{Sc} from interacting with PrP^{C} either because (1) the PrP^{Sc}/PrP^{C} interaction prevents superinfecting PrP^{Sc} from gaining access to PrP^{C} or (2) that blocking PrP^{Sc} conversion has depleted the available PrP^{C} for the superinfecting strain to convert.

14.3.2 The Role of PrP^{C} in Prion Strain Interference

Current evidence suggests that prion strain interference is not due to the blocking strain PrP^{Sc} converting all of the available PrP^{C} to PrP^{Sc}. In prion infected animals, the abundance of PrP^{C} and the levels of PrP^{C} mRNA are unchanged compared to that in uninfected controls, suggesting that PrP^{C} abundance is not rate limiting

(Meyer et al. 1986; Oesch et al. 1985). Alternatively, increased expression of PrP^C in transgenic animals results in a corresponding reduction of incubation period; however, it is not known if this is due to an increase in the rate of agent replication or the animals becoming more sensitive to prion-induced neurodegeneration (Fischer et al. 1996; Telling et al. 1996a). Even if the overall abundance of the PrP^C does not change as a result of prion infection, it cannot be excluded that changes in the cellular localization of PrP^C due to blocking strain replication may contribute to prion strain interference. In vitro studies have provided direct evidence of the role of PrP^C in strain interference. Recent results using PMCA have demonstrated that PrP^{Sc} accumulates to higher levels in HY TME seeded reactions compared with DY TME seeded reactions (Shikiya et al. 2010). If PrP^C were rate limiting in PMCA, the abundance of PrP^{Sc} should be similar between the two strains. These data indicate that in the DY TME seeded reactions, not all of the available PrP^C has been converted to PrP^{Sc}. Under these same PMCA conditions, DY TME can interfere with, or completely block, the emergence of HY TME (Shikiya et al. 2010). Recent studies have demonstrated that there is a direct correlation between the abundance of PrP^C and the amount of PrP^{Sc} produced in PMCA (Mays et al. 2009). Taken together, these observations are consistent with the hypothesis that strain interference is not due to a DY TME converting the available PrP^C to PrP^{Sc}, but rather that DY TME is sequestering PrP^C or another cofactor required for HY TME conversion.

14.3.3 Prion Replication Cofactors and Prion Strain Interference

Depletion or sequestration of a necessary prion conversion cofactor by the blocking strain may underlie prion strain interference. Polyanions such as RNA are important for the conversion of PrP^C to PrP^{Sc} in PMCA and colocalize with PrP^{Sc} in vivo (Deleault et al. 2003, 2007; Geoghegan et al. 2007). Incorporation of RNA into the growing PrP^{Sc} fibril or sequestration of RNA by the blocking strain PrP^{Sc} could lead to interference of superinfecting strains that require RNA for the conversion process (Gonzalez-Montalban et al. 2011). Glycosaminoglycans (GAGs) are another polyanion that is implicated in prion conversion and colocalize to cellular locations where prion conversion occurs (Caughey and Kocisko 2003; Snow et al. 1989). Similarly to RNA, prion strains may compete for GAGs as a limiting cofactor involved in prion conversion.

14.3.4 Prion Strain Interactions and Interconversion

Direct interaction between the blocking and superinfecting strain PrP^{Sc} may account for strain interference. In this model, blocking strain PrP^{Sc} binds to the PrP^{Sc} from the superinfecting strain. From this point two main outcomes are possible. First, the blocking strain may be able to convert the superinfecting strain's PrP^{Sc} to the blocking strain PrP^{Sc} strain-specific conformation. This will allow for greater accumulation of blocking strain PrP^{Sc} and a reduction in the PrP^{Sc} of the superinfecting strain, leading to strain interference. Prion strain interconversion has been suggested in vitro by strain-induced alterations in the migration of PrP^{Sc}, the sensitivity of PrP^{Sc} to PK digestion, and conformations switching within synthetic PrP fibrils (Makarava et al. 2009; Nishina et al. 2004; Wadsworth et al. 1999). The second possibility is that direct interaction of PrP^{Sc} from the blocking and superinfecting strain may lead to a hybrid PrP^{Sc} molecule that may have a diminished capacity to convert PrP^{C} to PrP^{Sc} from one or both strains. This model is consistent with the observation that a vast excess of the blocking strain is required for interference to occur. Evidence to support this hypothesis is the detection of hybrid PrP^{Sc} deposits in mice infected with two strains (Nilsson et al. 2010).

14.4 Prion "Vaccination" and Strain Interference

The concept of a prion "vaccine" was first proposed over 30 years ago (Dickinson and Prusiner 1979). The envisioned vaccine strain would not cause disease in the vaccinated host but would block subsequent infection with a pathogenic prion strain. This prion vaccine would not protect the host via an immune response to the agent as conventional vaccines do, but would instead occupy prion replication sites and prevent infection by a pathogenic prion strain. At the time of its proposal, all blocking prion strains eventually would kill the host and the difficulty, as outlined by Dickinson, was in identifying a strain that would not cause disease yet retain the ability to interfere. While a "vaccine" strain that completely protects the host has not been identified, there is an example of a prion strain that can interfere with a pathogenic strain yet does not cause disease by extraneural routes of infection.

Intraperitoneal or oral inoculation with the DY TME agent does not result in clinical disease within the lifespan of the host (Bartz et al. 2004, 2005). Additionally, in spleen, lymph nodes, PNS, and CNS, DY TME agent replication is not detectable by animal bioassay or PrP^{Sc} deposition as determined by Western blot or immunohistochemistry (Bartz et al. 2005; J. Bartz unpublished data). Interestingly, inoculation of the DY TME agent can modestly extend the incubation period of the HY TME agent following both i.p. and per os routes of inoculation (Bartz et al. 2004; Schutt and Bartz 2008). The mechanism underlying this observation is not known; however, several possibilities exist. First, the DY TME agent may be occupying replication sites in locations in the host that are used by the superinfecting strains for neuroinvasion that were not examined for DY TME agent replication. Second, the DY TME agent is blocking replication sites but the amount of DY TME agent in these tissues is below the sensitivity of animal bioassay. Of these two possibilities, the first one seems the most likely based on the large time interval required for DY TME to be able to interfere with superinfecting strains. It remains to be determined if other prion strains will be better "vaccine" candidates or if the DY TME agent can be modified to have a greater effect. Regardless, too little is currently known about prion strains to accurately address the frequency of reversion of a nonpathogenic vaccine strain to a pathogenic strain. For a prion vaccine to be safely used, it must not revert into a pathogenic strain or revert at a very low rate. Prion strains are thought to be due to strain-specific conformations of PrP^{Sc} (Bessen and Marsh 1994). However, it is not known if the PrP^{Sc} from a given prion strain contains a single conformation or many PrP^{Sc} conformations (i.e., prion quasispecies) (Collinge and Clarke 2007; Domingo et al. 1978; Li et al. 2009). A better understanding of strains is required before this concept can be fully explored.

Acknowledgments This work was supported by the National Center for Research Resources (P20 RR0115635-6 and C06 RR17417-01) and the National Institute for Neurological Disorders and Stroke (R01 NS052609).

References

Bartz JC, Bessen R, McKenzie D, Marsh R, Aiken JM (2000) Adaptation and selection of prion protein strain conformations following interspecies transmission of transmissible mink encephalopathy. J Virol 74:5542–5547
Bartz JC, Aiken JM, Bessen RA (2004) Delay in onset of prion disease for the HY strain of transmissible mink encephalopathy as a result of prior peripheral inoculation with the replication-deficient DY strain. J Gen Virol 85:265–273
Bartz JC, Dejoia C, Tucker T, Kincaid AE, Bessen RA (2005) Extraneural prion neuroinvasion without lymphoreticular system infection. J Virol 79:11858–11863
Bartz JC, Kramer ML, Sheehan MH, Hutter JAL, Ayers JI, Bessen RA, Kincaid AE (2007) Prion interference is due to a reduction in strain-specific PrPSc levels. J Virol 81:689–697
Basler K, Oesch B, Scott M, Westaway D, Walchli M, Groth D, McKinley M, Prusiner S, Weissmann C (1986) Scrapie and cellular PrP isoforms are encoded by the same chromosomal gene. Cell 46:417–428
Bessen R, Marsh R (1992a) Biochemical and physical properties of the prion protein from two strains of the transmissible mink encephalopathy agent. J Virol 66:2096–2101
Bessen R, Marsh R (1992b) Identification of two biologically distinct strains of transmissible mink encephalopathy in hamsters. J Gen Virol 73:329–334
Bessen RA, Marsh RF (1994) Distinct PrP properties suggest the molecular basis of strain variation in transmissible mink encephalopathy. J Virol 68:7859–7868
Bessen R, Kocisko D, Raymond G, Nandan S, Lansbury P, Caughey B (1995) Non-genetic propagation of strain-specific properties of scrapie prion protein. Nature 375:698–700
Bruce M, Will R, Ironside J, McConnell I, Drummond D, Suttie A, McCardle L, Chree A, Hope J, Birkett C, Cousens S, Fraser H, Bostock C (1997) Transmissions to mice indicate that "new variant" CJD is caused by the BSE agent. Nature 389:498–501
Castilla J, Saa P, Hetz C, Soto C (2005) In vitro generation of infectious scrapie prions. Cell 121:195–206
Caughey B, Kocisko DA (2003) Prion diseases: a nucleic-acid accomplice? Nature 425:673–674
Caughey B, Raymond G (1991) The scrapie-associated form of PrP is made from a cell surface precursor that is both protease- and phospholipase-sensitive. J Biol Chem 266:18217–18223
Caughey B, Raymond GJ, Bessen RA (1998) Strain-dependent differences in beta-sheet conformations of abnormal prion protein. J Biol Chem 273:32230–32235
Colby D, Zhang Q, Wang S, Groth D, Legname G, Riesner D, Prusiner S (2007) Prion detection by an amyloid seeding assay. Proc Natl Acad Sci USA 104:20914–20919
Collinge J, Clarke A (2007) A general model of prion strains and their pathogenicity. Science 318:930–936

Deleault NR, Lucassen RW, Supattapone S (2003) RNA molecules stimulate prion protein conversion. Nature 425:717–720

Deleault N, Harris B, Rees J, Supattapone S (2007) Formation of native prions from minimal components in vitro. Proc Natl Acad Sci USA 104:9741–9746

Dickinson A (1976) Scrapie in sheep and goats. Front Biol 44:209–241

Dickinson A, Prusiner S (1979) The scrapie replication-site hypothesis and its implications for pathogenesis. In: Prusiner SB, Hadlow WJ et al (eds) Slow transmissible diseases of the nervous system, vol 2. Academic, New York, pp 13–31

Dickinson A, Fraser H, Meikle V, Outram G (1972) Competition between different scrapie agents in mice. Nat New Biol 237:244–245

Dickinson AG, Fraser H, McConnell I, Outram GW, Sales DI, Taylor DM (1975) Extraneural competition between different scrapie agents leading to loss of infectivity. Nature 253:556

Domingo E, Sabo D, Taniguchi T, Weissmann C (1978) Nucleotide sequence heterogeneity of an RNA phage population. Cell 13:735–744

Fischer M, Rulicke T, Raeber A, Sailer A, Moser M, Oesch B, Brandner S, Aguzzi A, Weissmann C (1996) Prion protein (PrP) with amino-proximal deletions restoring susceptibility of PrP knockout mice to scrapie. EMBO J 15:1255–1264

Geoghegan JC, Valdes PA, Orem NR, Deleault NR, Williamson RA, Harris BT, Supattapone S (2007) Selective incorporation of polyanionic molecules into hamster prions. J Biol Chem 282:36341–36353

Gonzalez-Montalban N, Makarava N, Savtchenko R, Baskakov IV (2011) Relationship between conformational stability and amplification efficiency of prions. Biochemistry 50:6815–6823

Hirogari Y, Kubo M, Kimura KM, Haritani M, Yokoyama T (2003) Two different scrapie prions isolated in Japanese sheep flocks. Microbiol Immunol 47:871–876

Hunter N, Hope J, McConnell I, Dickinson A (1987) Linkage of the scrapie-associated fibril protein (PrP) gene and Sinc using congenic mice and restriction fragment length polymorphism analysis. J Gen Virol 68:2711–2716

Kimberlin R, Walker C (1978) Evidence that the transmission of one source of scrapie agent to hamsters involves separation of agent strains from a mixture. J Gen Virol 39:487–496

Kimberlin R, Walker C (1985) Competition between strains of scrapie depends on the blocking agent being infectious. Intervirology 23:74–81

Kimberlin R, Cole S, Walker C (1987) Temporary and permanent modifications to a single strain of mouse scrapie on transmission to rats and hamsters. J Gen Virol 68:1875–1881

Kimberlin R, Walker C, Fraser H (1989) The genomic identity of different strains of mouse scrapie is expressed in hamsters and preserved on reisolation in mice. J Gen Virol 70:2017–2025

Kocisko D, Come J, Priola S, Chesebro B, Raymond G, Lansbury P, Caughey B (1994) Cell-free formation of protease-resistant prion protein. Nature 370:471–474

Lasmezas C, Deslys J, Demaimay R, Adjou K, Lamoury F, Dormont D, Robain O, Ironside J, Hauw J (1996) BSE transmission to macaques. Nature 381:743–744

Li J, Browning S, Mahal S, Oelschlegel A, Weissmann C (2009) Darwinian evolution of prions in cell culture. Science 327(5967):869–872

Makarava N, Ostapchenko V, Savtchenko R, Baskakov I (2009) Conformational switching within individual amyloid fibrils. J Biol Chem 284:14386–14395

Manuelidis L, Lu ZY (2003) Virus-like interference in the latency and prevention of Creutzfeldt–Jakob disease. Proc Natl Acad Sci USA 100:5360–5365

Manuelidis L (1998) Vaccination with an attenuated Creutzfeldt–Jakob disease strain prevents expression of a virulent agent. Proc Natl Acad Sci USA 95:2520–2525

Manuelidis L, Yun Lu Z (2000) Attenuated Creutzfeldt–Jakob disease agents can hide more virulent infections. Neurosci Lett 293:163–166

Marsh RF, Burger D, Eckroade R, Zu Rhein GM, Hanson RP (1969) A preliminary report on the experimental host range of the transmissible mink encephalopathy agent. J Infect Dis 120:713–719

Mays CE, Titlow W, Seward T, Telling GC, Ryou C (2009) Enhancement of protein misfolding cyclic amplification by using concentrated cellular prion protein source. Biochem Biophys Res Commun 388:306–310

Meyer RK, McKinley MP, Bowman KA, Braunfeld MB, Barry RA, Prusiner SB (1986) Separation and properties of cellular and scrapie prion proteins. Proc Natl Acad Sci USA 83:2310–2314
Nilsson KPR, Joshi-Barr S, Winson O, Sigurdson CJ (2010) Prion strain interactions are highly selective. J Neurosci 30:12094–12102
Nishida N, Katamine S, Manuelidis L (2005) Reciprocal interference between specific CJD and scrapie agents in neural cell cultures. Science 310:493–496
Nishina K, Jenks S, Supattapone S (2004) Ionic strength and transition metals control PrPSc protease resistance and conversion-inducing activity. J Biol Chem 279:40788–40794
Oesch B, Westaway D, Walchli M, McKinley MP, Kent SB, Aebersold R, Barry RA, Tempst P, Teplow DB, Hood LE (1985) A cellular gene encodes scrapie PrP 27–30 protein. Cell 40:735–746
Parchi P, Gambetti P (1995) Human prion diseases. Curr Opin Neurol 8:286–293
Polymenidou M, Stoeck K, Glatzel M, Vey M, Bellon A, Aguzzi A (2005) Coexistence of multiple PrPSc types in individuals with Creutzfeldt–Jakob disease. Lancet Neurol 4:805–814
Prusiner S (1991) Molecular biology of prion diseases. Science 252:1515–1522
Schutt CR, Bartz JC (2008) Prion interference with multiple prion isolates. Prion 2:61–63
Shikiya RA, Bartz JC (2011) In vitro generation of high titer prions. J Virol 85(24):13439–13442
Shikiya RA, Ayers JI, Schutt CR, Kincaid AE, Bartz JC (2010) Coinfecting prion strains compete for a limiting cellular resource. J Virol 84:5706–5714
Sigurdson C, Nilsson K, Hornemann S, Heikenwalder M, Manco G, Schwarz P, Ott D, Rulicke T, Liberski P, Julius C, Falsig J, Stitz L, Wuthrich K, Aguzzi A (2009) De novo generation of a transmissible spongiform encephalopathy by mouse transgenesis. Proc Natl Acad Sci USA 106:304–309
Snow A, Kisilevsky R, Willmer J, Prusiner S, DeArmond S (1989) Sulfated glycosaminoglycans in amyloid plaques of prion diseases. Acta Neuropathol 77:337–342
Taylor D, Dickinson A, Fraser H, Marsh R (1986) Evidence that transmissible mink encephalopathy agent is biologically inactive in mice. Neuropathol Appl Neurobiol 12:207–215
Telling GC, Haga T, Torchia M, Tremblay P, DeArmond S, Prusiner S (1996a) Interactions between wild-type and mutant prion proteins modulate neurodegeneration in transgenic mice. Genes Dev 10:1736–1750
Telling GC, Parchi P, DeArmond S, Cortelli P, Montagna P, Gabizon R, Mastrianni J, Lugaresi E, Gambetti P, Prusiner S (1996b) Evidence for the conformation of the pathologic isoform of the prion protein enciphering and propagating prion diversity. Science 274:2079–2082
Wadsworth JD, Hill AF, Joiner S, Jackson GS, Clarke AR, Collinge J (1999) Strain-specific prion-protein conformation determined by metal ions. Nat Cell Biol 1:55–59

Chapter 15
Introduction to Yeast and Fungal Prions

Reed B. Wickner

Abstract Prions are infectious proteins, not requiring an accompanying nucleic acid for the transmission to a new individual. In 1994, we found that the long-known cytoplasmic genes [URE3] and [PSI+] were actually prions of Ure2p and Sup35p, respectively. These, and a variety of yeast and fungal prions found since then are based on self-propagating amyloids, but one prion based on a protease that self-activates shows that not all infectious proteins need be amyloids. The importance of chaperones in prion propagation, the involvement of many other cellular systems, and development of anti-prion measures—some potentially active against mammalian prions, have enriched the prion field. The in-register parallel architecture of yeast prion amyloids can explain how a single protein can faithfully propagate any of several structurally different prion variants/strains. Discovery of an array of new prions, and interesting new variants of old prions continues to expand our understanding of this phenomenon.

Keywords Ure2p • Sup35p • Rnq1p • HET-s • [PSI+] • [URE3] • [PIN+] • [Het-s] • Chaperones • Prion variants

15.1 Mysterious Non-Chromosomal Genetic Elements in Yeast

The non-chromosomal genetic elements [PSI+] and [URE3] were discovered in the 1960s and 1970s due to the pioneering work of Brian Cox (1965) and Francois Lacroute (1971) and their coworkers. [PSI+] enhances weak nonsense-suppressor tRNA mutations or can even be a nonsense-suppressor on its own, allowing growth

R.B. Wickner, M.D. (✉)
Laboratory of Biochemistry and Genetics, National Institute of Diabetes and Digestive and Kidney Diseases, National Institutes of Health, Bldg. 8, Room 225, NIH, 8 Center Drive MSC 0830, Bethesda, MD 20892-0830, USA
e-mail: wickner@helix.nih.gov

W.-Q. Zou and P. Gambetti (eds.), *Prions and Diseases: Volume 1, Physiology and Pathophysiology*, DOI 10.1007/978-1-4614-5305-5_15,

of cells with a premature translation termination mutation in an essential gene (Cox 1965; Liebman et al. 1975). In combination with a strong nonsense suppressor tRNA mutation, [PSI+] is lethal, as one would expect from excessive read-through of normal termination codons (Cox 1971). Mating a strain carrying a classical nonsense-suppressor mutation in a tRNA gene with another strain lacking the mutation produces diploids heterozygous for the suppressor mutation. When the diploid cells undergo meiosis, two of the spores in each tetrad have the suppressor mutation and two do not. This is 2+:2− meiotic segregation. However, [PSI+] segregated 4+:0, meaning that all of the meiotic segregants have the suppression-enhancing genetic element (Cox et al. 1988). This showed that [PSI+] was a non-chromosomal genetic element, assumed at the time to be a replicating DNA plasmid or RNA virus.

The [URE3] non-chromosomal genetic element was found in studies involving uracil biosynthesis that led to control of nitrogen source utilization (Lacroute 1971; Drillien et al. 1973). In the first step of uracil biosynthesis, aspartate is condensed with carbamyl phosphate to form ureidosuccinic acid (USA), a reaction catalyzed by aspartate transcarbamylase (*URA2*). On media with a rich nitrogen source, such as ammonia, yeast will not take up USA to feed a *ura2* mutant. However, cells growing on a poor nitrogen source, such as proline, or *ure2* mutants, can do so (Aigle and Lacroute 1975). One dominant "mutant" able to take up USA on ammonia-containing medium showed non-chromosomal segregation in meiosis (like [PSI+] above), and was designated [URE3] (Lacroute 1971).

15.2 Discovery of Yeast Prions and the Three Genetic Criteria

These cytoplasmic genes were long unexplained, but careful studies by Michel Aigle, with Lacroute, showed that the [URE3] cytoplasmic element required the chromosomal *URE2* gene for its propagation (Aigle and Lacroute 1975). Both *ure2* mutants and strains carrying the [URE3] genetic element have the *same* phenotype (Aigle and Lacroute 1975). In contrast, chromosomal *mak* mutants unable to propagate the killer factor (M dsRNA) have the *opposite* phenotype (non-killer) of strains carrying M dsRNA (killer), and *pet* mutants unable to propagate the mitochondrial DNA have the *opposite* phenotype (glycerol negative) of cells carrying mitDNA (glycerol positive). This was the first clue that led us to suggest that [URE3] and [PSI+] were prions (Wickner 1994) (Fig. 14.1). Nucleic acid replicons (viruses and plasmids) depend for their propagation on chromosomal genes, but the general pattern is that a mutant in the chromosomal gene has a phenotype opposite to that of cells carrying the replicon. If a prion produces a phenotype as a result of deficiency of the normal form, then the presence of the prion should give the same phenotype as mutation of the gene for the normal form (Wickner 1994). But a continuous supply of the normal form is necessary for the continued propagation of the prion. Thus we inferred that [URE] must be a prion of the Ure2 protein. The same relation had just been reported by Cox for [PSI+] and *sup35* (Cox 1993; Doel et al. 1994), from which we inferred that [PSI+] was a prion of Sup35p (Wickner 1994).

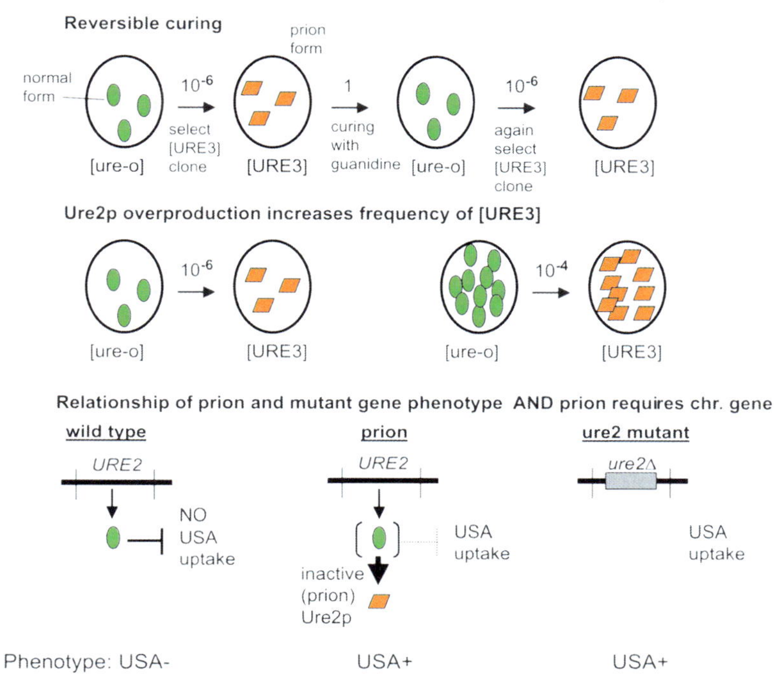

Fig. 15.1 Three genetic criteria for identification of prions in yeast and fungi (Wickner 1994)

We devised two other tests for a prion (Fig. 15.1). Curing a nucleic acid replicon leaves a cell that cannot re-acquire the element without its introduction from outside or from another cell. In contrast, prion curing should be reversible. The protein is still being made and could undergo the prion change (rarely) to produce prion-containing cells from those previously cured (Wickner 1994). We showed this to be true for [URE3] (Wickner 1994) as Lund and Cox had previously found for [PSI+] (Lund and Cox 1981), supporting our view that both were prions.

Overproducing a chromosomally encoded protein required for propagation of a plasmid or virus will not induce the appearance of either replicon. But overproduction of a protein able to form a prion should increase the frequency of prion formation. With more molecules of the protein present, the chances that a prion conversion will happen should increase, whatever the mechanism of prion formation. We showed that overproduction of Ure2p increased the frequency of [URE3] arising by ~100-fold (Wickner 1994), and Chernoff reported a similar result for overproduction of Sup35p inducing [PSI+] appearance (Chernoff et al. 1993). We inferred that [URE3] is a prion of Ure2p, and [PSI+] a prion of Sup35p (Wickner 1994) (Figs. 15.1 and 15.2).

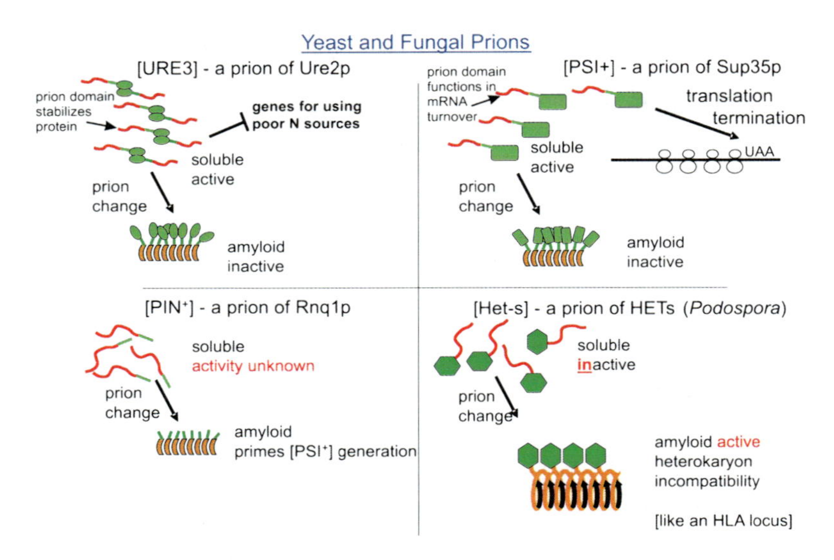

Fig. 15.2 The four most-studied prions of yeast and fungi

15.3 The [Het-s] Prion of *Podospora anserina*

When two colonies of a filamentous fungus grow toward each other, they fuse cellular processes to form, in effect, a single syncytium (a heterokaryon), allowing the exchange of nutrients [reviewed by Saupe (2000)]. However, this fusion process is limited to closely related strains, a limitation enforced by testing of identity of about a dozen polymorphic loci scattered about the genome. Nonidentity of alleles at even a single one of these loci produces death of the first fusing cellular processes and a barrier to further fusions, a process called heterokaryon incompatibility.

One of these loci is called *het-s* with alleles *het-s* and *het-S*. The proper incompatibility between *het-s* and *het-S* strains is only observed if the *het-s* cells have a non-chromosomal gene, [Het-s] (Rizet 1952), shown to be a prion of the HET-s protein (product of the *het-s* allele) (Coustou et al. 1997). [Het-s] has proven to be a very important system for many aspects of prion studies [reviewed by Saupe (2007)] (Fig. 15.2).

15.4 The [PIN+] Prion (Rarely) Seeds Other Prions

Although overproduction of Sup35p induced [PSI+] appearance (Chernoff et al. 1993), it was noted that in some strains, there was no such effect (Derkatch et al. 1997). Crossing strains in which overproduction of Sup35p induced [PSI+] generation with those in which it did not resulted in all meiotic segregants showing the induction, evidence of a non-chromosomal genetic element, that was named [PIN+], for [PSI]-inducibility (Derkatch et al. 1997). Using the genetic criteria above, it was shown that [PIN+] was a prion of Rnq1p (Derkatch et al. 2001), a protein rich in N and Q residues that had been shown to carry out a self-propagating aggregation (Sondheimer and Lindquist 2000).

In the course of showing that [PIN+] is a prion of Rnq1p, it was found that overexpression of any of a variety of QN-rich proteins had a [PIN+]-like effect, stimulating prion formation by Sup35p (Derkatch et al. 2001; Osherovich and Weissman 2001). In fact, several proteins detected in this screen were later found to form prions themselves, including Swi1p ([SWI+]) and Cyc8p ([OCT+]) (Du et al. 2008; Patel et al. 2009).

15.5 [ISP+], a Nuclear Prion of Spf1p

Starting with the observation of a non-chromosomal genetic element producing an anti-suppressor effect—the opposite of [PSI+] and hence the name [ISP+]—Moronova and coworkers have found a prion of the transcription factor Sfp1p that is largely confined to the nucleus and thus is poorly transmitted in cytoplasmic mixing experiments (cytoduction) (Volkov et al. 2002; Rogoza et al. 2010). The detailed mechanism of the anti-suppressor effect is not yet clear, but Sfp1p is known to regulate ribosomal protein biogenesis (Fingerman et al. 2003).

15.6 [BETA], an Enzyme-Based Prion

We shall see that most yeast prions are self-propagating amyloids, but one is simply an enzyme that, under certain circumstances, is necessary for activation of its own inactive precursor protein (Roberts and Wickner 2003). Vacuolar protease B is made as an inactive precursor that is normally activated by vacuolar protease A (Jones 1991). In the absence of protease A, the protease B can inefficiently activate its own precursor, but on the usual medium, this activation cascade quickly dies out (Zubenko et al. 1982). However, because protease B expression is glucose-repressed, growth of cells on glycerol medium results in the self-activation being indefinitely sustained. The active enzyme then acts like a prion (called [BETA]), showing that prions (infectious proteins) need not be amyloids (Roberts and Wickner 2003).

15.7 Amyloids as the Basis of most Yeast Prions

Restricted domains of Ure2p (Masison and Wickner 1995; Masison et al. 1997) and Sup35p (TerAvanesyan et al. 1994) are sufficient for propagation of the [URE3] and [PSI+] prions. These prion domains are the N-terminal Q/N rich parts of the respective molecules, although the prion domains of HET-s and Rnq1p are C-terminal (Balguerie et al. 2003; Vitrenko et al. 2007), and the HET-s prion domain is not Q/N rich.

Amyloid formation by prion domains (King et al. 1997; Taylor et al. 1999) and full-length prion proteins (Glover et al. 1997; Taylor et al. 1999), along with protease resistance of Ure2p in extracts of [URE3] strains (Masison and Wickner 1995) and aggregation of Sup35p in [PSI+] strains (Patino et al. 1996; Paushkin et al. 1996) first suggested that amyloid was the basis of [URE3] and [PSI+]. The [Het-s] system was the first in which prion infection by amyloid formed in vitro from recombinant protein was achieved (Maddelein et al. 2002). The key to this experiment was that the amyloid form of HET-s was infectious, but the soluble form or a nonspecific aggregate was not. Since overexpression of prion proteins dramatically increases the frequency of prion induction, it was critical to show that one was not simply increasing the supply of the prion protein in the transfected cells. Similar results were shown for [PSI+] (King and Diaz-Avalos 2004; Tanaka et al. 2004) and later for [URE3] (Brachmann et al. 2005) and [PIN+] (Patel and Liebman 2007).

As will be discussed in another chapter, the amyloids of the prion domains of Ure2p, Sup35p, and Rnq1p are in-register parallel beta sheets, multiply folded along the long axis of the fiber. This architecture can explain the ability of proteins to template any of several different structures, based on different locations of the folds and/or different extents of the beta sheet (see Wickner et al., Chap. 16).

15.8 Chaperones and Other Cellular Factors Affecting Prion Propagation

The finding that overproduction or deficiency of the disaggregating chaperone Hsp104 resulted in the loss of [PSI+] (Chernoff and Ono 1992; Chernoff et al. 1995) began a series of studies in which a host of chaperones were found to intimately affect the generation and propagation of yeast prions [reviewed in Reidy and Masison (2011)]. Hsp104 works with Hsp70s and Hsp40s in renaturing proteins (Glover and Lindquist 1998), and, indeed, cytoplasmic Hsp70s are needed for yeast prion propagation (Jung et al. 2000), and Hsp40s also affect yeast prions (Moriyama et al. 2000; Sondheimer et al. 2001). Each of the known yeast amyloid-based prions requires Hsp104 for its propagation (Derkatch et al. 1997; Moriyama et al. 2000; Du et al. 2008; Patel et al. 2009).

The primary means by which Hsp104–Hsp70–Hsp40 help prions propagate is by splitting amyloid filaments, thus producing two filaments where there was one (Paushkin et al. 1996; Ness et al. 2002; Kryndushkin et al. 2003) (Fig. 15.3). This constitutes prion replication, a process which must keep up with cell division if the prion is to be maintained. However, the mechanism by which overexpression of Hsp104 cures [PSI+] is unclear, and is clearly not simply resolubilization of Sup35 by extensive cleavage of filaments [reviewed by Reidy and Masison (2011)].

The Hsp70 family includes four soluble cytoplasmic members, Ssa1–Ssa4 and the two ribosome-associated chaperones, Ssb1 and Ssb2. The highly homologous

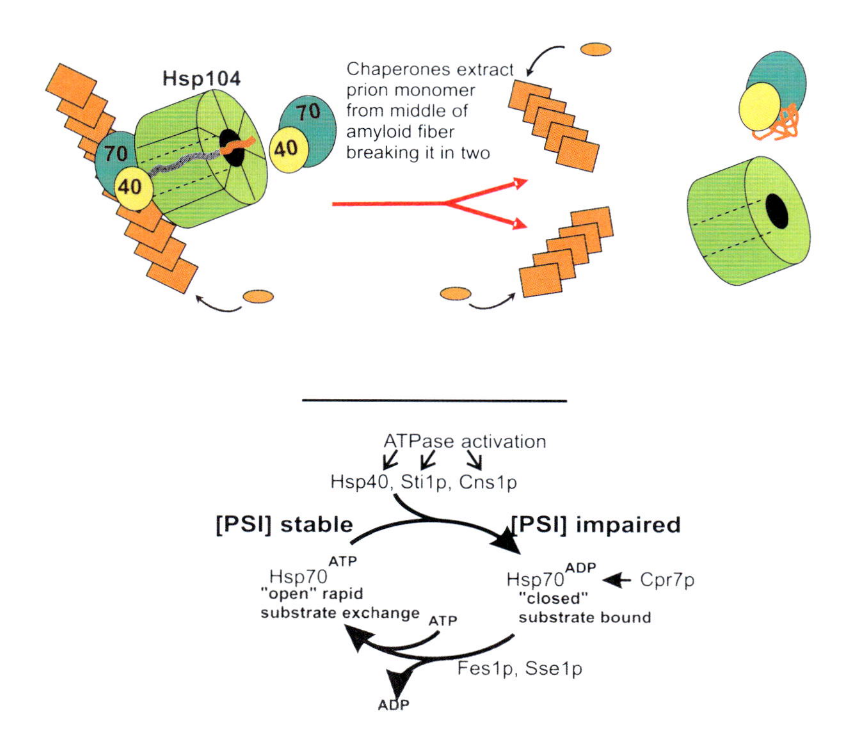

Fig. 15.3 Chaperones cleave prion amyloid filaments, generating new seeds (*Top*). Hsp104, Hsp70s, and Hsp40s extract a monomer from the middle of the filament, thus producing two filaments [reviewed by Reidy and Masison (2011)]. Chaperones (Liu et al. 2010; Reidy and Masison 2011) and Btn2p (Kryndushkin et al. 2008) may also function in prion segregation (not shown). Hsp40s, co-chaperones, and nucleotide exchange factors regulate the role of Hsp70s in prion propagation [reviewed by Sharma and Masison (2009)] (*Bottom*)

Ssa's show surprising specificity for promoting or inhibiting the [PSI+] and [URE3] prions (Schwimmer and Masison 2002; Sharma and Masison 2008). Overproduction of the Btn2 protein or its homolog can cure [URE3], apparently by drawing most or all of the Ure2p aggregates to a single place in the cell (Kryndushkin et al., 2008). It was suggested that this structure is analogous or homologous to the mammalian aggresome. The assymetrical retention of aggregates in the mother cell at a particular site (Liu et al., 2010) may be a related phenomenon.

15.9 Prion Variants and the Species Barrier

A striking characteristic of prions in nearly all systems (the exceptions are interesting!) is the ability of a single protein sequence to stably propagate any of an array of prion "strains" or "variants." Different prion variants are distinguished in mammals by the incubation period, the regions of the brain affected, and the disease signs

Fig. 15.4 Prion variants and the species barrier. As in animal systems, the facility of prion transmission across a species barrier depends on the prion variant (Edskes et al. 2009)

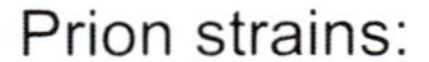

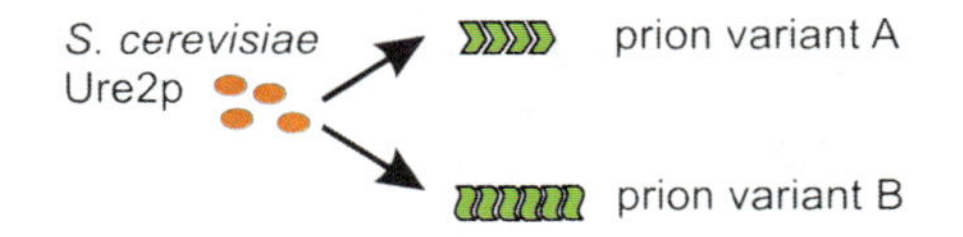

Species barrier:

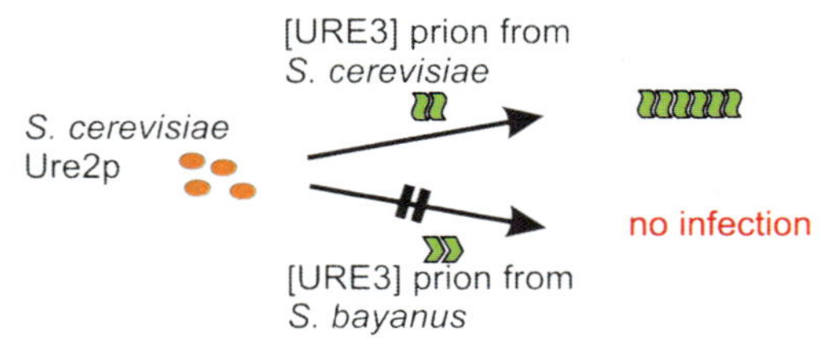

Prion strain affects species barrier:

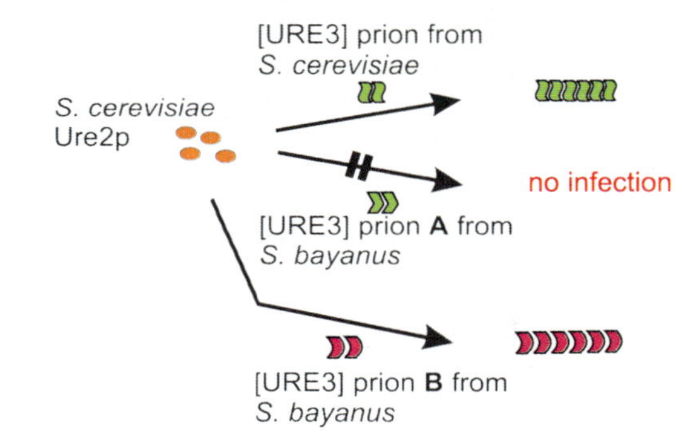

[reviewed by (Bruce 1993)]. In yeast, prion variants (Derkatch et al. 1996; Schlumpberger et al. 2001; Bradley et al. 2002) differ in the intensity of the prion phenotype, the stability of the prion, the response to overproduction or deficiency of various chaperones (Chernoff et al. 1999; Kushnirov et al. 2000), ability to cross species barriers (Edskes et al. 2009), and whether the prion is lethal or pathogenic or not (McGlinchey et al. 2011) (Fig. 15.4).

Prion variants are clearly due to faithfully propagated differences in amyloid structure (e.g., King 2001; King and Diaz-Avalos 2004; Tanaka et al. 2004), but the detailed structure of no prion variant is yet know. However, as detailed in the next chapter (Chap. 15), yeast prions have an in-register parallel architecture that suggests a detailed mechanism of variant information propagation. An important subject of future studies will be elucidation of the detailed nature of prion variant differences, and study of how they produce different pathologies.

15.10 Perspective

The yeast and fungal prion field has blossomed, becoming important for the overall understanding of prions and amyloid diseases in general. Yeast's traditional lead in genetic studies is being complemented with biochemical, cell biological and structural studies to produce a variety of insights important for all prion systems.

References

Aigle M, Lacroute F (1975) Genetical aspects of [URE3], a non-Mendelian, cytoplasmically inherited mutation in yeast. Mol Gen Genet 136:327–335

Balguerie A, Dos Reis S, Ritter C, Chaignepain S, Coulary-Salin B, Forge V, Bathany K, Lascu I, Schmitter J-M, Riek R, Saupe S (2003) Domain organization and structure–function relationship of the HET-s prion protein of *Podospora anserina*. EMBO J 22:2071–2081

Brachmann A, Baxa U, Wickner RB (2005) Prion generation in vitro: amyloid of Ure2p is infectious. EMBO J 24:3082–3092

Bradley ME, Edskes HK, Hong JY, Wickner RB, Liebman SW (2002) Interactions among prions and prion "strains" in yeast. Proc Natl Acad Sci USA 99:16392–16399

Bruce ME (1993) Scrapie strain variation and mutation. Br Med Bull 49:822–838

Chernoff YO, Ono B-I (1992) Dosage-dependent modifiers of PSI-dependent omnipotent suppression in yeast. In: Tuite MF, Brown AJP, McCarthy JEG (eds) Protein synthesis and targeting in yeast. Springer, Berlin, pp 101–107

Chernoff YO, Derkach IL, Inge-Vechtomov SG (1993) Multicopy SUP35 gene induces de novo appearance of psi-like factors in the yeast *Saccharomyces cerevisiae*. Curr Genet 24:268–270

Chernoff YO, Lindquist SL, Ono B-I, Inge-Vechtomov SG, Liebman SW (1995) Role of the chaperone protein Hsp104 in propagation of the yeast prion-like factor [psi$^+$]. Science 268:880–884

Chernoff YO, Newnam GP, Kumar J, Allen K, Zink AD (1999) Evidence for a protein mutator in yeast: role of the Hsp70-related chaperone Ssb in formation, stability and toxicity of the [PSI+] prion. Mol Cell Biol 19:8103–8112

Coustou V, Deleu C, Saupe S, Begueret J (1997) The protein product of the *het-s* heterokaryon incompatibility gene of the fungus *Podospora anserina* behaves as a prion analog. Proc Natl Acad Sci USA 94:9773–9778

Cox BS (1965) PSI, a cytoplasmic suppressor of super-suppressor in yeast. Heredity 20:505–521

Cox BS (1971) A recessive lethal super-suppressor mutation in yeast and other PSI phenomena. Heredity 26:211–232

Cox BS (1993) Psi phenomena in yeast. In: Hall MN, Linder P (eds) The early days of yeast genetics. Cold Spring Harbor Laboratory Press, Cold Spring Harbor, pp 219–239

Cox BS, Tuite MF, McLaughlin CS (1988) The Psi factor of yeast: a problem in inheritance. Yeast 4:159–179

Derkatch IL, Chernoff YO, Kushnirov VV, Inge-Vechtomov SG, Liebman SW (1996) Genesis and variability of [*PSI*] prion factors in *Saccharomyces cerevisiae*. Genetics 144:1375–1386

Derkatch IL, Bradley ME, Zhou P, Chernoff YO, Liebman SW (1997) Genetic and environmental factors affecting the de novo appearance of the [*PSI+*] prion in *Saccharomyces cerevisiae*. Genetics 147:507–519

Derkatch IL, Bradley ME, Hong JY, Liebman SW (2001) Prions affect the appearance of other prions: the story of *[PIN]*. Cell 106:171–182

Doel SM, McCready SJ, Nierras CR, Cox BS (1994) The dominant *PNM2*⁻ mutation which eliminates the [PSI] factor of *Saccharomyces cerevisiae* is the result of a missense mutation in the *SUP35* gene. Genetics 137:659–670
Drillien R, Aigle M, Lacroute F (1973) Yeast mutants pleiotropically impaired in the regulation of the two glutamate dehydrogenases. Biochem Biophys Res Commun 53:367–372
Du Z, Park K-W, Yu H, Fan Q, Li L (2008) Newly identified prion linked to the chromatin-remodeling factor Swi1 in *Saccharomyces cerevisiae*. Nat Genet 40:460–465
Edskes HK, McCann LM, Hebert AM, Wickner RB (2009) Prion variants and species barriers among *Saccharomyces* Ure2 proteins. Genetics 181:1159–1167
Fingerman I, Nagaraj V, Norris D, Vershon AK (2003) Spf1 plays a key role in yeast ribosome biogenesis. Eukaryot Cell 2:1061–1068
Glover JR, Lindquist S (1998) Hsp104, Hsp70, and Hsp40: a novel chaperone system that rescues previously aggregated proteins. Cell 94:73–82
Glover JR, Kowal AS, Shirmer EC, Patino MM, Liu J-J, Lindquist S (1997) Self-seeded fibers formed by Sup35, the protein determinant of [*PSI*+], a heritable prion-like factor of *S. cerevisiae*. Cell 89:811–819
Jones EW (1991) Three proteolytic systems in the yeast *Saccharomyces cerevisiae*. J Biol Chem 266:7963–7966
Jung G, Jones G, Wegrzyn RD, Masison DC (2000) A role for cytosolic Hsp70 in yeast [PSI+] prion propagation and [PSI+] as a cellular stress. Genetics 156:559–570
King CY (2001) Supporting the structural basis of prion strains: induction and identification of [*PSI*] variants. J Mol Biol 307:1247–1260
King CY, Diaz-Avalos R (2004) Protein-only transmission of three yeast prion strains. Nature 428:319–323
King C-Y, Tittmann P, Gross H, Gebert R, Aebi M, Wuthrich K (1997) Prion-inducing domain 2–114 of yeast Sup35 protein transforms in vitro into amyloid-like filaments. Proc Natl Acad Sci USA 94:6618–6622
Kryndushkin DS, Alexandrov IM, Ter-Avanesyan MD, Kushnirov VV (2003) Yeast [*PSI*⁺] prion aggregates are formed by small Sup35 polymers fragmented by Hsp104. J Biol Chem 278:49636–49643
Kryndushkin D, Shewmaker F, Wickner RB (2008) Curing of the [URE3] prion by Btn2p, a Batten disease-related protein. EMBO J 27:2725–2735
Kushnirov VV, Kryndushkin D, Boguta M, Smirnov VN, Ter-Avanesyan MD (2000) Chaperones that cure yeast artificial [*PSI*⁺] and their prion-specific effects. Curr Biol 10:1443–1446
Lacroute F (1971) Non-Mendelian mutation allowing ureidosuccinic acid uptake in yeast. J Bacteriol 106:519–522
Liebman SW, Stewart JW, Sherman F (1975) Serine substitutions caused by an ochre suppressor in yeast. J Mol Biol 94:595–610
Liu B, Larsson L, Caballero A, Hao X, Oling D, Grantham J, Nystrom T (2010) The polarisome is required for segregation and retrograde transport of protein aggregates. Cell 140:257–267
Lund PM, Cox BS (1981) Reversion analysis of [psi-] mutations in *Saccharomyces cerevisiae*. Genet Res 37:173–182
Maddelein ML, Dos Reis S, Duvezin-Caubet S, Coulary-Salin B, Saupe SJ (2002) Amyloid aggregates of the HET-s prion protein are infectious. Proc Natl Acad Sci USA 99:7402–7407
Masison DC, Wickner RB (1995) Prion-inducing domain of yeast Ure2p and protease resistance of Ure2p in prion-containing cells. Science 270:93–95
Masison DC, Maddelein M-L, Wickner RB (1997) The prion model for [URE3] of yeast: spontaneous generation and requirements for propagation. Proc Natl Acad Sci USA 94:12503–12508
McGlinchey R, Kryndushkin D, Wickner RB (2011) Suicidal [PSI+] is a lethal yeast prion. Proc Natl Acad Sci USA 108:5337–5341
Moriyama H, Edskes HK, Wickner RB (2000) [URE3] prion propagation in Saccharomyces cerevisiae: requirement for chaperone Hsp104 and curing by overexpressed chaperone Ydj1p. Mol Cell Biol 20:8916–8922
Ness F, Ferreira P, Cox BS, Tuite MF (2002) Guanidine hydrochloride inhibits the generation of prion "seeds" but not prion protein aggregation in yeast. Mol Cell Biol 22:5593–5605

Osherovich LZ, Weissman JS (2001) Multiple Gln/Asn-rich prion domains confer susceptibility to induction of the yeast *[PSI+]* prion. Cell 106:183–194
Patel BK, Liebman SW (2007) "Prion proof" for [PIN+]: infection with in vitro-made amyloid aggregates of Rnq1p-(132–405) induces [PIN+]. J Mol Biol 365:773–782
Patel BK, Gavin-Smyth J, Liebman SW (2009) The yeast global transcriptional co-repressor protein Cyc8 can propagate as a prion. Nat Cell Biol 11:344–349
Patino MM, Liu J-J, Glover JR, Lindquist S (1996) Support for the prion hypothesis for inheritance of a phenotypic trait in yeast. Science 273:622–626
Paushkin SV, Kushnirov VV, Smirnov VN, Ter-Avanesyan MD (1996) Propagation of the yeast prion-like [*psi*$^+$] determinant is mediated by oligomerization of the *SUP35*-encoded polypeptide chain release factor. EMBO J 15:3127–3134
Reidy M, Masison DC (2011) Modulation and elimination of yeast prions by protein chaperones and co-chaperones. Prion 5(4) 245–249
Rizet G (1952) Les phenomenes de barrage chez *Podospora anserina*: analyse genetique des barrages entre les souches s et S. Rev Cytol Biol Veg 13:51–92
Roberts BT, Wickner RB (2003) A class of prions that propagate via covalent auto-activation. Genes Dev 17:2083–2087
Rogoza T, Goginashvili A, Rodionova S, Ivanov M, Viktorovskaya O, Rubel A, Volkov K, Mironova L (2010) Non-mendelian determinant [ISP+] in yeast is a nuclear-residing prion form of the global transcriptional regulator Sfp1. Proc Natl Acad Sci USA 107:10573–10577
Saupe SJ (2000) Molecular genetics of heterokaryon incompatibility in filamentous ascomycetes. Microbiol Mol Biol Revs 64:489–502
Saupe SJ (2007) A short history of small s: a prion of the fungus *Podospora anserina*. Prion 1:110–115
Schlumpberger M, Prusiner SB, Herskowitz I (2001) Induction of distinct [URE3] yeast prion strains. Mol Cell Biol 21:7035–7046
Schwimmer C, Masison DC (2002) Antagonistic interactions between yeast [PSI+] and [URE3] prions and curing of [URE3] by Hsp70 protein chaperone Ssa1p but not by Ssa2p. Mol Cell Biol 22:3590–3598
Sharma D, Masison DC (2008) Functionally redundant isoforms of a yeast Hsp70 chaperone subfamily have different antiprion effects. Genetics 179:1301–1311
Sharma D, Masison DC (2009) Hsp70 structure, function, regulation and influence on yeast prions. Protein Pept Lett 16(6):571–581
Sondheimer N, Lindquist S (2000) Rnq1: an epigenetic modifier of protein function in yeast. Mol Cell 5:163–172
Sondheimer N, Lopez N, Craig EA, Lindquist S (2001) The role of Sis1 in the maintenance of the [RNQ+] prion. EMBO J 20:2435–2442
Tanaka M, Chien P, Naber N, Cooke R, Weissman JS (2004) Conformational variations in an infectious protein determine prion strain differences. Nature 428:323–328
Taylor KL, Cheng N, Williams RW, Steven AC, Wickner RB (1999) Prion domain initiation of amyloid formation in vitro from native Ure2p. Science 283:1339–1343
TerAvanesyan A, Dagkesamanskaya AR, Kushnirov VV, Smirnov VN (1994) The *SUP35* omnipotent suppressor gene is involved in the maintenance of the non-Mendelian determinant [psi+] in the yeast *Saccharomyces cerevisiae*. Genetics 137:671–676
Vitrenko YA, Pavon ME, Stone SI, Liebman SW (2007) Propagation of the [*PIN*$^+$] prion by fragments of Rnq1 fused to GFP. Curr Genet 51:309–319
Volkov KV, Aksenova YA, Soom MJ, Sipov KV, Svitin AV, Kurischko C, Shkundina IS, Ter-Avanesyan MD, Inge-Vechtomov SG, Mironova LN (2002) Novel non-Mendelian determinant involved in the control of translation accuracy in *Saccharomyces cerevisiae*. Genetics 160:25–36
Wickner RB (1994) [URE3] as an altered *URE2* protein: evidence for a prion analog in *S. cerevisiae*. Science 264:566–569
Zubenko GS, Park FJ, Jones EW (1982) Genetic properties of mutations at the *PEP4* locus in *Saccharomyces cerevisiae*. Genetics 102:679–690

Chapter 16
Yeast Prions Are Pathogenic, In-Register Parallel Amyloids

Reed B. Wickner, Herman K. Edskes, David A. Bateman, Amy C. Kelly, and Anton Gorkovskiy

Abstract Most yeast prions are self-propagating amyloids of normally non-amyloid proteins. The prion domains of Ure2p, Sup35p, and Rnq1p each form highly infectious in-register parallel β-sheet amyloids. This architecture can explain perhaps the most mysterious prion phenomenon: the stable propagation of any of several prion variants ("strains") by a single amino acid sequence. We have thus proposed a detailed model for the mechanism of templating of protein conformation by amyloid filaments. The yeast prions [URE3] and [PSI+] are diseases of yeast, with different variants differing in the degree to which they deter cell growth or viability, but even the most mild forms not being found in wild strains. Sequence conservation of the prion domains reflects the important non-prion function of these domains, not conservation of prion-forming ability, which does not require sequence conservation and is, in fact, not conserved. Upon infection with a prion, cells undergo induction of Hsp70s and Hsp104, indicative of a stress response: the cells know that prion infection is not a good thing.

Keywords Lethal prions • Solid-state NMR • Templating of protein conformation • Prion structure

It is now well established in both mammalian and yeast systems [ref to Chap. 1] that a single prion-forming protein can support the faithful propagation of any of several (perhaps many) different prion "strains" or "variants" [reviewed by Derkatch et al. (1996); Bruce (2003)]. It is also clear that different prion variants are based on different amyloid conformations (Bessen and Marsh 1992; Caughey et al. 1998; Toyama et al. 2007). This means that having assumed a particular amyloid conformation,

R.B. Wickner, M.D. (✉) • H.K. Edskes, Ph.D. • D.A. Bateman
• A.C. Kelly, Ph.D. • A. Gorkovskiy, Ph.D.
Laboratory of Biochemistry and Genetics, National Institute of Diabetes and Digestive and Kidney Diseases, National Institutes of Health, Bethesda, MD, USA
e-mail: reedw@helix.nih.gov

W.-Q. Zou and P. Gambetti (eds.), *Prions and Diseases: Volume 1, Physiology and Pathophysiology*, DOI 10.1007/978-1-4614-5305-5_16,

a prion protein can instruct a new molecule joining the end of the amyloid filament to assume the same conformation as those already in the filament. How does this work? This is the central mystery of the prion phenomenon. Our proposed mechanism based on our demonstration of the in-register parallel architecture of yeast prion proteins (Wickner et al. 2007, 2008a, b, 2010) appears to be the only candidate explanation.

A second leading issue in yeast prions is their biological role. The [Het-s] prion of *Podospora anserina* is necessary for a normal physiological function of this organism, heterokaryon incompatibility (Coustou et al. 1997; Saupe 2007). This led us to state that this was the first functional prion (Wickner 1997). Because the yeast prion variants usually studied are relatively benign, it was suggested that yeast prions actually helped the host (Eaglestone et al. 1999). Claims of an advantage of [PSI+] or [URE3] (True and Lindquist 2000) have not been reproducible (Namy et al. 2008), and we will review the evidence that these prions are, in fact, diseases of yeast.

16.1 Yeast Prion Variants

Prion variants (called strains in mammals) were first recognized in scrapie transmitted to mice, where the different isolates produce dramatically different incubation times, different distributions of brain lesions, and different species barriers (Bruce 2003). In yeast, prion variants were first observed by Derkatch and Liebman (Derkatch et al. 1996) as different phenotype intensities and different stabilities of independent [PSI+] isolates. Variants of [URE3] and [PIN+] have also been observed (Schlumpberger et al. 2001; Bradley et al. 2002). Yeast prion variants differing in their response to overproduction or deficiency of chaperones (Chernoff et al. 1999; Kushnirov et al. 2000a, b), or in their transmission to other species (Edskes et al. 2009) are also well documented. Most recently, it has been found that common variants of [PSI+] or [URE3] can kill or severely impair the growth of the host, unlike the usual mild variants that have been studied in the past (McGlinchey et al. 2011).

Different prion variants are apparently due to different amyloid structures. Bessen and Marsh showed that the protease-resistant domain of PrP differed in the Hyper and Drowsy variants of transmissible mink encephalopathy studied in mice (Bessen and Marsh 1992). Studies from the laboratories of King and of Weissman have shown that amyloids of Sup35p fragments seeded by different [PSI+] variants (King 2001; King and Diaz-Avalos 2004; Chang et al. 2008) or with differing variant spectra on infection (Tanaka et al. 2004, 2006; Toyama et al. 2007) involve different extents of the Sup35NM region. However, these studies do not deal with the question of what the actual structure is, or how structural information is passed from prion protein molecules already in the fiber to molecules joining the fiber.

16.2 Shuffled Prion Domains of Sup35p or Ure2p Can Still Be Prions

To determine if there were specific sequences in the Ure2p or Sup35p prion domains that were needed for prion formation, we randomly shuffled these domains, and tested five shuffled sequences for prion formation. Surprisingly, we found that each of the five shuffled sequences of each prion domain could form prions (Ross et al. 2004, 2005a, b), showing that, for at least these prion domains, sequence was not critical and that prion formation depended more on amino acid content. The degree to which different residue types contribute to prion formation has been further examined as well (Toombs et al. 2010, 2011). Because the Sup35p prion domain, in common with PrP, has oligopeptide repeat sequences, many authors have proposed that these sequences are important. Indeed, deletion or further duplication of these repeats do indeed affect prion propagation and generation (Liu and Lindquist 1999; Shkundina et al. 2006), but such manipulations also affect the length and composition of the prion domain. Our finding that shuffled sequences (lacking the repeats) (Ross et al. 2005a, b) and results of Toombs et al., that shuffling just the repeats, do not impair generation or propagation of prions (Toombs et al. 2011) imply that the repeats are not critical. It is possible that the repeats are significant for the mRNA turnover role of the Sup35p prion domain (Hoshino et al. 1999; Hosoda et al. 2003; Funakoshi et al. 2007) (see below).

That prion forming ability was impervious to shuffling the amino acid sequence also implied that the prion structure must be an in-register parallel sheet (Ross et al. 2005a, b). The well-known sequence dependence of prion *propagation*, the "species barrier," seemed to be at odds with our finding that prion formation did not require any specific sequence. However, the sequence specificity for propagation simply means that there are specific interactions between amino acid side chains in the process of molecules adding to the end of an amyloid filament. If these specific interactions are complementary interactions, like the A–T and G–C interactions of DNA strands, shuffling the sequence would surely destroy the complementarity. However, if the specific interactions were between identical amino acid residues, then shuffling the sequence would still allow the same interactions, but they would occur in a different order. We thus predicted that the Ure2p and Sup35p prion domains would have an in-register parallel structure in their infectious amyloids (Ross et al. 2005a, b). As we describe in the following section, we verified this inference over the next few years.

16.3 Solid-State NMR Shows In-Register Parallel Architecture of Yeast Prion Amyloids

Meredith and coworkers were the first to demonstrate an in-register parallel amyloid structure (a peptide fragment of Abeta in this case) and used a solid-state NMR approach (Benzinger et al. 1998). Using singly carbonyl ^{13}C-labeled peptides, they showed a uniform ~5 Å distance between the labeled atoms, essentially the distance between strands in a beta-sheet. Because the molecules were singly labeled in each case, this could only be explained by an in-register parallel structure (Benzinger et al. 1998). Indeed, detailed studies have shown that the full-length Abeta amyloid has this architecture (Antzutkin et al. 2000; Balbach et al. 2000).

We have used a similar approach, but because the yeast prion domains are too long to synthesize, we used molecules labeled with a single carbonyl-^{13}C amino acid, at each of the (usually several) sites it occurs in the sequence. We found that the nearest neighbor labeled amino acid was generally about 5 Å away (Shewmaker et al. 2006; Baxa et al. 2007; Wickner et al. 2008a, b). Because there were several labeled residues in each molecule, it was critical to show that the nearest neighbor labeled atom was indeed in another molecule. This was done by diluting labeled molecules with unlabeled molecules and showing that the nearest neighbor distance was increased to the extent predicted based on the degree of dilution. Confirmation of the in-register parallel structure of the Ure2p prion domain has come recently from electron spin resonance studies (Ngo et al. 2011).

Amyloids of the Ure2 or Sup35 prion domains made for these NMR experiments generally produced a mixture of prion variants on transformation into yeast (King and Diaz-Avalos 2004; Tanaka et al. 2004; Brachmann et al. 2005). Correspondingly, two-dimensional ^{13}C–^{13}C solid-state NMR experiments show broad peaks indicative of microheterogeneity of sample conformations (Shewmaker et al. 2006; Baxa et al. 2007; Wickner et al. 2008a, b). Growing Sup35NM filaments at 4C or 37C produces amyloid that on infection in yeast gives largely homogeneous [PSI+] prion variants (Tanaka et al. 2004). Interestingly, hydrogen–deuterium exchange showed different extents of the slow-exchange regions in these preparations (Toyama et al. 2007). We found that each of these variant amyloid preparations showed the in-register parallel architecture (Shewmaker et al. 2009). However, it is not clear that these amyloid preparations are homogeneous, since the H–D exchange does not show single-exponential kinetics (Toyama et al. 2007).

Electron micrographs of amyloid formed from Ure2p or Sup35p prion domains show diameters of roughly 5 and 12 nm, respectively (Glover et al. 1997; Taylor et al. 1999). However, if the structures were single unfolded beta-sheets, they would be about 23 and 40 nm wide. Thus, the sheets must be folded along the long axis of the filaments. We suggest that prion variants may differ in the location of these folds (Wickner et al. 2008a, b).

Melki and collaborators have proposed that the core of Ure2p amyloid is composed of the C-terminal domain, with the N-terminal domain playing a peripheral role, and with no change to beta-sheet conformation (Bousset et al. 2002, 2003; Loquet et al. 2009). This model clearly does not apply to the Ure2p prion amyloid filaments, since (1) the N-terminal domain is necessary and sufficient for propagation of the [URE3] prion (Masison and Wickner 1995; Masison et al. 1997); (2) amyloid filaments of the prion domain alone or fused to various other proteins can efficiently transmit the [URE3] prion to cells on transformation (Brachmann et al. 2005); (3) the prion domain is unstructured in the native form (Pierce et al. 2005), but infectious amyloid of the Ure2p prion domain has beta-sheet structure by CD, by solid-state

NMR, by Raman spectroscopy, by electron diffraction, and by X-ray fiber diffraction (Taylor et al. 1999; Baxa et al. 2003, 2005, 2007); (4) mass per length measurements of the infectious fibrils show approximately one molecule per 4.7 Å (Baxa et al. 2003), consistent with the in-register parallel model, but inconsistent with the Melki model; (5) solid-state NMR data show the C-terminal domain essentially unchanged on filament formation by full-length Ure2p (Loquet et al. 2009), but the prion domain changing to beta-sheet structure (Kryndushkin et al. 2011).

Lindquist and coworkers have proposed a beta-helix model with head-to-head and tail-to-tail junctions for the prion domain of Sup35p (Sup35NM) (Krishnan and Lindquist 2005; Dong et al. 2010). This conclusion was based on the failure to find interaction between molecules of a large probe with orientation-dependent fluorescence (pyrene) attached at mutant cysteine residues. It is possible that the large probe affected the structure of the amyloid formed. Moreover, the beta-helix model is ruled out by (1) mass per length measurements of infectious Sup35N or Sup35NM filaments (Diaz-Avalos et al. 2005; Chen et al. 2009), which gave one molecule per 4.7 Å, consistent with the in-register parallel architecture, while the beta helix model predicts less than half a molecule per 4.7 Å; (2) deletion of the "tail" region is fully compatible with transmission of various [PSI+] variants (Bradley and Liebman 2004; Shkundina et al. 2006); and (3) the solid-state NMR data described above (Shewmaker et al. 2006, 2007, 2008, 2009) are incompatible with this model.

16.4 In-Register Parallel Architecture Explains Protein Templating of Conformation

A model of the in-register parallel structure is shown in Fig. 16.1. There is a line of each amino acid residue along the long axis of the filaments. What holds the molecules in-register in the yeast prion amyloid structure? The main chain hydrogen bonds between the amide H and the amide carbonyl of the peptide bond are the primary beta-sheet hydrogen bonds between molecules, and are oriented along the long axis of the filament, but are not sequence specific. It is interactions between the amino acid side chains that must be maintaining the structure in-register. If aligned, glutamine side chains can form hydrogen bonds as first suggested by Perutz for Huntingtin (Perutz et al. 1994). Aligned asparagine side chains can form a similar line of hydrogen bonds as can serines or threonines. Alignment of hydrophobic residues will likewise be favored by hydrophobic interactions of their side chains. Only charged residues will not want to be aligned because it brings identical charges close together, but charged residues are strongly underrepresented in the yeast prion domains.

At least for Sup35p amyloid filaments, elongation occurs by the addition of monomers to the ends of the filament (Collins et al. 2004). The prion domain of at least the native Ure2p is unstructured (Pierce et al. 2005). Formation of these amyloids is a change from unstructured to parallel in-register beta-sheet, with the sheet folded length-wise at specific sites (Fig. 15.1). We proposed that the same side chain—side chain bonds that hold the molecules in the filament in register direct the molecule

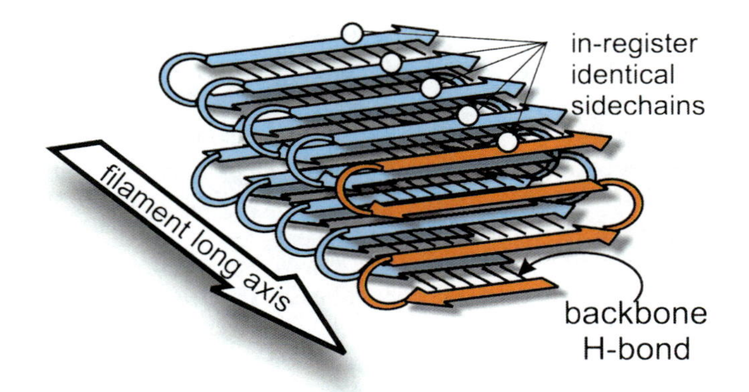

Fig. 16.1 In-register parallel beta-sheet architecture of the yeast prion amyloids [modified from Shewmaker et al. (2006)]. The side chains of a given residue form a line along the long axis of the filament. It is favorable interactions among such identical aligned side chains that keep the chains in-register. Electron microscopic measurements of filaments imply that the sheets must be folded along the long axis of the filaments as shown here

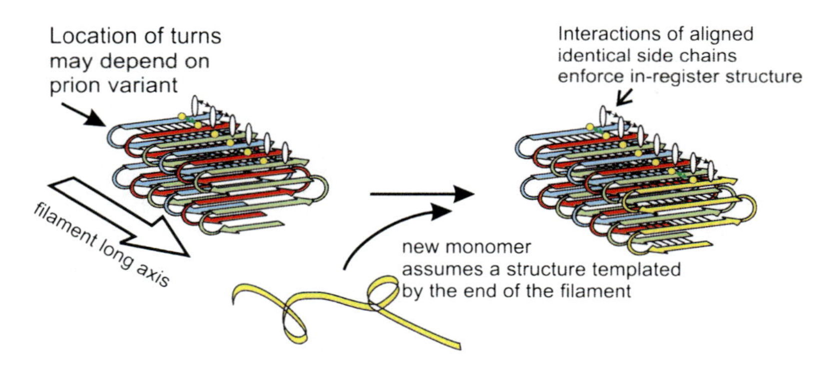

Fig. 16.2 The in-register parallel beta-sheet architecture suggests that prion variants differ in the location of the folds of the sheet, and implies a mechanism by which conformation can be inherited (Wickner et al. 2007, 2008a, b, 2010). The same favorable interactions among identical side chains that keep the structure in-register direct a monomer joining the end of the filament to assume the same conformation as molecules already in the filaments. Thus, the protein templates its own conformation in the same way that a DNA molecule templates its sequence

joining the end of the filament to assume the same conformation as the molecules already in the filament (Wickner et al. 2007, 2008a, b, 2010) (Fig. 16.2). Thus, just as DNA templates sequence, a protein can template conformation. Different protein conformations (=different prion variants/strains) can be faithfully propagated, and so proteins can act as genes.

16.5 Biology of Yeast Prions

Because de novo generation of prions in yeast, as in mammals, is a stochastic process, it is not well suited as an adaptive measure, and likely to be simply an infectious disease. Ure2p is a regulator of nitrogen catabolism, turning off the genes encoding transporters and enzymes needed for assimilation of poor nitrogen sources when the cells have available a good nitrogen source (Drillien et al. 1973; Courchesne and Magasanik 1988). When Ure2p is converted to amyloid in [URE3] cells, it loses its flexibility and is locked in the off position. Sup35p is a subunit of the translation termination factor, and it seems unlikely that cells will regulate translation at the termination step. Moreover, inefficient translation termination must produce read-through of many or most mRNAs, resulting in a wide variety of pathologic proteins.

16.5.1 [Het-s]: Benefit and Detriment

[Het-s], a prion of the filamentous fungus *P. anserina*, is necessary for heterokaryon incompatibility, a normal process in which the fungus recognizes self/nonself, and avoids fusing with colonies not very closely related (Coustou et al. 1997; Saupe 2000). The *het-s* locus has alleles *het-s* and *het-S*, differing at 13 amino acid residues in the 289 residue protein, and found at about equal frequency in wild strains (Dalstra et al. 2003). Only *het-s* cells can have the prion form, and when *het-s* [Het-s] cells fuse with *het-S* cells, the fused cells die and build a barrier to further fusions. We suggested that this was the first prion to have a function for the host, rather than being a disease (Wickner 1997). However, the [Het-s] prion also is involved in a meiotic drive phenomenon (much like the t locus of mice or segregation distorter in *Drosophila*), where an allele of a gene promotes its inheritance, not by benefiting the organism, but by cheating on meiosis, killing germ cells with other alleles. When female *het-s* [Het-s] cells mate with male *het-S* cells, most meiotic segregants with the *het-S* allele are killed (Dalstra et al. 2003). Thus, [Het-s] might be viewed as a disease of *Podospora* and the heterokaryon incompatibility, a secondary phenomenon.

16.5.2 Proposed Benefits of Yeast Prions

Eaglestone and Tuite were the first to suggest that yeast prions might benefit their hosts when they reported that [PSI+] had a general protective effect against heat or elevated ethanol concentrations (Eaglestone et al. 1999). True and Linquist then explored a large array of conditions in several isogenic pairs of [PSI+] and [psi-] strains and failed to reproduce the reported protective effects against heat or ethanol; in fact, there was no condition under which [PSI+] was protective in all cases

(True and Lindquist 2000). In most strains under most conditions, if there was a difference, [psi-] proved to be healthier than [PSI+] (True and Lindquist 2000). Nonetheless, the authors proposed that [PSI+] was helping yeast evolve by, in some cases, protecting cells from adverse conditions (True and Lindquist 2000).

This hypothesis is reminiscent of Lindquist's earlier proposal of an "evolutionary capacitor" role of Hsp90s in *Drosophila* (Rutherford and Lindquist 1998). Inhibition or mutation of Hsp90 resulted in the frequent appearance of morphological changes in the offspring. It was proposed that Hsp90 prevented the expression of accumulating mutations, until a stress condition, by otherwise occupying the Hsp90, allowed their expression, providing a diversity of phenotypes for evolution to operate on (Rutherford and Lindquist 1998). This type of "selection for evolvability" may be impossible in principle (Sniegowski et al. 2000). Moreover, it is now clear that Hsp90 is part of the "piwi" anti-transposon system operating in the germ line of *Drosophila* (Specchia et al. 2010; Gangaraju et al. 2011), and that inactivation of Hsp90 results in transposon-induced mutations, some of which have morphological consequences. No evidence in the original report tested whether the mutations producing the morphological changes were preexisting or not (Rutherford and Lindquist 1998).

If the phenotypes produced by yeast prions were to aid evolution, they would have to be beneficial, at least occasionally. Thus, it is particularly damaging to the evolvability model for yeast prions that Namy et al. (2008) were unable to reproduce the favorable phenotypes reported by True and Lindquist (True and Lindquist 2000), even though they used the same strains.

It has been reported that under certain stress conditions, the frequency of [PSI+] arising increases, and this is interpreted as an adaptive response (Tyedmers et al. 2008). However, the authors could not detect this effect with the normal Sup35 protein sequence, and only found it with an artificial construct that converts to [PSI+] with dramatically higher frequency. In addition, the authors reported that of four of the six conditions producing more frequent [PSI+], acquisition of the prion was detrimental to the cells (Tyedmers et al. 2008). These results actually argue against the "prion as evolvability factor" model. It is also argued that prion-forming ability is conserved across evolution, but we will see (in the following section) that this is not the case, with close homologs of prion proteins of *S. cerevisiae* being unable to form prions.

16.6 Evidence that Yeast Prions Are Diseases

Although it is impossible to test all possible conditions or to know what conditions represent what portion of the yeast natural habitat, there is a way to take a sort of integral over all conditions. The infectivity of yeast prions means that even if they are a net detriment to yeast, they may be found at some frequency in the wild. For example, the uniformly fatal chronic wasting disease of deer and elk is found infecting ~10% of animals in Wyoming and Colorado (Williams 2005). An infectious element that was beneficial to its host would spread rapidly, because effect on the host and infectivity would be working in the same direction instead of in opposition. Thus, an infectious element that is not found in the wild must be detrimental to its host. We surveyed 70 wild strains, and found each of the known parasitic nucleic acid replicons, including the L-A and L-BC dsRNA viruses, the 20S and 23S single-stranded RNA replicons, and the two micron DNA plasmid. However, neither [PSI+] nor [URE3] was present in any of the wild strains (Nakayashiki et al. 2005). This indicates that the overall effect of these prions is detrimental. In more limited surveys, other groups also found [PSI+] absent from wild strains (Chernoff et al. 2000; Resende et al. 2003). We did, however, find the [PIN+] prion at a frequency comparable to that of the parasitic nucleic acid replicons.

The partial conservation of sequence of the prion domains of Ure2p and Sup35p has been advanced as an argument that prion formation must be a benefit to yeast (Harrison et al. 2007). However, we have shown that prion formation is not determined by the prion domain sequence for either Ure2p or Sup35p (Ross et al. 2004, 2005a, b). Rather it is amino acid composition that is critical (Toombs et al. 2010, 2011). The sequence conservation probably reflects the normal non-prion functions of the prion domains. The Sup35p prion domain is necessary for the general mRNA turnover system, linking translation termination to the mRNA decay process by interactions with the polyA binding protein and the polyA RNAses (Hoshino et al. 1999; Hosoda et al. 2003; Kobayashi et al. 2004). The Ure2p prion domain stabilizes the protein against decay in vivo (Shewmaker et al. 2007). Thus, the presence of these domains across evolution and their conservation of sequence probably reflect the importance of these functions, and do not provide evidence for value of the prions to the host.

Although several homologs of Sup35p and Ure2p have been shown capable of being prions (Chernoff et al. 2000; Kushnirov et al. 2000a, b; Santoso et al. 2000), there are notable exceptions. The Ure2p's of *Saccharomyces castellii* (Edskes et al. 2009), *Candida glabrata* (Edskes et al. 2011), and *Kluyveromyces lactis* (Safadi et al. 2011). *Candida glabrata* is closely related to *S. cerevisiae*, and cannot form a [URE3] prion, but *C. albicans*, which is farther away, forms a [URE3] with properties similar to those of [URE3] of *S. cerevisiae* (Edskes et al. 2011).

Yeast cells (like other cells) react to a variety of stresses by inducing the production of heat shock proteins. Yeast induce both Hsp104 and Hsp70s on infection with the [URE3] and/or [PSI+] prions, indicating that the yeast cell's view of prion infection is unfavorable (Jung et al. 2000; Schwimmer and Masison 2002).

The prion domains of Sup35p and Ure2p change far more rapidly in evolution than do the non-prion parts of the same molecules (Kushnirov et al. 1990, 2000a, b; Chernoff et al. 2000; Santoso et al. 2000; Edskes and Wickner 2002; Baudin-Baillieu et al. 2003). Many of these changes produce barriers to transmission, species barriers that prevent the spread of the prions among the inter-mating *Saccharomyces* species (Chen et al. 2007; Edskes et al. 2009). In analogy with the protection afforded by the 129M/V PrP polymorphism in humans, it is likely that these mutations were selected to protect cells against infection by a prion from a more common Sup35p or Ure2p allele.

Table 16.1 Evidence that [PSI+] and [URE3] prions are diseases

[PSI+] and [URE3] are not found in wild strains	Chernoff et al. (2000), Resende et al. (2003), Nakayashiki et al. (2005)
Prion domains of Sup35p and Ure2p have important non-prion functions	Hoshino et al. (1999), Hosoda et al. (2003), Kobayashi et al. (2004), Shewmaker et al. (2007)
Cells mount a stress reaction when infected with [PSI+] or [URE3]	Jung et al. (2000), Schwimmer and Masison (2002)
Prion domains change more rapidly than non-prion domains, producing prion transmission barriers	Chen et al. (2007), Edskes et al. (2009)
Lethal and extremely toxic prion variants of [PSI+] and [URE3] are common	McGlinchey et al. (2011)
Prion-forming ability is not well conserved even among close relatives of *S. cerevisiae* Sup35p and Ure2p	Edskes et al. (2009, 2011), Safadi et al. (2011)

In spite of this array of data that [PSI+] and [URE3] are detrimental to yeast (Table 16.1), these prions are frequently cited in reviews as functional/beneficial amyloids (Shorter and Lindquist 2005; Chiti and Dobson 2006), probably because cells carrying the usually studied yeast prion variants do not seem particularly sick in the laboratory. If there were a [PSI+] variant that adsorbed all of the cell's Sup35p in the filaments, the cells would be dead because Sup35p is essential. To isolate such a "suiψdal [PSI+]," we prepared a strain with a normal full-length chromosomal *SUP35*, and carrying a counter-selectable plasmid with a doxycycline-repressed *SUP35C* gene, lacking the prion domain. The plasmid-encoded *Sup35C* cannot be incorporated into the amyloid filaments because it lacks the prion domain, and its expression was adjusted so that it was sufficient by itself to keep cells alive, but so low that cells would be Ade + because of increased read-through of the *ade1-14* nonsense mutation. [PSI+] clones were then tested for growth on medium that required loss of the plasmid carrying *SUP35C*. Eight percent of [PSI+] was totally unable to grow after plasmid loss, and 46% grew extremely slowly (McGlinchey et al. 2011). The existence of "suiψdal" and sick [PSI+] show that maintaining the ability to become [PSI+] does not come without a severe price.

Ure2p is not essential to yeast, and in many strains, *ure2*Δ does not even slow growth. However, we found that frequently [URE3] isolates grew extremely slowly, forming only tiny colonies on rich media (McGlinchey et al. 2011). These prion variants are not slowing growth by producing deficiency of Ure2p, since deletion of the *URE2* gene in this background is harmless. The prion must be having some toxic effect on the cell, perhaps adsorbing some essential component or interacting in a detrimental way with some other cellular component. Further work will be required to understand the nature of these toxic actions.

16.7 Perspective

Because it forms a single amyloid structure—corresponding to a single prion variant—the HET-s amyloid structure has been solved in a series of elegant studies (Ritter et al. 2005; Siemer et al. 2006a, b; Wasmer et al. 2008). [Het-s] was evolved to be a prion, and so forms only a single amyloid structure with the selected properties. It will be necessary to develop a method to obtain substantial amounts of yeast prion amyloid in a single conformation in order to obtain more detailed structural information. The in-register parallel architecture represents what is common among the structures, but the material used in these studies has, unavoidably, represented a mixture of structures. Future work on the biology of yeast prions will include studies of the mechanisms by which they produce harm to the cells, mechanisms that go beyond the mechanism known since the first studies of yeast prions (Wickner 1994) of simple depletion of active prion protein by conversion to the prion form.

Acknowledgment This work was supported by the Intramural Program of the National Institute of Diabetes and Digestive and Kidney Diseases.

References

Antzutkin ON, Balbach JJ, Leapman RD, Rizzo NW, Reed J, Tycko R (2000) Multiple quantum solid-state NMR indicates a parallel, not antiparallel, organization of beta-sheets in Alzheimer's beta-amyloid fibrils. Proc Natl Acad Sci USA 97:13045–13050

Balbach JJ, Ishii Y, Antzutkin ON, Leapman RD, Rizzo NW, Dyda F, Reed J, Tycko R (2000) Amyloid fibril formation by A beta 16–22, a seven-residue fragment of the Alzheimer's beta-amyloid peptide, and structural characterization by solid state NMR. Biochemistry 39:13748–13759

Baudin-Baillieu A, Fernandez-Bellot E, Reine F, Coissac E, Cullin C (2003) Conservation of the prion properties of Ure2p through evolution. Mol Biol Cell 14:3449–3458

Baxa U, Taylor KL, Wall JS, Simon MN, Cheng N, Wickner RB, Steven A (2003) Architecture of Ure2p prion filaments: the N-terminal domain forms a central core fiber. J Biol Chem 278:43717–43727

Baxa U, Cheng N, Winkler DC, Chiu TK, Davies DR, Sharma D, Inouye H, Kirschner DA, Wickner RB, Steven AC (2005) Filaments of the Ure2p prion protein have a cross-beta core structure. J Struct Biol 150:170–179

Baxa U, Wickner RB, Steven AC, Anderson D, Marekov L, Yau W-M, Tycko R (2007) Characterization of β-sheet structure in Ure2p1-89 yeast prion fibrils by solid state nuclear magnetic resonance. Biochemistry 46:13149–13162

Benzinger TL, Gregory DM, Burkoth TS, Miller-Auer H, Lynn DG, Botto RE, Meredith SC (1998) Propagating structure of Alzheimer's beta-amyloid(10–35) is parallel beta-sheet with residues in exact register. Proc Natl Acad Sci USA 95:13407–13412

Bessen RA, Marsh RF (1992) Biochemical and physical properties of the prion protein from two strains of the transmissible mink encephalopathy agent. J Virol 66:2096–2101

Bousset L, Thomson NH, Radford SE, Melki R (2002) The yeast prion Ure2p retains its native α-helical conformation upon assembly into protein fibrils in vitro. EMBO J 21:2903–2911

Bousset L, Briki F, Doucet J, Melki R (2003) The native-like conformation of Ure2p in fibrils assembled under physiologically relevant conditions switches to an amyloid-like conformation upon heat-treatment of the fibrils. J Struct Biol 141:132–140

Brachmann A, Baxa U, Wickner RB (2005) Prion generation in vitro: amyloid of Ure2p is infectious. EMBO J 24:3082–3092

Bradley ME, Liebman SW (2004) The Sup35 domains required for maintenance of weak, strong or undifferentiated yeast [PSI+] prions. Mol Microbiol 51:1649–1659

Bradley ME, Edskes HK, Hong JY, Wickner RB, Liebman SW (2002) Interactions among prions and prion "strains" in yeast. Proc Natl Acad Sci USA 99(Suppl 4):16392–16399

Bruce ME (2003) TSE strain variation: an investigation into prion disease diversity. Br Med Bull 66:99–108

Caughey B, Raymond GJ, Bessen RA (1998) Strain-dependent differences in beta-sheet conformations of abnormal prion protein. J Biol Chem 273:32230–32235

Chang H-Y, Lin J-Y, Lee H-C, Wang H-L, King C-Y (2008) Strain-specific sequences required for yeast prion [PSI+] propagation. Proc Natl Acad Sci USA 105:13345–13350

Chen B, Newnam GP, Chernoff YO (2007) Prion species barrier between the closely related yeast proteins is detected despite coaggregation. Proc Natl Acad Sci USA 104:2791–2796

Chen B, Thurber KR, Shewmaker F, Wickner RB, Tycko R (2009) Measurement of amyloid fibril mass-per-length by tilted-beam transmission electron microscopy. Proc Natl Acad Sci USA 106:14339–14344

Chernoff YO, Newnam GP, Kumar J, Allen K, Zink AD (1999) Evidence for a protein mutator in yeast: role of the Hsp70-related chaperone Ssb in formation, stability and toxicity of the [PSI+] prion. Mol Cell Biol 19:8103–8112

Chernoff YO, Galkin AP, Lewitin E, Chernova TA, Newnam GP, Belenkiy SM (2000) Evolutionary conservation of prion-forming abilities of the yeast Sup35 protein. Mol Microbiol 35: 865–876

Chiti F, Dobson CM (2006) Protein folding, functional amyloid and human disease. Annu Rev Biochem 75:333–366

Collins SR, Douglass A, Vale RD, Weissman JS (2004) Mechanism of prion propagation: amyloid growth occurs by monomer addition. PLoS Biol 2:1582–1590

Courchesne WE, Magasanik B (1988) Regulation of nitrogen assimilation in *Saccharomyces cerevisiae:* roles of the *URE2* and *GLN3* genes. J Bacteriol 170:708–713

Coustou V, Deleu C, Saupe S, Begueret J (1997) The protein product of the *het-s* heterokaryon incompatibility gene of the fungus *Podospora anserina* behaves as a prion analog. Proc Natl Acad Sci USA 94:9773–9778

Dalstra HJP, Swart K, Debets AJM, Saupe SJ, Hoekstra RF (2003) Sexual transmission of the [Het-s] prion leads to meiotic drive in *Podospora anserina*. Proc Natl Acad Sci USA 100:6616–6621

Derkatch IL, Chernoff YO, Kushnirov VV, Inge-Vechtomov SG, Liebman SW (1996) Genesis and variability of [*PSI*] prion factors in *Saccharomyces cerevisiae*. Genetics 144:1375–1386

Diaz-Avalos R, King CY, Wall JS, Simon M, Caspar DLD (2005) Strain-specific morphologies of yeast prion amyloids. Proc Natl Acad Sci USA 102:10165–10170

Dong J, Castro CE, Boyce MC, Lang MJ, Lindquist S (2010) Optical trapping with high forces reveals unexpected behaviors of prion fibrils. Nat Struct Mol Biol 17:1422–1430

Drillien R, Aigle M, Lacroute F (1973) Yeast mutants pleiotropically impaired in the regulation of the two glutamate dehydrogenases. Biochem Biophys Res Comm 53:367–372

Eaglestone SS, Cox BS, Tuite MF (1999) Translation termination efficiency can be regulated in *Saccharomyces cerevisiae* by environmental stress through a prion-mediated mechanism. EMBO J 18:1974–1981

Edskes HK, Wickner RB (2002) Conservation of a portion of the *S. cerevisiae* Ure2p prion domain that interacts with the full - length protein. Proc Natl Acad Sci USA 99(Suppl 4):16384–16391

Edskes HK, McCann LM, Hebert AM, Wickner RB (2009) Prion variants and species barriers among *Saccharomyces* Ure2 proteins. Genetics 181:1159–1167

Edskes HK, Engel A, McCann LM, Brachmann A, Tsai H-F, Wickner RB (2011) Prion-forming ability of Ure2 of yeasts is not evolutionarily conserved. Genetics 188:81–90

Funakoshi Y, Doi Y, Hosoda N, Uchida N, Osawa M, Shimada I, Tsujimoto M, Suzuki T, Katada T, Hoshino S (2007) Mechanism of mRNA deadenylation: evidence for a molecular interplay between translation termination factor eRF3 and mRNA deadenylases. Genes Dev 21: 3135–3148

Gangaraju VK, Yin H, Weiner MM, Wang J, Huang XA, Lin H (2011) *Drosophila* Piwi functions in Hsp90-mediated suppression of phenotypic variation. Nat Genet 43:153–158

Glover JR, Kowal AS, Shirmer EC, Patino MM, Liu J-J, Lindquist S (1997) Self-seeded fibers formed by Sup35, the protein determinant of [*PSI+*], a heritable prion-like factor of *S. cerevisiae*. Cell 89:811–819

Harrison LB, Yu Z, Stajich JE, Dietrich FS, Harrison PM (2007) Evolution of budding yeast prion-determinant sequences across diverse fungi. J Mol Biol 368:273–282

Hoshino S, Imai M, Kobayashi T, Uchida N, Katada T (1999) The eukaryotic polypeptide chain releasing factor (eRF3/GSPT) carrying the translation termination signal to the 3'-poly(A) tail of mRNA. J Biol Chem 274:16677–16680

Hosoda N, Kobayashii T, Uchida N, Funakoshi Y, Kikuchi Y, Hoshino S, Katada T (2003) Translation termination factor eRF3 mediates mRNA decay through the regulation of deadenylation. J Biol Chem 278:38287–38291

Jung G, Jones G, Wegrzyn RD, Masison DC (2000) A role for cytosolic Hsp70 in yeast [PSI+] prion propagation and [PSI+] as a cellular stress. Genetics 156:559–570

King CY (2001) Supporting the structural basis of prion strains: induction and identification of [*PSI*] variants. J Mol Biol 307:1247–1260

King CY, Diaz-Avalos R (2004) Protein-only transmission of three yeast prion strains. Nature 428:319–323

Kobayashi T, Funakoshi Y, Hoshino S, Katada T (2004) The GTP-binding release factor eRF3 as a key mediator coupling translation termination to mRNA decay. J Biol Chem 279: 45693–45700

Krishnan R, Lindquist S (2005) Structural insights into a yeast prion illuminate nucleation and strain diversity. Nature 435:765–772

Kryndushkin DS, Wickner RB, Tycko R (2011) The core of Ure2p prion fibrils is formed by the N-terminal segment in a parallel cross-β structure: evidence from solid-state NMR. J Mol Biol 409:263–277

Kushnirov VV, Ter-Avanesyan MD, Didichenko SA, Smirnov VN, Chernoff YO, Derkach IL, Novikova ON, Inge-Vechtomov SG, Neistat MA, Tolstorukov II (1990) Divergence and conservation of *SUP2 (SUP35)* gene of yeasts Pichia pinus and *Saccharomyces cerevisiae*. Yeast 6:461–472

Kushnirov VV, Kochneva-Pervukhova NV, Cechenova MB, Frolova NS, Ter-Avanesyan MD (2000a) Prion properties of the Sup35 protein of yeast *Pichia methanolica*. EMBO J 19:324–331

Kushnirov VV, Kryndushkin D, Boguta M, Smirnov VN, Ter-Avanesyan MD (2000b) Chaperones that cure yeast artificial [*PSI*[+]] and their prion-specific effects. Curr Biol 10:1443–1446

Liu JJ, Lindquist S (1999) Oligopeptide-repeat expansions modulate "protein-only" inheritance in yeast. Nature 400:573–576

Loquet A, Bousset L, Gardiennet C, Sourigues Y, Wasmer C, Habenstein B, Schutz A, Meier BH, Melki R (2009) Prion fibrils of Ure2p assembled under physiological conditions contain highly ordered, natively folded molecules. J Mol Biol 394:108–118

Masison DC, Wickner RB (1995) Prion-inducing domain of yeast Ure2p and protease resistance of Ure2p in prion-containing cells. Science 270:93–95

Masison DC, Maddelein M-L, Wickner RB (1997) The prion model for [URE3] of yeast: spontaneous generation and requirements for propagation. Proc Natl Acad Sci USA 94:12503–12508

McGlinchey R, Kryndushkin D, Wickner RB (2011) Suicidal [PSI+] is a lethal yeast prion. Proc Natl Acad Sci USA 108:5337–5341
Nakayashiki T, Kurtzman CP, Edskes HK, Wickner RB (2005) Yeast prions [URE3] and [*PSI*+] are diseases. Proc Natl Acad Sci USA 102:10575–10580
Namy O, Galopier A, Martini C, Matsufuji S, Fabret C, Rousset C (2008) Epigenetic control of polyamines by the prion [*PSI*+]. Nat Cell Biol 10:1069–1075
Ngo S, Gu L, Guo Z (2011) Hierarchical organization in the amyloid core of yeast prion protein Ure2. J Biol Chem 286(34):29691–29699
Perutz MF, Johnson T, Suzuki M, Finch JT (1994) Glutamine repeats as polar zippers: their possible role in inherited neurodegenerative diseases. Proc Natl Acad Sci USA 91:5355–5358
Pierce MM, Baxa U, Steven AC, Bax A, Wickner RB (2005) Is the prion domain of soluble Ure2p unstructured? Biochemistry 44:321–328
Resende CG, Outeiro TF, Sands L, Lindquist S, Tuite MF (2003) Prion protein gene polymorphisms in *Saccharomyces cerevisiae*. Mol Microbiol 49:1005–1017
Ritter C, Maddelein ML, Siemer AB, Luhrs T, Ernst M, Meier BH, Saupe SJ, Riek R (2005) Correlation of structural elements and infectivity of the HET-s prion. Nature 435:844–848
Ross ED, Baxa U, Wickner RB (2004) Scrambled prion domains form prions and amyloid. Mol Cell Biol 24:7206–7213
Ross ED, Edskes HK, Terry MJ, Wickner RB (2005a) Primary sequence independence for prion formation. Proc Natl Acad Sci USA 102:12825–12830
Ross ED, Minton AP, Wickner RB (2005b) Prion domains: sequences, structures and interactions. Nat Cell Biol 7:1039–1044
Rutherford SL, Lindquist S (1998) Hsp90 as a capacitor for morphologic evolution. Nature 396:336–342
Safadi RA, Talarek N, Jacques N, Aigle M (2011) Yeast prions: could they be exaptations? The *URE2*/[URE3] system in *Kluyveromyces lactis*. FEMS Yeast Res 11:151–153
Santoso A, Chien P, Osherovich LZ, Weissman JS (2000) Molecular basis of a yeast prion species barrier. Cell 100:277–288
Saupe SJ (2000) Molecular genetics of heterokaryon incompatibility in filamentous ascomycetes. Microbiol Mol Biol Rev 64:489–502
Saupe SJ (2007) A short history of small s: a prion of the fungus *Podospora anserina*. Prion 1: 110–115
Schlumpberger M, Prusiner SB, Herskowitz I (2001) Induction of distinct [URE3] yeast prion strains. Mol Cell Biol 21:7035–7046
Schwimmer C, Masison DC (2002) Antagonistic interactions between yeast [PSI+] and [URE3] prions and curing of [URE3] by Hsp70 protein chaperone Ssa1p but not by Ssa2p. Mol Cell Biol 22:3590–3598
Shewmaker F, Wickner RB, Tycko R (2006) Amyloid of the prion domain of Sup35p has an in-register parallel β-sheet structure. Proc Natl Acad Sci USA 103:19754–19759
Shewmaker F, Mull L, Nakayashiki T, Masison DC, Wickner RB (2007) Ure2p function is enhanced by its prion domain in *Saccharomyces cerevisiae*. Genetics 176:1557–1565
Shewmaker F, Ross ED, Tycko R, Wickner RB (2008) Amyloids of shuffled prion domains that form prions have a parallel in-register β-sheet structure. Biochemistry 47:4000–4007
Shewmaker F, Kryndushkin D, Chen B, Tycko R, Wickner RB (2009) Two prion variants of Sup35p have in-register β-sheet structures, independent of hydration. Biochemistry 48:5074–5082
Shkundina IS, Kushnirov VV, Tuite MF, Ter-Avanesyan MD (2006) The role of the N-terminal oligopeptide repeats of the yeast Sup35 prion protein in propagation and transmission of prion variants. Genetics 172:827–835
Shorter J, Lindquist S (2005) Prions as adaptive conduits of memory and inheritance. Nat Rev Genet 6:435–450
Siemer AB, Arnold AA, Ritter C, Westfeld T, Ernst M, Riek R, Meier BH (2006a) Observation of highly flexible residues in amyloid fibrils of the HET-s prion. J Am Chem Soc 128: 13224–13228

Siemer AB, Ritter C, Steinmetz MO, Ernst M, Riek R, Meier BH (2006b) 13C, 15N resonance assignment of parts of the HET-s prion protein in its amyloid form. J Biomol NMR 34:75–87
Sniegowski PD, Gerrish PJ, Johnson T, Shaver A (2000) The evolution of mutation rates: separating causes from consequences. Bioessays 22:1057–1066
Specchia V, Piacentini L, Tritto P, Fanti L, D'Alessandro R, Palumbo G, Pimpinelli S, Bozzetti MP (2010) Hsp90 prevents phenotypic variation by suppressing the mutagenic activity of transposons. Nature 463:662–665
Tanaka M, Chien P, Naber N, Cooke R, Weissman JS (2004) Conformational variations in an infectious protein determine prion strain differences. Nature 428:323–328
Tanaka M, Collins SR, Toyama BH, Weissman JS (2006) The physical basis of how prion conformations determine strain phenotypes. Nature 442:585–589
Taylor KL, Cheng N, Williams RW, Steven AC, Wickner RB (1999) Prion domain initiation of amyloid formation in vitro from native Ure2p. Science 283:1339–1343
Toombs JA, McCarty BR, Ross ED (2010) Compositional determinants of prion formation in yeast. Mol Cell Biol 30:319–332
Toombs JA, Liss NM, Cobble KR, Ben-Musa Z, Ross ED (2011) [PSI] maintenance is dependent on the composition, not the primary sequence, of the oligopeptide repeat domain. PLoS One 6:e21953
Toyama BH, Kelly MJ, Gross JD, Weissman JS (2007) The structural basis of yeast prion strain variants. Nature 449:233–237
True HL, Lindquist SL (2000) A yeast prion provides a mechanism for genetic variation and phenotypic diversity. Nature 407:477–483
Tyedmers J, Madariaga ML, Lindquist S (2008) Prion switching in response to environmental stress. PLoS Biol 6:e294
Wasmer C, Lange A, Van Melckebeke H, Siemer AB, Riek R, Meier BH (2008) Amyloid fibrils of the HET-s(218–279) prion form a beta solenoid with a triangular hydrophobic core. Science 319:1523–1526
Wickner RB (1994) [URE3] as an altered *URE2* protein: evidence for a prion analog in *S. cerevisiae*. Science 264:566–569
Wickner RB (1997) A new prion controls fungal cell fusion incompatibility. Proc Natl Acad Sci USA 94:10012–10014
Wickner RB, Edskes HK, Shewmaker F, Nakayashiki T (2007) Prions of fungi: inherited structures and biological roles. Nat Rev Microbiol 5:611–618
Wickner RB, Dyda F, Tycko R (2008a) Amyloid of Rnq1p, the basis of the [*PIN*$^+$] prion, has a parallel in-register β-sheet structure. Proc Natl Acad Sci USA 105:2403–2408
Wickner RB, Shewmaker F, Kryndushkin D, Edskes HK (2008b) Protein inheritance (prions) based on parallel in-register β-sheet amyloid structures. Bioessays 30:955–964
Wickner RB, Shewmaker F, Edskes H, Kryndushkin D, Nemecek J, McGlinchey R, Bateman D, Winchester C-L (2010) Prion amyloid structure explains templating: how proteins can be genes. FEMS Yeast Res 10:980–991
Williams ES (2005) Chronic wasting disease. Vet Pathol 42:530–549

Index

W.-Q. Zou and P. Gambetti (eds.), *Prions and Diseases: Volume 1, Physiology and Pathophysiology*, DOI 10.1007/978-1-4614-5305-5,